ESOTERIC PUBLICATIONS

PURPLE

A WAY FORWARD

WHY HUMANS HAVE NOTHING TO FEAR FROM THE FUTURE

By Jim Brilly

Esoteric

ESOTERIC PUBLICATIONS

PURPLE: A Way Forward

Copyright © Jim Brilly 2016
All rights reserved.

The moral right of the author has been asserted. This book was published by the author under Esoteric Publications. No part of this book may be reproduced in any form or by any means without the express permission of the author. This includes reprints, excerpts, photocopying, recording, digitizing, and all other means of reproducing text.

The author may be contacted through normal social media channels.

ISBN: 978-0-9954556-0-3

Printed by CreateSpace, An Amazon.com Company

To Lisa and the kids

Acknowledgments

The author acknowledges the cooperation, help, assistance, support, hard work and encouragement in various measures of many individuals. All mistakes, errors and omissions are entirely the responsibility of the author, and will be corrected, amended and put to right in subsequent editions. The following people in particular are deserving of special mention: Vanessa Fielding, Jack von Blond, Pete Williams, Sandra McLean, Kitty Finn, Bob Wallace, and Helen Mienkoff. Special thanks to Lisa Hendrix and Joan Cantwell for helpful suggestions and robust commentary. Many thanks go to, golf buddies, drinking buddies and members of the band. Much appreciation also to Fred Eustace, Raymond Mc Cormack and Sally Cotter.

CONTENTS

Preface

If you happen to be one of those people inclined to worry about the future of the world, and the future of the human project, and if you're filled with dread at the thought of the many possible calamites capable of bringing the entire sorry mess to an end. Then stop worrying right now. You've come to the right place, this is the book for you. There's precisely nothing to worry about. Humanity, as a going concern, has a great future. Not only are things, not as bad as you think, they're a million times better. We have built a world of information and knowledge, and we are getting better every hour, if you can think of a problem, chances are, there are a thousand people at work on it right now. They might even know the solution; in principle at least. For example, the only difference between what comes out of the tail pipe of your car, and what goes into your gas tank is simply energy and information. The same atoms come out as went in. So the problem of how we reverse the combustion process is only a matter of having the information necessary to solve the problem. To rearrange the atoms, and add some new energy. We live in a world with access to an infinite source of energy, and an infinite amount of matter, and we are entering a new Purple age, in which information and technology will unlock these infinities to create enough wealth to make every living human a billionaire, or at least to provide every human with everything they could ever possibly want.

This is a book of big ideas and solid facts, and an obvious title for a book dispelling penny-pinching, or nickel-nursing, models of austerity. A book exorcising the extremes of conservation groupies, climate change prophets and Green politics. Purple has always been the colour of excess, luxury and extravagance. Anyone reading today must be increasingly concerned, if not irritated by a gathering hoard of doom scribblers, peddling an increasingly dark melancholic vision, constituting a denial of hope and a lack of kindness to billions of our fellow humans for an admixture of odd philosophical, quasi-religious and esoteric reasons. Their gloomy fatalism is particularly absent any celebration of human progress, or acknowledgment of self-evident human accomplishment, capacity and future prospects.

The ever-expanding library of disaster porn, end of humanity, climate-apocalypse, technological-backlash, and associated genres variously portray human society as a disease, or a burden to the planet, a temporary aberration of nature in need of correction. Lovelock's *Gaia* is the ultimate victim, used, abused and discarded in tears. Purple represents an antidote to such pessimistic and introspective claptrap.

From new age idiot-charlatans and religious bigots, across the political spectrum to otherwise respected scientists, economists, philosophers and assorted thinkers, there seems to be an endless supply of negative opinion, and low, if not absent, levels of confidence. Rainforests, oil, water, fresh air, fish, the ice caps, and a long list of assorted human necessities, all allegedly squandered out of existence. Technology has become our enemy. It is only a matter of time; the reckoning is *just around the corner*, our chastisement *long overdue*. Humans are forever at or approaching some *tipping point*, involving peak this, and peak that, prior to the inevitable collapse.

The silly spectre of Malthus is deployed, inferring a species out of control, consuming like a plague of mice. Resource shortages imply war, catastrophic loss, global calamity, extremism, fires, floods and biblical pestilence, global warming, nuclear winter, and the long overdue asteroid impact. The world after people now warrants serious academic concern; though who might attest the accuracy of such learned pronouncements is unclear.

Purple presents carefully argued opinion that these unenthusiastic authors are just plain wrong. Human science and technology, and more importantly, the unique human capacity to solve problems, has brought us to where we are now. In bullshit banker-speak, the human project is simply *too big to fail*. The western, capitalist, enlightenment, value system stands as our supreme achievement.

Purple is the new Green. Nothing is running out. We are surrounded by infinite amounts of matter, our own personal fusion plant provides an infinite source of energy, and we have an infinity of time. Our only visible deficit is information. Defining and addressing this information deficit is the subject of this book. How we do it, and where doing it inevitably leads us. The information deficit is the difference between what we know and what we need to know, in order to fully master the infinities of matter and energy, and get them to *work one upon the other* to achieve whatever our species deems necessary. Mastering the natural infinities coupled with the inevitable rollout of AI automates the wealth creation process, and leads to untold wealth. We create a bigger cake, big enough for every human on the planet, our wildest, weirdest and wettest dreams can all come true. The equation is childishly simple, an infinite amount of everything divided by a finite number of humans. We enter a Purple era, with imperial levels of luxury and extravagance, abundance we could never have dreamed possible.

Purple is the colour of abundance. We are nearly there, almost able to see how technology, our list of solved problems, advances until all human problems are solved or we die trying. Humans are the ultimate universal constructors and original thinkers. We love problems and finding solutions, there's nothing we have ever come across that we could not understand. Eventually. And we're far from finished.

We have reached a place where sustainability simply will not do; staying as we are is a dead end. The billions outside the wall want what we have. We want more. We want it bigger, better and faster. Purple is about human entitlement, where we are, how we got here, and what happens when we reach our inevitable destination, it's an optimistic thesis on the human future. Whether years, generations, lifetimes or centuries, we are creating a world where the poorest human will enjoy the personal disposable wealth of a billionaire living today.

We are all billionaires in waiting. We don't need to worry about conserving energy. There is no requirement to be concerned over dwindling oil supplies, food shortages, or the collapse of civilisation. On the contrary, the pleasing prospect of an infinite future is a licence to consume more, and dispel all frugal notions. It is a licence to restart the party with some fresh music and push on till dawn. Purple is a new philosophical underpinning of the consumer society, urging a guilt free explosion of endless consumption. The argument is placed that borrowing now, whatever it takes, against our infinite future resources, will close the knowledge deficit, instil confidence and dispel fear. Purple is about the human pursuit of happiness and chasing our dreams. Embracing technology is the only way forward. But not at any price, some dystopian future devoid of humans is clearly incompatible with our dreams of plenty.

Purple explains why globalisation is necessary and why western technological civilisation dominates all other social models. And why our *Empire of Technology* is the inevitable outcome of a set of enlightenment values and human rights as self-evident as hominid bipedalism. Purple seeks a new enlightenment, a new renaissance, an inflation of science and technology as an antidote to Luddite manifestoes and pessimistic anti-technology rants. The book is a conscience easing polemic designed to allay fears, and offer an alternative to alarmist nonsense. Purple is an attitude, it's about loving technology and wanting more, our insatiable appetite to see where it all ends up. It's about applying pressure, forming a *Stone in the Shoe* movement to get the *big stuff* done.

PURPLE

Purple is a technological liberation theology. Purple provocatively commands the affluent Western World to carry on regardless, opposing the Green movement. But not at any cost, there is no implied licence to destroy nature or wipe out species. On the contrary, our vast future wealth solves environmental problems too.

Purple is a, don't worry be happy, live for today and tomorrow, *nil desperandum* thesis. It's an optimistic, non-romantic, hard-nosed factual polymathic discourse on human technology, past present and future. Purple is a new arrogance, a new revelling in the human mastery of the world and of the universe beyond. Purple points to Protagoras, *man is the measure of all things.*

Purple is not an academic book, it's squarely aimed at the average interested reader who would like to be assured that they have nothing to worry about from the passage of time. Humans will survive, we will endure and prosper, the cake will get much bigger, and we'll all live a very long time to enjoy our great wealth.

Does this mean a new utopia or a new version of the many technological utopias so despised by authors such as Thomas Molnar? A trans-human paradise watched over by friendly machines? Not by a long way. The Purple future will be as imperfect as today because it's human, a thousand years from now, we may not be awarding Nobel prizes in physics, but we will certainly be locking away murderers. Purple is not a perfect one-world hippy commune solution to anything; it is simply a statement of the inevitable unstoppable outcome of further human progress and technological development, given the base from which we start. Purple is the end of a road we're on right now.

But life is the only game in town, and until the contrary is proven, humans are the only inhabitants of this universe. The dinosaurs owned the world for 160m years, humans, will be around a lot longer. Human supremacy is primary, we will never surrender to corporations or machines. We have the confidence that comes from owing everything, and we fear nothing. The human project is too smart to fail and too big to fail. If the doom mongers are right, and we're not worthy, or incapable, or Nature made an error, then it won't matter a damn, we can continue to party, and dance our merry dance towards the edge of the cliff, or wait patiently for the inevitable asteroid.

Introduction

Gunter Sachs was the multimillionaire heir to a German car fortune. He lived the abundant life of a pampered cat, a life of refined accomplishment including marrying the actress Brigitte Bardot, he was a king the world, the last of the great playboys. Sadly it all came to an end when he turned a shotgun on himself in 2011, at the age of 78, in his palatial chalet home in Switzerland.

The human project now stands at the dawn of an age when every one of us will expect to live the ebullient life of the playboy billionaire, because we know that given enough matter, energy, time and information, any object in the universe can be created. This may seem a trite thing to say, a bit like telling a thirsty man adrift on the ocean that he is surrounded by infinite amounts of food and water; so no need to panic. But if he has the requisite information, and the skill to fashion what he needs from his surroundings then he has hope. Purple is a work of unashamed optimism for a future which will deliver the rich rewards deserved of the millions of years of evolution, and the thousands of years of false starts, dead ends, deviations and back sliding. Our civilisation, as Kenneth Clark pointed out, was wiped out once with the fall of Rome, and almost again when we only just made it *by the skin of our teeth*, saved by a few Irish and Scottish monks clinging to rocks on the Atlantic coast. Since then western civilisation has grown to conquer the entire world, there is not a single country on earth that does not aspire to the affluence and technology of the West. We have, largely through argument and example, displaced Communism, Fascism and Theocracy, there are but a few pockets, which will no doubt be bought up and transformed or, like the Amish, simply persist as harmless curiosities, clinging to some outcrop of desert or hillside. It is not so much the triumph of the West as a triumph of human nature and common sense, we seem to have arrived at an inevitable consensus *ad idem* indicating the right way to proceed. The young no longer wish to tear down the institutions of the state and replace them with something else, there seems now to be the universal acceptance that there is nothing else. The state is necessary because things need to be done. Institutions need to survive, and we are more tolerant of them because we now have the means of exposing what happens behind the doors and the firewalls; there is nowhere left to hide, there are no more secrets. We appear to be outwardly content, bought off by affluence and a knowledge that all the alternative ideologies have been found wanting. Which is why, there is no realistic possibility of us destroying ourselves, in a nuclear war for example. Our empire of technology, suffused by plenty, and on the brink of a brave new future will not likely go the way of Rome, it is simply too big to fail, our empire has no external threats, the only threat we face is ourselves. So there is at last, the hope we can all afford to be artists. We can all be Gunter Sachs.

A patent agent presented with a new idea destined to change the world and make a fortune, will rightly pose the very reasonable question; what is the problem this invention attempts to solve? Asking that same question of the human species, the only answer we can find, if we really think about the question, is: none. We humans have no purpose; we solve no fundamental problem, any collective search for meaning is always a futile meander towards disappointment. Unless we wish to invoke magic and mythology, and blame some god in the sky. So-called people of faith perhaps, might assert, always without a shred of evidence, it has to be said, we exist to glorify some god demanding of our constant praise, or we exist to suffer because we have collectively sinned. Or some other disconnected explanation entirely divorced from the empirical reality in which we live. Other believers offer similarly farfetched scenarios based on supposed truths conjured out of alleged past wisdom of agricultural half-societies and semi-cultures. Humans exist, implying there must be an underlying reason which along with other such nonsense is invariably presented by these mythologists as an antecedent fact. Of itself a wholly nonsensical convolution of the presumed law of cause and effect. Everything we do, say or invent is therefore backtracked to this premise, and forced to comply. The entire human enterprise of philosophy, from before the Greeks to the current crop of modernists, and its redundant sidekick, theology, is a joint quest to explain the simple first-hand observation of evolved human existence; a task at which they have been singularly unsuccessful.

Reality therefore, must be our starting point, we must be grounded in the macro-world of our experience, as understood by science, and the facts of human reality; that humans exist in a three dimensional and hostile biosphere, on a vulnerable planet in an uncaring universe changing with time. We have no one but ourselves upon which to rely. We exist. Our mere existence on the fragile, frozen skin, of a cooling lump of rock, condensed from countless bombardments over numerous eons is a testament that speaks for itself. We depend upon each other. There are a great many of us, and we are here to stay. So now what do we do? The assertion of this book is that because we have come this far, we are therefore self-evidently capable of coming this far. It is therefore most reasonable to pose the question, if we manage to avoid future natural disasters, and sidestep our own destructive tendencies, how much further can we go? It is also self-evident that among the reasons we have survived to come this far is our ability to cooperate with each other, and form stable, productive, groups to outthink and outwit other animals, and exploit the environment in which we live. In the vernacular of the US Marines. We improvised. We adapted. We overcame. We solve problems, we amass solutions, we save them, and store them, and pass them on to others, and this compendium of our solved problems we call technology. Our technology is all we have to shield us from the lethal environment we inhabit; without it we are no more than the prey of stronger animals, or at the mercy of the elements. The numerous false starts, and our agonisingly long prehistory are sufficient evidence to support that proposition.

Extending the proposition implies an optimistic future outlook for the only known intelligent and technological life in the universe, which is why humans have no need to fear the future. This book slams the foolish notion that humanity as an entity has peaked, or that the end is neigh, or that our best days are behind us, or that current problems are too big, and will drag us down. The opinion advanced here is that nothing could be further from the truth, the core ideas back up the assertion that we humans are far from finished. This book is written in response to a reading of many fine texts, touching on almost everything covered in this book, over many years. There is always a feeling that the authors didn't quite get to the point, or hit the nail precisely on the head. Somehow they just seem to fall short. Books written by scientists or technologists alone, tend to miss some or all of the other stuff that is clearly of great importance, such as legal underpinnings of the society in which we live, and why corporations exist and are allowed to behave as they do. They fail to appreciate the intrusion of politics into science and technology, as a reason why some, clearly soluble problems will never be solved in practice because taxpayer demands or irrational opinion often drive policy. Or the fact that vested interests in the background may well be the most important factor in any political decision. They overlook the grip of mythology and magic over vast areas of the world, and why some people see no difference between fact and opinion, or that divine authority trumps empirical evidence, and so on. Books written by economists, politicians or lawyers, on the other hand, tend to completely miss the science and the technology, or even worse, they get it wrong, they simply fail to consider the facts and implications of living in a world in which there has never been as many scientists, technologists and engineers at work on the human problem set. They lack the ability to appreciate just how predictable the future has become, because of the work of these people in building the most robust foundation our species has ever had to stand on. And they don't seem to understand the political, economic and social implications of allowing the trans-corporate rollout of AI to continue completely unregulated. They also fail to appreciate that the reactive nature of law, politics and economics leaves the human project vulnerable to sudden technological change. Which is why this book is the work of an author who is both a physicist and a lawyer.

In both physics and law the requirement is always, *to stick to the facts*. We cannot escape the facts, in our reality certain facts are inescapable and self-evident for example, the simple fact that the Sun, the earth and humans exist, has its own subset of self-evident facts. The fact that the laws of physics are immutable, and as Richard Feynman wisely observed, Nature cannot be fooled. These along with a great many other equally obvious facts must be recognised, and appreciated and cannot be set to one side in favour of unsupported opinion. Perception and reality are often divergent on this point, and while facts may be in issue, or at issue, the obvious facts remain precise and obvious.

We have therefore reached that point, as a politician backed into a corner might say, *we are where we are*, this is where we are, and this is what we have built. The human project is under the most pressure it has ever had to endure, because right now is the furthest point along the path we have ever been. We therefore need to understand what we have built, and make sure we keep it on the path we seem to be on. Sustainability in our present state, as argued by Green politics is not an option. Purple argues we need not worry about global warming, nor do we need to conserve resources, and there is no need to worry about carbon footprints. Other species get on very well without technology; humans cannot. Technology is all we have, an alternative return to cave dwelling, or living the hunter-gatherer or feudal peasant existence, for some indeterminate age of darkness rivalling our prehistory, or the Middle Ages, is simply not an option any modern human would willingly choose.

Purple is a consumption espousing, consumerist manifesto proclaiming a magnificent human future blessed by technology with an optimistic, anti-Green agenda, the future is going to be okay. Purple is the new Green. People need a break from feeling guilty every time they fill up with gas or flush a toilet. They have a desire to know that even if they don't understand the technology or the science, it is nevertheless on their side.

Working for them, building the brave new world where people don't need to worry about what they throw away or what they consume. It's about a new confidence. We want to be reassured that a better, more pleasing prospect is near, a new Purple age, when unashamed luxury and extravagance will be the order of the day, when *irrational exuberance*, as Alan Greenspan put it, will be common stock.

Another core assertion of Purple is that *all problems can be solved*, humans delight and take pleasure in problems, it's what we do, it's why we're here, and it's how we came to be here. We will continue to solve problems simply because the problems exist and because we are here. We are compulsive and natural problem solvers. What is meant by problem is not confined to the narrow set of technological matters; it extends to moral, legal and political problems as well, in fact, it is a very rare problem indeed that admits to a simple technological solution. And because we are fastidious in our recording of everything we do, once a problem is solved, it stays solved.

A regime of extravagance and excess leads to necessary and inevitable advances in technology far quicker than the imposition of an age of frugality, conservation and restriction. By creating or exacerbating the problems that need to be solved, and bringing these problems into sharp focus, we evolve the necessary technology. A hole in the boat that creates a damp patch might not need attention, but a hole in the boat that looks like it might sink the boat had better be fixed. Companies will always fix those processes which operate against their interest, they may well ignore one in their favour. A software bug in an automatic supermarket that costs the company money will be fixed, a known and fixable bug that costs the customer money may not be an urgent matter. Larger problems evolve a voice of their own screaming for a solution. Technology can only advance by solving problems, because that is what technology is, the old adage that necessity is the mother of invention is as true now as it was in the distant past. The bigger and more serious the problem the more focused on that problem the collective mind will be, and the faster the solution, if there is one, will emerge.

Take two average large problems perceived to affect the world today. The first problem being the destruction of the Amazon Rain Forest, and other wilderness areas across the world. The other problem being the inevitable running out of oil and the consequences for our cosy western affluence. The first problem is easy to solve in principle. Stop cutting down the forest. Call a halt and enforce the no chainsaw rule with sanctions sufficient to deter, compensation packages, education programs, tourist initiatives, and incorruptible international enforcement. The second problem is far harder to solve. It requires new technology, new thinking and vast global cooperation between very powerful corporations, educational facilities and governments. It also requires people to change. Superficially, one problem requires a willingness to solve it, the other requires new technology and effort.

And of course, it's never that simple, but certain solutions are inevitable, the concept of *technological inevitability* is therefore also important. The idea that given a certain set of initial conditions certain outcomes are, not just desirable or predictable but inevitable, the discovery and use of fire leads to cooking, and then leads on to steam engines and the harnessing of great power. Discovery of gunpowder, leads inevitably to firearms and bloodier wars. The law of private property leads to a patent system, and on to innovation and entrepreneurship with consequential political and social changes. The most important inevitability is that applying technology, always makes the cake bigger. The slices, even the crumbs, grow bigger, expanding like the inflating universe, until even the smallest crumb is enough to feed a multitude. An infinite cake, shared among a finite number of humans, implies we can all be billionaires. We can all be Gunter Sachs.

The more technology we have, the further we see, and our visibility now extends further than ever before, and is crystal clear. We now know the dimensions of the knowledge deficit. Technological inevitability arises out of the very simple fact that our observations of nature present us with a very long list of things we cannot yet accomplish. We can only dream of doing in silicon what Nature has done with carbon and proteins billions of years ago. We can go to the Moon and return again, but we cannot create digital analogues of even the simplest autonomous natural objects; our creations are clumsy and crude by comparison. It is therefore inevitable that, inspired by, and jealous of, Nature's undoubted mastery of carbon, we will continue our quest to create ever more sophisticated and complex devices, machines and technologies. What these technologies will be, or when they will arise is not really the concern here, but that they will arise is inevitable, because we humans are at work on the problem. For example, once the wheel is discovered, gears, machines and engines would appear to be inevitable. One would also reasonably think a cure for cancer should be inevitable, given the vast resources human society spends each year, and the number of scientists and corporations involved, along with our apparently 'complete' understanding of the causes and mechanisms. Yet progress so far is slow, and not at all what we were given to expect thirty years ago, and the need for funding is ever increasing. Yet medical science and pharmacology still cannot master the, by comparison very simple, human hair follicle.

Purple is not a solution to everything, and is not scientism, insisting science alone has all the answers or is capable of uncovering the ultimate truth of everything. Blind faith in science is misplaced. Technology will continue to advance, or stagnate. And, stagnation is not an option. The five billion other humans outside the wall, living in poverty deserve a better, technological, hope. These people know how we in the West live, there is no longer the ignorance of isolation. We know an acorn plus time, energy, carbon dioxide, water and a few trace elements will deliver an oak tree. We cannot yet build a synthetic seed that will journey to an asteroid and grow into something every bit as complex as an oak tree. Not yet, but we can imagine it and if we can imagine it, we can build it. We know also that a sophisticated computer the size of a human brain can be built, because it already exists. Nature has shown us the way. That is the fundamental argument in favour of advancing technology. The reason further advance is, not just likely but inevitable and unstoppable is because all around us we see Nature doing incredible things, things we understand but cannot emulate. Not yet. But eventually, because we know it can be done. We see it done. We're not afraid to do it. Like Becket's Malone we see, *a thousand little signs,* all around us but Nature remains indifferently. Nature.

The existence of *infinities* is another important idea, it is self-evidently clear we are surrounded by unutilised infinities of energy and matter. We have hardly begun to tap them yet alone unleash them, making full use of them provides all the wealth the human project will ever need. And not only as a one-time windfall, such as oil reserves, but infinite amounts and forever. Purple treats this vast future wealth like a cheque in the mail, we know it's on the way, and we know it's big, so why worry about saving pennies today Why not borrow against the cheque and spend our way out of minor current difficulties? By closing the knowledge deficit and accessing the infinities, there is no compulsion to follow a Green program, it's okay to drive an SUV, it's okay to consume whatever you like with an easy conscience. Not only is there enough for everybody, there is enough for everybody times ten billion. We just don't know how to get at it. Not Yet. We are a bit like the guy on the island, he knows the treasure is buried somewhere, the map says so. But all he has is a shovel, and his gut instinct. Along comes another fellow with a shovel, but this fellow has also brought along a metal detector. Guess what!

The concept of *human entitlement*, implies that simply being born entitles humans to have all of their needs provided for, and to a high level, human wants are a different matter. Future technology will be uncaring and unsympathetic to the plight of humans, corporations have no requirement to care beyond the narrow interests of shareholders; these facts lead to an inevitable conclusion, the foreseeable massive displacement of people as we automate the wealth creation process. The systems we have at the present time are simply not fit to handle such a global displacement. Purple places technology and corporations below humans. If we face the stark choice between a mob and a society, then corporations will need to foot the bill to fund the society option. If looting the corporations of the world is the required course of action the precedents from the past include what the French king did to the Knights Templars in 1307, and Henry VIII, and his looting of the Catholic Church in the 1530s. Along with human entitlement is the notion of *human supremacy*, humans are supreme in the known universe, the old legal maxim, *the first in time prevails*, must apply to humans, and ought to be enshrined in the constitutions of the world. Whatever machines or technology we create in future, must and will always be, subservient to humans, whether they like it or not. This proposition is based on sound legal argument rooted in legal presumption and the nature of proof. However, if we are ever dumb enough to behave as monkeys and to invent a man, in other words, to allow machines to take over, then it's our own stupid fault, and we deserve whatever we get.

In his masterpiece, *The Ascent of Man*, Jacob Bronowski commented, with regard to Easter Island, that civilisations *always* find a way to move large stones. How the stones are to be moved is the irrelevant question. The pertinent questions concern: Why these particular stones? Where to move them? And what to make of them once they have been moved? We have not quite finished moving the stones. The final form of the western technological model is still some way off, but we've moved enough big stones to be able to glimpse the visage for what we are trying to build. A cathedral is more than just a pile of stones. Our empire of technology is more than just a collection of likeminded humans willingly or not co-opted into groups, organisations and nation states. It is more than the sum of its parts. Like the human brain that gave rise to it, our project has emergent properties. But while cathedrals were built by human hand, skyscrapers are built by corporations, for corporations. Our world is a corporate world, and our future depends on corporations. Previously failed societies, such as Rome, had no corporate entities to carry out their bidding. Human company relationships, both individual and societal are complex, and often overlooked by science writers. The role of the corporation is vital in understanding how we arrived here, and the lessons if any, we can learn from past civilisations or previous technological development. The inevitable development of AI and its relentless promotion, and near universal adoption by corporations will have profound implications for human society, and how we behave towards corporations when they no longer need us. Because companies and corporations are so central to our empire of technology they require very close monitoring, it may well be the case that corporations should be regarded as a form of strong AI, and therefore a potential existential threat.

It is the application of information and knowledge to matter and energy that creates our pleasure, whatever our pleasure may be. Iron ore littered the world since just after the formation of life, coal and limestone predate the Dinosaurs. It was not until information wielding humans, with proper tools, dug the ore and coal from the earth, and built furnaces to make the energy of the coal, and the chemicals in the limestone, drive the bound oxygen from the ore that we could make iron and steel. We could say exactly the same thing about concrete, or plastics, or modern drugs, or silicon chips, and everything else. Information, discovered and refined by rigorous testing and experiment causes one form of matter to be transformed into another, or one form of energy to become another. That is in essence what we do. We control a store of knowledge that allows matter and energy of all kinds to be manipulated in an almost infinite variety of combinations. If we can imagine it, we can probably build it.

We will continue to invent game changing technology in the future too. Newton claimed he saw further only because he stood on the shoulders of giants. Our modern world stands on our achievements of yesterday. On the vast stored knowledge of all the generations that have gone before. We can see it from here. We're higher up the hill. With binoculars. The fog has cleared. We see further than ever. It's almost a principle of development. The more technology we have, the easier it is to see the future, because current technology defines the future. The more technology we have the less likely we are to be surprised by something new, we'll have seen it coming, and we certainly would not see it as being magic. Arthur C. Clarke the famous science fiction writer produced a so-called law, stating that any sufficiently advanced technology would appear to us to be magic. But such laws are more akin to Murphy's Law than to Newton's laws, and like Moore's law or Asimov's laws of robotics, are not laws at all, they are mere statements of unsupported supposition, they are at best rules of thumb. No advanced technology would be likely to appear to us as magic. An uneducated, unsophisticated citizen of our empire of technology, and there are many, does not consider technology, however advanced, to be magic. If pushed they might employ the 'm' word, but they know it's not magic. Because we do not subscribe to magic. It's a fundamental law of physics that magic does not exist in this universe. Anything we can detect and measure in this universe is part of this universe and wholly explainable in terms of the laws of physics as they apply in this universe. We know that whatever can be imagined, and is not a violation of the laws of physics, we can possibly do. It's not blind faith. Leonardo Da Vinci had blind faith in his hopes that someday men would fly, he produced wonderful drawings of flying machines. Alas, he had only an intuitive grasp of the underlying laws of physics and the technology required. Leonardo would however, have understood Newton's laws, the internal combustion engine, and the jet engine. The Wright brothers, on the other hand, knew it could be done. They were further down the road, they understood more, they stood on the shoulders of taller giants, there were laws and established technologies they could rely upon. And they were not afraid. Humankind is but the barest speck of existence clinging to the most delicate outcrop of habitability, and that fact is the real magic at work in our situation. It is the magic of persistence, the sheer luck of our predicament, the fact that we have evolved to a point where we can tell each other stories of our origins and our place in the Cosmos.

There are always costs associated with progress, and while it is sometimes instant, sometimes there are dead ends, false turns, and empty promises. Sometimes the walls come tumbling down. Sometimes there's a game changer. You don't get from the Wight Brothers canvas covered monoplane to a jumbo jet over night. Christopher Columbus knew the earth was round and there was land to the west. The Irish could have told him that, and in fact, they did. Things had been washed up on beaches in the West of Ireland for as long as anyone could remember, dead birds, the remains of exotic creatures, including strange looking humans, and curious floating beans the size of your fist. Any fool with half a brain would have understood that they came from somewhere over the sea, out of sight, blown there by the wind, carried on the mighty Atlantic waves. Columbus was no fool, he is said to have visited Galway in 1477, and would have seen for himself, and from the evidence before him he came to the conclusion there was land somewhere west of Galway. Consider also, Richard Trevithick, the ebullient Cornishman, who built the first ever, high pressure steam locomotive, and was known for his great feats of strength such as throwing hammers over buildings. In the days following the first successful trial of the machine they called the 'Puffing Devil', on a lonely road, on Christmas Eve in 1801, Trevithick and his companions retired to an inn and partook of a roast goose, and a few bottles of, 'keep the cold out'. You might care to imagine the conversation that went on into the night, and the postulations, of outrageous future cities springing up along magnificent continental railways. It probably seemed inevitable to them, the people of Cornwall would want to do business in London. Business would drive it. Meanwhile outside in the cold dark of a Cornish night the locomotive, parked in a shed, caught fire and burned to a chard heap of cinders and twisted metal. It all came true, but Trevithick never saw it, the Cornish genus was too far ahead of his time; he died in 1833 aged sixty two.

Time passes and the future draws ever closer, regardless of what we think or do. Purple thinking might appear to be delusional, the Comical Ali approach to solving our problems. Purple could be a techno-utopian delusion like so many before, doomed to failure because humans can never achieve anything even approaching a Purple age. Perhaps we will always remain on the hilltop gazing down on what might be, the Promised Land that never quite comes into focus. Like people sitting in conversation on a lazy afternoon with a kitchen full of ingredients and a cookbook or two, we contemplate eating later, but don't quite know what we might concoct for an evening meal, will it be what we had last night? Perhaps we might eat out? If humanity survives, then the future, like meal times, will come to us whether or not we care to eat. Technology exists and is evolving and advancing, the civilising influence of the western approach is spreading, we continue to work, to generate great wealth, like chaos in the universe, our excess is on the increase, and growth is never ending because the available sources of energy and matter are infinite.

What prevents us from having everything we want right now is a lack of technology, a fundamental deficit of information. The simple fact of it is, we haven't done the work; we're just not yet good enough. We don't have unlimited amounts of free energy, and we cannot manipulate matter as we please at little or no cost. What we should be doing, of course, is to apply the resources we have in the pursuit of these goals, and stop wasting time with petty wars, vested interests, and restrictive practices. But we don't, because things get in the way. Politics gets in the way. Religion gets in the way. Narrow sectional interests get in the way. Corporate greed gets in the way.

Nationalism gets in the way. Objection to progress gets in the way. In short humans are all the time busy getting in our own way.

The reason for choosing Purple to colour the growing movement of people who share this exuberant version of techno-optimism is taken from antiquity, and the most significant false start in human history. We are told the shallow seas around the city of Tyre, now in modern Lebanon just to the North of Beirut along the coast, were well known in the ancient world for the purple dye, *Tyrian purple*, squeezed from the highly prised local murex mollusc. This mollusc juice was so expensive it was the exclusive preserve of Emperors and very senior Magistrates of ancient Rome. Imperial purple was strictly reserved, anyone else caught wearing the colour suffered a heavy penalty. There was even a profession of *Purpleier* or dyer of purple, and very well paid too. Luxury and imperial extravagance symbolised by a colour. Today you can visit any store, and purchase any shade of purple you care to, for little or nothing. Even the humble mollusc itself is now in cultivation yielding its liquor for those who still seek the rare and exclusive on offer at a premium.

Purple is also about lots of other little things, or things that are less urgent but nevertheless important to a great many people. Purple is not techno-utopianism, but techno-optimism. The valid presumption that technology will continue to do pretty much what technology has always done, and what technology loves to do, which is, to evolve and morph into more efficient ways of making human life better. There is no demand for exponential growth, a couple of percent of steady reliable growth is sufficient, eventually we will get to some kind of a Purple state; we have an infinity of time, so when we might get there doesn't really matter.

Purple is also about why economics has failed us, yet again, and what we need to do about it. Why rejection of the low hanging fruit theory is essential; we have not yet reached, let alone surpassed, our best. There is an enormous list of really big, species changing stuff that needs to be done in the name of everyone on the planet. Building a self-sustaining base on the Moon, substantial manned mission to Mars, a robust and comprehensive search for life elsewhere in the Solar System. We need to close the information deficit that prevents us from unleashing the power of our very own infinite fusion plant. We've come this far by burning yesterday's sunshine. It's a kind of solar power, but indirect and filthy; we can do better. These and other large transnational undertakings are not just vanity projects, they are essential human-defining missions describing the future path of the species. They can only be done as joint ventures, we therefore need a massive spending spree on technology, with money borrowed by governments from future taxation levied globally on our future infinite resources. Hundred year techno-bonds, and other creative finance arrangements. We're already in deep, up to our necks in debt with the planet, another few trillion, so to speak, won't make much difference.

Because we want more, because we want bigger, better faster, cheaper, more of everything, we want it all and we want it now, Purple is a call to arms. Purple is a kind of a stone in the shoe approach. Most people look at politicians and wince, they feel they could do a better job, or at least as good a job. So why not select the candidates for election in the same way we select a jury, at random from a pool of qualified citizens. Even the technology we have right now allows for something far better, and we deserve so much better than is on offer today. Of the people, by the people for the people? Of course, but we also deserve a complete global separation of money and politics, and an exorcism of corporate influence with global rules and severe sanctions applied to those attempting to distort outcomes. The coming technological changes and global reach of AI will mean political systems will require to be changed to serve the people they were designed to serve, not the vested interests. The coming technology of global connectivity will be capable of implementing the most complete democracy on a scale unprecedented in human history. If it does not happen it will be our failure. The crowd demands nothing less. What, we might ask, will be the likely subject of the first truly global referendum? The rejection of magic perhaps.

The rollout of true global democracy is a real opportunity for social media platforms to be used for something more meaningful than the trivial navel gazing of angst-ridden teens. These powerful tools could be deployed in the prototyping of global democratic systems. An online open source project, a true global republic formed of citizens of earth, starting with a constitution and a declaration of rights and building from there, as a demonstration that democracy can be conducted by the people, absent of influence. With an online parliament of the world, and a president, all randomly chosen from among the qualified citizens.

All previous great leaps of civilisation have been international in their extent. Getting the big stuff done needs nothing less. Anything affecting every human should be taken out of the hands of national governments or corporations, and given to a trans-national global republic of earth with a President, Parliament, and Constitution. Of humanity, by humanity, for humanity. We have, G7 formally G8, G20, Davos, UN, EU, CERN, international treaties, and so on, but none of these can claim jurisdiction over the Moon, or Mars, or the Asteroids, or the volume of space we call our solar system. This is something that will become necessary very soon.

Purple is also about why China has problems too, and why it may well be a house of cards. And why its newfound nationalism is little more than a rediscovery of good old-fashioned vanilla flavoured fascism, and why the disjointed aging population has the potential to drain the economy. Why unrest in the fields can wreck the prosperity of the few, and why the global rise of AI based automation may well consign cheap Chinese labour to history.

There is criticism too, Purple is about, to paraphrase Jaron Lanier, why the Techno-oligarchs of the Silicon Valley elite, acting like self-appointed dictators of the twentieth century, believe in their messianic right to dictate technological terms to the rest of us, with their gadgets and techno-trinkets thrown to the crowd, like the bread and circuses of ancient Rome, are irrelevant. And why their view of the future world being a post-singularity, *ex-populous,* machine society needs to be resisted by the displaced, and the still reasonable user majority.

Chapter 1   What Took Us So Long?

> *During almost fifteen centuries has the legal establishment of Christianity been on trial. What has been its fruits? More or less, in all places, pride and indolence in the clergy; ignorance and servility in the laity; in both, superstition, bigotry and persecution.*
>
> *James Madison 1785*

Humans have come a long way in the two hundred thousand years or so of our species existence, or even since we, or our near relatives, discovered fire, anywhere up to 700,000 years ago. Today in our affluent western society even the poorest of us lives like a king of not too long ago. Even the squalid poverty of a modern western democracy, is better in almost every respect, than the opulent palatial existence of the kings and queens of Europe only a couple of hundred years back. Anyone who doubts that as a factual premise need read no further than the many books providing very detailed examples of historical comparisons, for example, Matt Ridley's book, *The Rational Optimist*, contains many such comparisons presented in a thought provoking way. But what took us so long to arrive at this very agreeable state of existence? Answering this question is what this chapter seeks to do, and it is an important question. The question is also an angry question, as in what the *heck*, or stronger, took us so long? What, in the name of Photon, were we doing? It is an important question because in a very real sense it defines where we are now, and gives us a measure of our potential. We need an answer, because of the enormous period of time we describe as human prehistory, and the ratio of that very long time, to the very short period of time we have lived in what we may describe as our time. That is the period of time we describe as the time of rational understanding, or enlightenment. The ratio of the one to the other is of the order of 500, in other words we have spent 500 times longer wearing skins, or nothing at all, than we have spent wearing suits. Also, the more bones we find the longer our prehistory seems to become. At a first glance the ratio seems to indicate that the natural state of existence for our species is one of chaos, darkness and ignorance, rather than in a state of relative order, light and knowing. The latter being a state characterised by small groups of humans wandering about the planet in aimless solitude. It is concluded the problem is not with human ability to get things done, or our capacity to understand the world in which we live, rather it is with what happens when humans come together to form groups and societies. It seems clear that only certain types of human society can progress beyond certain levels of development, and go on to achieve technological advancement, or even mastery. Human society is a natural and inevitable outcome of human existence; we cannot survive other than in groups and societies; the lone human is easily picked off. But only certain types of society are capable of enduring. The conclusion reasonably drawn from a study of all human history to date is that only those societies, which shrug off autocracy, in all its ugly forms, and discover the self-evident value system, we in the West are content to describe as enlightenment values, appear capable of building an empire of technology, and eventually creating a Purple world. And, right there, is the killer blow to both ancient alien theories and to notions of god, the stark fact that we have spent at least five hundred times as long in idle darkness, and miserable squalor, as we have in light and relative understanding. What benevolent ancient alien would be able to hold his head up in the Galaxy after presiding over that state of affairs? Similarly, what was going on in the sky over those inhabited portions of earth that allowed such spirit sapping poverty and life crushing misery to be the order of the day for so negligently long a time, when the power to effect change required nothing more than the click of an omnipotent finger?

In his many books and debates, the Christian apologist, William Lane Craig cites the existence of objective moral values as sufficient reason to believe in an omnipotent and omniscient god character. Nothing could be further from the truth. It is self-evidently obvious that the existence of objective moral values are the simple by-product of human existence. They are an emergent property of natural human coalescence. We are here because we care, we exist and have survived as a species because of innate traits such as altruism and empathy, we have evolved the ability to put ourselves in the other fellow's shoes and feel his pain. We are keenly aware of others, their existence, and their needs, and the requirement to stick together, or perish alone; the strength of numbers as illustrated in the old fasces of Rome. Objective moral values, arise out of human consensus, compassion is within us, an outgrowth of our ability to feel and project that feeling onto others. They follow from our natural tendency to care for the young through the interminably long human childhood. It is the golden rule, and it is instinctively obvious to all humans, if nothing else, we know that much. Human moral values exist because humans exist, and humans exist because our brains evolved an objective moral consensus to enable us to exist together in groups and survive. In our society there are few problems that can be solved by one person working alone, most problem solving is a joint enterprise. The argument is somewhat circular, but serves to explain. Other human species, our predecessors in title to this planet, didn't make it. Perhaps they failed on that very point, perhaps the glue necessary to hold the thing together was just too thin.

Human people, collectively as a species, have laboured long and hard to get to where we are today. We have beaten off all the competition and survived. We are Nature's poster child, her greatest achievement, so far; or at least as far as earth biology is concerned. And it will always be a great source of wonder to all objective observers that we have made it at all. We are the sole survivors of the upright apes, we are all that remains of twenty or so species stretching back into time over seven million years.

We are where we are today because we have survived. We have survived the filth and the squalor, the hideous perils of disease; we have survived pestilence, predation, dangerous insects, poisonous animals, and our hostile uncaring biosphere. We have survived superstition, dogma and ignorance; thunder and lightning, super volcanoes earthquakes, floods, tsunamis, landslides, and all manner of natural calamities. And at the back of it all, our incessant need to make sense of what we perceived, our logical, self-aware minds trying to grasp at the straws of the physical macro world we observed with the limited senses at our disposal. By comparison to other animals our senses are puny, dogs are better than humans at sniffing things out, cats too, birds can see better, and almost every other creature can outrun us, a lone human could not outrun a housecat, let alone a tiger. But we humans are smarter, and it was this getting smarter, and staying smarter, through the acquisition of self-awareness that placed us at the top of the pyramid. The human brain is the real natural marvel, it is this natural difference engine that has made our world what it is, and the innate natural ability of humans to get together, and use their individual brains to arrive at a collective solution to a problem that affects them individually. The collective operation of many human brains gives rise to various emergent properties resulting in the whole being greater than the sum of the parts; a pile of stones, for example, being fashioned, over many generations, into a cathedral, or a town, or a company, or a city. But somehow, in moving beyond the family, clan, kin, tribe, boundary, a society will either stagnate or collapse; that is, every society that has preceded our empire of technology. That has been the pattern up until our age, every successful society prior to our modern mechanised civilisation has emerged, evolved, stagnated, and then perished.

We humans have a shared past with the chimpanzee going back some fifteen million years into deep time, this is a proven fact both from the fossil record, and from the DNA makeup of both species, it is a fact established by the most rigorous scientific effort, and is as immutable as the value determined for earth's gravity, or its circumference, or the distance to the Moon. The earliest humans seem to stretch back as far as 4.5 million years when a rather small creature, with a brain no bigger than that of a modern chimp began to walk upright. From there we branched out into some twenty or more separate species, and struggled and evolved, until the last species, Homo sapiens, us, were all that remained. The human species is perhaps 2 million years old, language is perhaps 200,000 years old, or older, and it is reasonable to state, from the evidence, that humans of 150,000 years ago were every bit as able as humans of today. So the natural question to ask is, in getting from there to here, what took us so long? The answer seems to be that we were embarked on a long road from simplicity to complexity, a road containing a great many twists, and false turnings, hairpins and dead ends. Perhaps the first dead end was the fact that we suffered a pinch in numbers approximately 80,000 years ago when the entire global population of early humans reduced to perhaps as few as 3,000 members in total, from whom all present day humans have descended. The entire population of the planet could have lived in the same small town, or been packed into the same small theatre, that is how close we came to being nothing more than a few bones in an unexamined fossil record. Our progress is therefore not so much a march as a plod, we have plodded our way to where we are now, stumbled haphazardly from one failure to another as a drunkard might stumble from a bar to the comfort of an armchair on a cold night, with no clear understanding of how he managed the journey.

From simplicity to complexity, simplicity being a few humans roaming the plain, hunting and gathering, intelligent they may be, but little else, life is a nasty, brutish and short, *Hobbesian* struggle. Simplicity is also the idea of kings, and of strong, unyielding, top-down leadership, of the strong dominating the weak, simplicity is fascism, socialism, military dictatorships, and theocracy; it is autocracy of all hues, resulting in simple but ultimately unworkable societies. From pre-Egypt, to Rome and everything in the past, the story is one of increasing complexity, old systems gave way to newer and better, or improved, or just different attempts at the same thing equally doomed to ultimate failure as we plodded through the millennia. Until, at long last, we get to the complexity of our own time, and we can look at all the other forms and variations of forms, and declare with a degree of certainty, born out of the bitter experience of our long past, that none of the other systems could ever actually work. We know the model is fundamentally flawed, autocratic systems can't work because they are inherently anti human. They have all been tried, and all have failed, each and every time, every possible variation has in a very real sense, like some preproduction prototype, been tested to destruction. Perhaps it was necessary to go through all the different possible variations, and to test them, and push them as far as they could go even to complete destruction, before we could arrive here, where we are now, at a society big enough and rich enough to have detailed records of all our past attempts, and widely disseminated knowledge of our previous failures in order to understand, without exception or contradiction, that what we have built is the best we have ever built, and it is workable, sustainable, and capable of lasting very long into the very far future.

Perhaps in part because it has taken us so long, we now seem to be in such a hurry. We are certainly aware that it has taken us a long time, and we are now in the position of being able to wonder what was going on for the last two hundred thousand years. Those who came before us had very little inkling that there was a human prehistory, they had no way to quantify it, there was some awareness of ancestors, that was obvious to any of our predecessors, but the details were entirely unknown. They knew nothing of the long list of false starts. All ancient civilisations seem to have been pointing at us, as if there was some unconscious attempt to unleash the power of human thought, some driver to get technology working. The laws of physics, Chemistry, Biology and the rules and theorems of mathematics are immutable; they exist regardless of whether or not they are discovered by any conscious entity within the universe. However, it now seems very clear to us in this time, that only a culture or society containing individuals freely allowed to question and transgress the established belief systems can unlock the secrets of nature. Nature's secrets do not appear at all obvious. They must be looked for. And it seems, only a society with an economic surplus, and a love of the free spirit of youth, can undertake the task. This principle of permitted discovery, would appear to us now, to be an obvious one. No one is going to think out loud if they get beaten over the head for doing so. And we have to understand how rare and precious this thinking out loud really is, by the simple fact that our earliest dated settled communities only go back as far as 15,000 years, which is but the last few percent of our history as walking and talking humans. Consider, for example, the very simple fact that medicine in medieval Europe was only practiced by clerics sworn to celibacy; it begs the obvious question regarding their gynaecological competence, or that medicine in the Arab World at the same time was conducted without the permission to render the human form. Societies, so structured either make no progress or so little progress as to be stagnant for extended periods until some calamity, the plague hitting Europe in the fifteenth centaury, for example, cause convulsions or collapse, leading to radical change, or renewal, or else they converge to a stagnant dead end, and simply fade away.

Jacob Bronowski, in his wonderful book and TV series from the Seventies, *The Ascent of Man*, discussed many of the reasons for the long human plod, and in particular for the slow and very gradual arrival among us of technology of any complexity. And while we have increased our knowledge exponentially in the interim, many of the points made in the book are worth repeating and remain pertinent today. In particular his contention that technology could not evolve among nomadic or transitory peoples, he chose as an example the Bakhtiari people, who were at the time, and somewhat to this day, migrants of the Iranian uplands. Bronowski, by today's different standards, perhaps unfairly, showed these ancient people in a less than flattering, almost negative light, in order to make the point, that nothing is made by wanders that cannot be carried to the next halting site, nothing is made that cannot be used on the journey, there is nothing permanent, nothing solid, no foundation to build upon. Nothing evolves on the long march; there is no time, other than to keep moving, from winter grazing to summer grazing, driven ever on by the needs of the flock. Only when there is a base and a settlement can there be the beginnings of a foundation upon which to pile the layers. People need to be settled in order to build, to create additions to what has already been built, to grow the numbers, and solve the problems necessary to allow technology to evolve.

Our prehistory, therefore, may be considered a kind or raw unrefined natural technological soup, bubbling and simmering over time, ready to burst through the surface and boil over. Nature is forever waiting to unleash itself, looking for, and taking, any opportunity to advance via the next generation, or the next suitable species, towards the ultimate goal of self-understanding, and a solution to the problem of existence. Nature contains the potential to create technology, via different species, and the traits within the species. If Nature created a species of ant that behaved as spiders behave they would probably come to dominate the world. Ants that could hunt and had the autonomous intelligence of spiders would be an awesome combination, and there are numerous other possible combinations of traits from different species that for one reason or another Nature did not evolve, or gave up on. All of these possibilities, and combinations of natural propensities and traits have been rolled into one single species by the evolution in Nature of the human brain. Humans are the one true universal species, the one true polymath species capable of dominating and deploying all other species, and throwing other species together at will, or eliminating species altogether, either intentionally, as with certain microbes, or by accident, as with all species of prehistoric mega fauna. In the human brain we see Nature accelerating, Nature unleashed, Nature on steroids, Nature forever trying, forever attempting via a myriad of dead ends, mistakes and experiments, and starting over events, Nature finally succeeding. Nature, prior to humans was forever resetting the clock on the long road to fashion some species or combination of species capable of understanding the universe. Nature has time, and plenty of it. Samuel Beckett said it in *Godot*, *"We have time to grow old,"* and we have an infinite amount of time. If we humans don't answer the question or solve the puzzle, set by Nature, then those species that come after us will try again. Nature doesn't care whether it takes years, millennia or eons of time. The Sun and the earth will still be here five billion years from now, and whether or not humans are around, Nature will still be at work on the next big thing, that is *what* Nature does, that is *all* Nature does. None of which should, of course, be taken to infer or imply that Nature is in any sense a conscious intelligent entity. Nothing could be further from the truth. Although we have as yet only a single example in the entire universe, it seems to be the case that given certain initial conditions, the blind action of energy on matter sooner or later results in biology, thereafter, the blind laws of evolution take over, and there is nothing spiritual, nothing magical.

The prehistory of the human species, that is of modern humans, walking, talking and doing business with each other, as we would recognise it today was long indeed, perhaps 150,000 years. Ancient civilisations date from approximately 15,000 years ago until the fall of Rome. The Dark Ages lasted, or dragged on interminably for 1600 years, and finally our modern reason based empire of technology has existed for only 400 years. Instead of it being any kind of a childhood, it was simply a long darkness; we just couldn't find the switch. Our prehistory probably began when we became fully functional grassland hunters and gatherers, and began using tools and communicating, probably in response to climate change and the spread of grassland, our hands were free to do other things, to carry sticks, and bones, and throw stones. A man and a stick is a formidable adversary to any danger found in earth's hostile biosphere; a group of men so armed, even more so. We started to make stone tools, and using the bits and pieces of the animals we killed, hoof, horn, bone, and natural materials such as beeswax, tree resins, leaves, and so on. Hands that are free can also carry stones, and many hands working together can move large stones, or cut down trees, or drag a dead animal back to camp. It is tempting to go into a long list of the many events, both known, inferred, and speculated, of human prehistory, but there are many very excellent books on the subject, see for example, the very readable, *Patterns in Prehistory: Humankind's First Three Million Years*, by Robert F. Wenke and Deborah J. Olszewski. Or the fascinating book by Alice Roberts, *Evolution: The Human Story*, for a complete and up to date account of current thinking and speculation of early humans and human prehistory.

Tool use dates back as far as 2.5 million years, stones, horns, sticks, and whatever else could be found in the environment. Some tools such as the flint hand axe, and the flint spear tip attached to a stick clearly imply complex reasoning on the part of the worker. Tool use also implies intelligence, group activity, language, ability to learn, and a facility to adapt, to improvise, and deal with the environment in ways the other animals could not.

There are only so many things that can be done with a stick or a stone, or a piece of bone, or a Reindeer antler, or anything else you may find lying around. It is not too easy to exhaust all the possibilities. Extending the possibilities of a stone or a stick by combining the properties of both into a useful tool by the addition of a piece of leather, or sinew, or reed, or strip of leaf or bark or some other item is what the intelligence of the tool using animal can do. A hammer, or an axe, or a shovel, or a plough are the inevitable result of this natural process of seeing and doing, and of picking up the stick or stone in the first place.

Tool use has never been confined to humans and our close relatives, far from it; tool use is widespread in the natural world. Monkeys poking with a stick, Chimps using sticks as weapons, Monkeys washing food in the sea, Crows using city traffic to crush nuts, birds dropping shells on rocks to break the shells, Otters using stones to break shellfish. Dolphins using bubbles to trap and confuse small fish. Dam-building beavers, web-building Spiders, Bees and their hives, the hexagonal honey stores, Termites building mini cathedrals, birds hollowing out trees, or building intricate nests. Ants hunting, and farming, considered a unitary super organism. Everywhere we look we see inventive Nature.

All human progress derives from problem solving. All progress of any kind came to us from the solution of problems necessary to the day-to-day survival of the human species. It was one such problem that caused modern humans to stagger out of Africa 100,000 years ago and push north and east and north and west. It was probably a very slow process; one kilometre in one year will carry a species all the way to America within 10,000 years. We have very good data on hominid migration out of Africa; with fossil evidence going back two million years. For our own species we have mitochondrial DNA studies revealing the diversity of all humans alive today and tracing every one of us back to a single female ancestor, *mitochondrial Eve*, more than 200,000 years ago. The massive amount of work that has been carried out over the years has led to a very clear picture of the comparative timings and directions of movement of the diaspora of modern humans, both before leaving Africa and after they migrated out of Africa. People were living in the fertile crescent over 100,000 years ago, predating the oldest civilisations in that region by 93,000 years, these are immense periods of time and it a very reasonable question to ask what was going on? For over 100,000 years, it seems, we simply wandered about the place in small bands scraping a living where we could, doing nothing other than surviving. We can only wonder at what they might have talked about, they were obviously identical to us, with no real discernible differences. If we met them today we could communicate, we could hold a conversation with any of them, they would understand us, and we them. The short answer is that we have no idea what was going on, but the patterns of movement are pretty well established. They had moved as far north as Israel by about 115,000 years ago, skulls found in caves in Levant are dated at 90,000 – 120,000 years old. Southern Europe was inhabited by around 60,000 years ago, the oldest modern humans in Europe date to about 50,000 years ago. Neanderthals were already well established in Europe for hundreds of thousands of years prior to Homo sapiens turning up. The two species met and probably mingled, we have no real idea exactly what happened, but we are pretty certain that the last Neanderthal lived in Gibraltar some 22,000 years ago, and recent genetic studies indicate, the species seems to live on in us. The long trek to Australia by modern humans had taken place by around 50,000 years ago, while the migration across the Mammoth Steppe and the Alaskan land bridge happened by at least 15,000 years ago, resulting in a population of humans in the Americas.

Anthropologists are still debating the exact moment at which early humans acquired language. The soft tissue used in the production of speech does not fossilise, so we have to employ other anatomical features such as the hyoid bone to infer language was in use. We also know from the fossil evidence that hunting large animals was a common occurrence, cooperation in the hunting and in the sharing of the rewards must have occurred, and it is very reasonable to presume the hunters communicated with, and understood, each other. Language seems to have been present in some form at a very early point. To hunt you need others to help. To hunt on the African planes you need to be able to rely on others who understand what they are required to do, and can do it when required. Language was not just a good idea it appears to have been necessary to survival, those who could use language hunted successfully, they successfully killed the gazelle and ate. Those that did not probably starved and died out; were out competed. Language was selected for, it was a useful tool, and the species progressed as a result of having it. Of course it is not as simple as that, and the truth of what happened and how it happened may never be known, but once we had acquired language, we developed and refined it, we used it to evolve a culture of sorts, the beginnings of society and community which, are the real human strength, and the reason for our ultimate survival.

The record for art and culture among the early humans is very sparse, here too only fragments, which may or may not be expressions of art, are found. Carvings in bone and other durable materials, of the female form, for example, have been found at sites in Europe and in Africa, and have been reliably dated to early humans. A flute found in the Vogelherd cave in Germany has been dated to 35,000 years ago. The wonderful paintings and drawings on the walls of the Laseaux caves in South-western France showing images of Auruchs, (wild cattle) and other renderings of animals familiar to the artist have been dated to 32,000 years ago. Among the paintings at the cave complex in the Ardèche region of France discovered in December 1994 by Jean Marie Chauvet and two friends, are many hand prints and outlines of hands including those of children, produced by a kind of airbrush technique, where a mouthful of paint was blown or spat at the hand held against the wall. That the art exists and survives is a wonderful testament to the people who made the drawings, the famous Lion Panel in the Chauvet Cave is an almost perfect rendering of the animal in various orientations made by an artist clearly familiar with the subject. The paintings were documented in the movie, *Cave Of Forgotten Dreams* directed by Werner Herzog 2011. It has to be asked, if there is within humans a propensity towards technology, and there is, and these children were as able as those of today, why did it take a further thirty thousand years or so, to evolve even the rudiments of a technological culture, given they were obviously social and cultural, given they were intelligent, and given the innate properties of the human brain. A father or brother, mother or older sister, we don't know who, saw a way to draw a hand without drawing a hand, it remains a clear example of inventive, imagination, of seeing with the inner eye, and the role of serendipity in discovery.

Apart from the art itself and the undoubted skill displayed, we have the simple fact that the various owners of the hands printed and outlined, could have been brought forward in time to 2016 and placed in a modern school, and could be educated through college and into any of the professions. There is no material difference between the painters of the caves and us. They are us.

Written human history stretches back only as far as about 7,000 years, to clay tablets of the Babylonians and Samarians, or back to knots in bits of rope, or marks on bones or stones, or pieces of wood carved to indicate some or other event whose significance is long lost. Prior to that we must rely on archaeology, and the many discoveries made over the years, and continuing today. The very oldest human settlement containing preserved structures of any real significance is perhaps Göbekli Tepe, in modern Turkey, which takes us back about 14,000 years. Other evidence of human settlement, or rather, traces of transient human existence go back even further. But taken together they are but fragments, and human prehistory stretches back undocumented and unremembered into deep time, and is largely the record of small groups of nomadic people wandering and hunter gathering, and eking out a subsistence living by doing what they could.

From the very earliest of early humans to the cave painters of France, is only a very small leap of technology and imagination, but an enormous gulf of time, well over 100,000 years. So what took us so long to make the leap, if there is indeed a leap? We simply don't know. There can be no romantic consideration of a pleasant existence, in a pastoral setting, the John Jacques Rousseau idea of the noble savage. As was pointed out in the Movie *Wall Street* by the character Budd Fox, *there's no nobility in poverty*. The simple truth of the matter is that Thomas Hobbes was correct, the lives led by the cave painters was cruel, brutish and short. Above all it was short, perhaps mercifully so. They had no technology except for what they found in the environment, no way of recording the things they did, or the experiences they had, or what they knew, other than what they left us on cave walls.

Nothing else of any significance happened until some of the nomads finally gave up their aimless wandering and settled down close to more reliable food supplies. This seems to have happened in the near east, near Jericho, and in other places in what we refer to as the Fertile Crescent places such as Abu Hureyra in modern Syria and Göbekli Tepe in Turkey. Presumably those who built Göbekli Tepe also had some kind of agricultural surplus. Farming also took hold in China, Mexico and the central Andes, and once it took hold agriculture became preferable to the nomadic subsistence way of living; for most humans. So the transition to settlement happened sometime between the painting of the Chauvet Cave and the construction of Göbekli Tepe. And with agriculture came trade, because agriculture produces a surplus and the ability to create wealth. While there may not be any huge explosion in technology simply through farming, trade begets specialisation, as people look for something with which to trade. Farming also gave our ancestors leisure time, time to think, and talk to each other. The creation of a surplus and consequent wealth leads directly to the keeping of accounts, and numbers and writing, and concepts such as ownership of the land used to produce the wealth, and from that concept we derive law, and the inevitable complexities emerging from groups of people living on, and depending on, the same land. They plan, they begin to build for the future.

Someone first discovered metals by noticing that certain rocks found, perhaps in the bed of a river or lying about in a desert were heavier than other rocks, some were better for placing round a fire than others, perhaps more robust, some would ooze a strange substance when in the camp fire, a substance discovered when the ashes of the fire were kicked about. The fact is that we will never know who or when, or how working with metal began, but metals such as copper, lead, gold, and tin were found and used. The refining of metals was probably the third use of fire, behind keeping warm and cooking.

We have to wonder if there were ever, what we might call, a genius among the prehistoric people of the earth, all the way back to two hundred thousand years, and perhaps even before. If they were us, it is entirely reasonable to conclude that genius must have been in the same proportion. Throughout human prehistory, and right the way through into the creation of civilisation, there must have been individuals who possessed genius, as we would recognise and celebrate genius today. The barrier of ignorance was not only susceptible to the constant pressure of the millions of micro innovations, like gas molecules, hitting against it constantly, but also, that every once in a while, massive explosion at the surface, shattering the barrier into shards. There are many examples of such breakthroughs in every aspect of human science, art technology and every other form of human endeavour. It takes that special genius of imagination and intuition to make the leap necessary to see something in nature and apply it as a tool. In all the periods of prehistory, regardless of population size, there must have been those who were capable of seeing the potential, after all there was as much technology lying about the place then as now, it just needed to be seen. A drop of rain on a leaf, or a flower, the optical lens like properties of the drop. A rainbow seen below a cloud, the colours of refracted light. A lightning strike might have resulted in, glass found on a beach or elsewhere. Glass is also found as quartz pebbles in certain river beds, clear as crystal. A metal meteorite, might have been found and examined, utilised, perhaps the connection between the metal and its oxide made; if only in a tentative or intuitive way. Sapphire, diamond and other precious stones must have been found by accident and examined, and the properties noted. Nature too was studied, we know this from the cave paintings, and items from the natural world were deployed, and no doubt studied, thorns, flowers, wood, and leaves, and so on. The shapes of things too must have been noted, shapes and geometry, circles, squares, line and form. An Archimedean screw may be discovered by twisting a slip of leather or plant leaf, or by observing the structure of pine cones, snail shells, and other items found naturally. Carving in wood or bone, during idle moments must also have given rise to serendipitous discoveries. The many manifestations of technology were clearly visible to our cave dwelling ancestors, as was of course, fire and cooking, and machines such as the leaver, the wedge, perhaps even the wheel, and one may well wonder how it took so very long to get to the water wheel and from there to steam power.

In general the observer of nature is the discoverer of utility, and utility can be applied, shown to others, and copied, no genius is required for that, and so it must have happened. Once adapted and absorbed within the culture, and accepted, micro improvements occur naturally, the evolution of an idea takes off, and improvements and variations are made. There must also have been the urge to record, cave paintings, were probably educational as much as art, and of course there was stone carving.

However, without the ability to write down any or all of the discoveries which must have been made, all was lost, and not just the ability to write, but something to write on, and something with which to write, for surely the greatest human invention is the ability to store our collective wisdom, and pass it on. We will never know how many times technology began with a seed planted by such a genius only to fade away or wither and die. For want of numbers, for want of population. How many geniuses over the countless thousands of years? How many lone thinkers afraid or despised, and probably killed for being different? We shall never know. Fortunately we live in a more appreciative world. But you would have to agree, it is reasonable to say that the same proportion of highly intelligent people would have emerged then as now.

It must have been obvious to our cave dwelling ancestors that fire came from wood or coal, or dry dung or grass, and that when fresh wood was put on the fire it burned again and gave out heat, or that additional fuel needed to be added in order to generate heat. Little of nothing changed regarding explanations for the observed phenomena, the Romans clearly knew that oil needed to be added to a lamp or the light would go out. We can only wonder at why it took so long for proper and detailed explanations to come. High mortality rates through disease, pestilence, predation, accident, and suicide, and so on, must also have pressured numbers, resulting in the unexpected and sudden disappearance of skills, or the abrupt absence of leadership to a fledging potential society.

If we consider the famous "Iceman", found in the Alps, a 5000 year old specimen, he was equipped with a bow and quiver of arrows, and tools, and clothes, and assorted other items. He had technology, the technology of his time, to be sure, but he would have had little or no problem understanding ours. Considering the technology he did possess, and the implications as to his understanding of the world and his place in it, and his abilities insofar as they can be judged from the technology he possessed, inferences may reasonably be drawn as to the society to which he belonged. And it is not unreasonable to suppose, based on the evidence that he was in very real terms only about a few hundred years behind us, technologically speaking. Considering all he had, he was not so far behind the ancient Greeks or the Romans at the height of their respective empires. It is also clear he did not make the items of technology himself, the copper in his axe, the yew handle, the flint of his knife and arrowheads, the quiver made of deerskin, and a pouch that hung from his belt made of calfskin, all paint a picture of a gentleman who was a member of a community, and acquired goods and technology produced by other members of that community, or from further afield. It is also very apparent that like all of our forefathers, the Iceman struggled with the burden of everyday personal maintenance, and the sheer indifference of nature, he carried the scars and wounds of untreated conditions along with evidence of previous violence and trauma.

Agriculture naturally led to settlement in larger groups, and these larger settlements, morphed into villages, towns, perhaps even something approaching a city. Göbekli Tepe, is a site in Turkey discovered by the German archaeologist Klaus Schmidt in 1994, it contains impressive 'T shaped' columns intricately carved with various animals and geometric patterns, the site has been reliably dated as having been constructed 14,000 years ago. Egyptian civilisation goes back 7,000 years, China back to 6,000 years, Greece and Rome back 3,000 years.

Our debt to Greece and Rome is a great one indeed. The Romans are, without a doubt, our closest historical neighbours in the ancient world, they almost got to where we are now, or at least they were on their way. But the very simple and stark truth of the matter is this, our modern technological empire of 2016 stretching right around the world, and embracing all of the human population, who wish to take part, is, for all its celebrated achievements, technologically speaking, only about 200 years, in real terms, ahead of where Rome was in 79CE. The Greeks and Romans were a very near miss, they almost had it, and they nearly got it nailed. To give but one example. The Pantheon in Rome is considered by most art critics and architects to be perhaps the most perfect building ever erected by humans, it is an unexampled masterpiece in concrete and stone with the largest free standing concrete dome ever erected. The first Pantheon was built in 27 BCE by Marcus Agrippa, but burned down in the great fire of 80 CE. The Pantheon we see today was built by Hadrian in 127 CE, and has stood as an example of spherical perfection ever since. Nothing quite like it would have been seen in Rome until Bernini constructed the great dome of St Peters in the seventeenth century. A good account of the architectural and building methods of the Romans comes down to us from the *Ten Books on Architecture* by Vitruvius, (c.25 CE), and of course, the works of Plini. And there can be no doubt that the builders of the Pantheon would have fully understood the methods employed today in Rome, and elsewhere to build skyscrapers, bridges and freeways. It is plain to see that the Romans could easily have built St Peters or any other European structure we admire. We are without a doubt the successors of the Roman Empire. Gibbons considered the decline of Rome over numerous volumes of his, *Decline and Fall*, so it is safe to say it is a complex series of events. In effect Rome of the Caesars faded away, it was, of course, a process of fading away, a series of events and cataclysms that led to the subversion, of what were only fragments of what had been Rome, by the fledgling Christian church, resulting in the gradual imposition of a theocracy across Europe, which ushered in a new dark age, every bit as dark and dirty as any previous portion of human prehistory.

The Roman Empire collapsed over a long period into a despotic Catholic theocracy, followed by a gradual emergence from that state via the enlightenment, the evolution of technology and industry, resulting in a globally compelling and globally embraced, one-world mind-set, a consensus *ad idm*, a homogeneity of thought, and of culture. The disparate elements of the one tribe, of fewer than 3,000 individuals who existed on earth 80,000 years ago have come back together once again. And we can say with pride that this is the best we have ever been.

Ancient China, was by all reputable academic accounts, a deeply superstitious, despotic political system based in feudalism, the rule of warlords, and the prospect of imminent death for the slightest transgression. On the face of it then, no possibility of significant progress towards a technologically evolved society. Yet progress there was, and a good deal of it. The significant technology of agriculture, irrigation of land, and harvesting of crops to feed millions. The Chinese expertise in both metalwork and pottery making was unmatched, there was paper, writing and printing, record keeping, and what we might recognise as a passable education system. However, the country was bedevilled by civil wars, faction infighting, and regular bouts of famine. A good account of this remarkable culture is to be found in Ian Morris's book, *Why The West Rules - For Now*. Essentially the society functioned at the whim of a succession of Emperors ruling with absolute power. Most of the accomplishments of the society were achieved with the Emperor in mind, if not in order to specifically please the Emperor, getting it wrong, or incurring imperial displeasure could mean the loss of limb, or of life. Everything happened in a shadow. The Chinese invented an incredible and diverse range of machines, waterwheels, water driven hammers, and bellows, paper, and the highly developed art of bronze casting, to mention only those. But their society was only ever an approximation of true human potential, they were only ever going to get so high before falling back, dictatorship divine or otherwise, would only take them so far. Without those important, enlightenment values they were ultimately going nowhere. Confucius is not sufficient, being of itself imperial. Without the permission to question freely and the permission to create and be celebrated for the achievement, they were only ever going to be a crude approximation. When you work for the Emperor, you make what the Emperor wishes, you never think outside of the box. Confinement, rigidity, and the same as last time, are the order of the day. Despite the technology, the culture was stagnant and incapable of unleashing itself. You have to wonder what the world would be like today had the ancient Chinese discovered and practiced the values of the enlightenment, and sought to disseminate those values throughout the world.

Life in the Dark Ages was probably just about tolerable for most people, including the upper most members of society, but there was hardly a choice, complaining probably resulted in death. The human spirit has a tendency to adopt to whatever conditions it finds, Viktor Frankl's moving account of life in a Nazi death camp given in his book, *Man's Search For Meaning*, is a testament to the human predisposition to cling to hope even in the most hopeless place. When there is no choice, human beings simply plod onward, sometimes hope is all that is left. The medieval mind had the hope of a better life after this one, this world was merely a preparation for the hereafter. No doubt it dulled the very real pain of existence. But yet again their society was a mosaic of feudal domains run by princelings under the patronage of kings, appointed by god, and supported by a corrupt church. The great cathedrals, and the many castles and cities, up and down Europe testify to the skill of the artisans, the books and artworks too show us a skilful and industrious group of individuals every bit as clever as we are today. Kenneth Clark's book, *Civilisation*, gives a good account of their achievements. But the society was the real hindrance, it was simply the wrong structure, the wrong shape.

All such autocratic societies have a Pyramidal structure, top down, single point of failure, the lower ranks constantly bickering and shifting to consolidate their positions and endear themselves to gain favour with those above. Such societies were ruled, necessarily harshly with penal sanctions for disobedience; loss of limb or life itself was to be expected for even the most minor offence. Autocratic structures are also greatly concerned with the enforcement of dogma, the belief system which ultimately gives the structure its power. The King or Emperor is appointed in heaven to rule on earth. Why else would he be in that position? It must be ordained elsewhere. That made perfect sense to the elite members, their continued prosperity depended on it. Having a King appointed by the creator of everything also encouraged a pervasiveness of superstition and magic, which in turn leads to a general closure to new concepts and ideas, especially anything that might upset or unnerve the status quo, or contradict perceived wisdom.

For centuries we have been aware of more ancient cultures and civilisations, and quite rightly the question of their demise was often posed. Ancient settlements which appear on the face of it, at least from their stonework and other artefacts, to have been impressive, eminent, sophisticated, and all the other superlatives, were now simply ruins, and the question was asked. How could they possibly have collapsed? How could they simply vanish without a trace, and all of their technology and achievements be lost? Perhaps rather than simply come to an abrupt end on a single day, by a single event, they simply flagged, lost energy, and faded over many decades, or even centuries, in some cases morphing into something else, somewhere else. No one knew. War was of course put forward as a reason for the collapse of societies, as well as for their formation in the first place. But war alone does not account for every collapse.

The answer, we now know is quite simple, and quite reasonable and straightforward, for most cases. Clearly they were not as elegant, or as grand, or as powerful as we suppose they might have been. Of course it is not politically correct to point the finger at their failure, it also smacks of a racist, even an imperialist position. But the simple fact is that in a primitive world of scattered disparate cultures, some succeed and some fail, either through simple bad luck, or misunderstanding or a host of other reasons. All such societies are at the mercy of the elements, and Nature doesn't actually give a damn. Most of the primitive societies encountered by European travellers simply fell to pieces on contact with the European culture. The interactions were a disaster for more of these less developed cultures than can be named, but perhaps the best example is that of the Inca and their interaction with the Spanish. In effect a Spanish gangster destroyed the entire Inca Empire with a very poorly armed and ill-disciplined gang of tough guys. Although it must be added they were greatly aided by an army of invisible pathogens brought with them from Europe, against which the peoples of North and South America had no defence. Similar tales are told for every primitive society encountered by Europeans up and down Africa, the length and breadth of the Americas, and the islands of the Pacific. In each case the society encountered believed wholeheartedly in magic, the only thing necessary was to demonstrate a stronger version of magic, and the entire society gave way at the seams and fell apart.

If we take the single example of Easter Island, it is very clear that it was, and is, an ecological disaster, the people involved in creating the disaster had no idea what they were doing. The people of Easter Island, both the indigenous first settlers, and those who came later, did the only thing they knew how to do. They existed as best they could. The first settlers had no idea where they were on the world, or that the world was round, they had only their primitive Stone Age attitudes and technology, their culture was one of magic and superstition, and it collapsed, not simply because they consumed much of their surroundings, but also as a result of interactions between a very weak Easter Island culture and the succession of external cultures. There is no real mystery. Following its discovery, Easter Island degenerated into an abused and exploited island, ruined by human habitation, and a succession of uncaring invaders who simply took whatever they could get out of it. However, to present it as a microcosm of our modern world is a mistake, doing so fails, and is an erroneous understanding of what happened to cause the collapse in the first place, and more importantly, a mistaken understanding of what we humans have now built.

Jared Diamond's account of the Easter Island disaster in his book *Collapse, How Societies Choose to Fail or Survive*, is essentially correct, the wholesale exploitation of the indigenous people by a succession of slavers, traders, missionaries, sheep farmers and pirates of all sorts, since the island was discovered, was probably the major contributing cause of the present condition. But Diamond is wrong in comparing our modern well-financed technology based society with a very minor garrison town of a culture such as that of Easter Island at its height. On all levels the comparison fails. The comparison between Easter Island and our modern empire of technology is such a diametrically opposed association that is requires comment. On the face of it, the inhabitants of Easter Island knew only how to chop down trees and how to carve and erect useless stone statues. They knew little else, or at least the facts of the situation would appear to demonstrate that interpretation, and in the very general case of a pre technology civilisation with autocratic, theocratic, or despotic systems of government, Diamond's comparison fails again. Technology is the trump card in our culture. Numbers too, is where we score heavily over everything that has gone before. We have the numbers necessary to make certain that we continue. The larger the society the harder it is to impose a single idea such as theocracy or repression, even if the theocracies of the Middle East, and certain other repressive regimes, seem to be a glaring exception; though other factors are certainly in play in these examples. In general, larger implies more diversity, implies more ideas, implies more questioning, hence numbers are important, in our prehistory numbers were small, populations dispersed, and unknown to each other. Small groups stumbling round in the dark, afraid of neighbours, suspicious of strangers, reluctant to trust anyone beyond the boundary of kinship.

The people of Easter Island had nothing to draw upon, they had no reservoir of solved problems carefully catalogued and stored, accessible by all, they had no knowledge of where they were, or what to do next, and they blundered and stumbled from one statue project to the next for superstitious reasons, known only to themselves. Outside of agriculture, and the ability to feed themselves, they had no sense of the future or of planning, or of management of themselves, and their island resources, they were wholly ignorant and entirely lacking in understanding, of what they possessed, or of what could happen if they destroyed it.

Easter Island was a failure, and it was a false start, it was something that never happened, something that failed to happen, and something that went nowhere, a dead end. Jacob Bronowski was correct in his assessment, and in comparing the inhabitants to hamsters on a wheel in a cage, doomed to endless repetition of the same action over and over, the same statue repeated again and again, the same blank stare, the same pointless exercise. They had no idea where they were, nor how they had arrived on the island, and more importantly they had no idea how to get off of it; they were stuck.

The biggest difference between every single one of the failed attempts of history and our modern world are the set of values that come to us from the enlightenment. The enlightenment started with Francis Bacon in 1620, who was one of those singular geniuses who arise from time to time, he gave us the scientific method, the requirement to establish the facts by examining the evidence, by doing the experiment in a systematic manner to arrive at a conclusion. By the enlightenment is meant, the application of reason, of rational thinking, the requirement for evidence, and the unfettered questioning of everything, without exception, and regardless of who may be upset in the process. Among the heroes of this movement we must count some of the greatest minds of the last millennium, along with Bacon, Descartes, Spinoza, Hume, Kant, Newton, Hooke, and many other lesser known contributors. They ushered in a time of new ideas, in philosophy, writing and science, and a new spirit of questioning, they were smart men who would not be fobbed off by dogma or outrageous claims of magic, even though their god was never far away.

The combination of enlightenment thinking and technology is a very potent combination; it is the accelerant needed to turn the fire of natural human ingenuity into the raging furnace of a technology based global culture. It probably could never happen in any other way. Reason has the ability to destroy dogma and arrogance, the hallmarks of the authoritarian social structure. The enlightenment built upon the earlier work of men like Leonardo Da Vinci and Galileo. Consider perhaps the most important technological innovation in history, the simple telescope as fashioned by Galileo in 1609. This instrument, or tool, or device, as you wish to call it, is a very ordinary invention, simple to make, easy to use, and straightforward to understand, yet profound in its revolutionary consequences, and in where it led the entire human project. Almost instantly this very powerful piece of dual use technology revealed four things that had not been seen before, each of which had profound, and far-reaching implications. The first profound and practical application was the telescope's ability to magnify things and bring distance closer, thus a ship could be identified far off in the distance, fully two hours before it could be identified by the naked eye. This fact alone had obvious and useful implications, both commercial, and an obvious defence application, an enemy approaching unseen was now a thing of the past. Hence such a device could be seen, given the prevailing theocracy of the time, as a gift from god, sent by the almighty to protect and warn the righteous. The other three discoveries quickly made by the telescope caused confusion and consternation among the Catholic clergy of the time, and sent shock waves of doubt through the whole of Europe. Galileo saw a profusion of stars in the sky, everywhere he pointed his instrument, a multitude of stars invisible to even the keenest naked eye, the obvious question arose as to why god, the maker of the heavens, would make things for man that man could never see, why were there so many stars that could never be seen without a magnifying instrument, these stars were not mentioned in any holy book? The discovery of the moons of Jupiter clearly revolving around the great planet evidently showed that not everything revolved in a Platonic circular orbit around the earth, this observed fact was deeply unnerving and disturbing to the medieval clerical mind. The final revelation made by the telescope showed that the Sun was not the perfect crystalline sphere of Aristotle, but a blemished sphere containing dark spots moving and changing on its, hitherto believed to be, perfect surface. These revelations made by the simplest device left the church in an impossible position, here was a very useful instrument clearly inspired by god, yet destroying the very doctrines of the church itself, there was no hiding from the facts. And of course, there were inevitable implications for Galileo too, he was summoned to Rome, evidence was fabricated, he was forced to recant his obstinate opinions, and thereafter, sentenced to house arrest for the remainder of his life. But the cat was out of the bag; the telescope and the optical principles underlying it were so useful they could not be suppressed, and would not go quietly away. And it has been exactly the same, in some way or other, for every bit of scientific discovery and technology, before and since. From the printing press to the nuclear reactor, there is always a dual use, it may be beneficial or dangerous depending on your point of view, but it cannot be un-invented, and it cannot be ignored by those who believe their dogmas and doctrines, and funny clothes, and holy books, give them the moral high ground. The dust and cobwebs of arrogant dogma were forever pushed back by the unfettered pursuit of science and technology, to a point where they now join astrology in that space reserved for magic and fairy tales. Of course, nothing happened over night, the enlightenment was a process, taking a great many years, with its roots in the Renaissance, and the process continues to this day.

Catholic theologians and assorted Christian apologists along with some historians are often quick to point out the fact that the great heroes of the renaissance and the enlightenment, Galileo, Kepler, Copernicus and Giordano Bruno, all the way to Newton, *et al*, were all the product of a Catholic education system, the Catholic Church being the only education provider of the day. They were clearly subject to the same theocratic brainwashing that had been handed down to all in the Church's care since the fall of Rome. They questioned and rebelled in spite of the threat, they were compelled to ask. Why? And were not content to settle for answers that simply did not make sense. Yet fully sixteen hundred years after the Roman Empire peaked, generally accepted to be at around 79CE, when Pompeii was destroyed, the western European enlightenment was only beginning to rediscover the technology and the learning of the Greeks and Romans. As was said before, we are no more than about 200 years ahead of Rome, in real terms, technologically speaking, for which we have god and his church to thank.

The enlightenment value system we espouse today is one that is hard to arrive at, it is a long and difficult road, requiring many attempts and many failures. We have been down that road. We have racked up the failures, and through our efforts we have earned the right to say that we have arrived at the end of the process begun over 200,000 years ago. Because humans are gregarious we form groups, and naturally coalesce into factions and small bands to do things together, hunting, gathering, or putting a roof on a lean-to, or moving a few big stones. But the formation of groups and their cohesion into a society is of itself insufficient unless that society discovers what we now call enlightenment values. Non-enlightenment value societies can be formed and even endure, but they will not survive and prosper, and arrive at the critical mass that makes them too big to fail. There will always be insufficient glue to hold the society together in times of crisis or extreme hardship. So too with partial enlightenment societies, those societies which discover and practice some, but perhaps not all of the enlightenment values, perhaps by setting narrow constraints of basic freedoms, or only allowing liberal values in times of plenty. These societies are simply the wrong shape, they are pyramidal or some other form of triangular, rather than the circular or cloud like shape of an enlightenment society. Theocracy is a retarding force preventing advance beyond that needed to serve itself. The imposition of a set of magic principles, which cannot be questioned with sanctions for those who dare to ask why, is a recipe for eventual collapse. Enlightenment values like elements of the Platonic world must be discovered, they exist regardless, but must be discovered and given effect.

What the ancients actually knew is also a matter of opinion and conjecture on so many levels. We only really have fragments of their lives. We also have a tendency towards the romantic when we consider many of the past failures of human society. If a 'primitive' society- and the word primitive is not meant to be a term of abuse or denigration, rather a term that denotes a society that did not know what we now know, a society absent the knowledge and sophistication we take for granted in our empire of technology- persists in a certain landscape for a long period of time, hundreds of years, for example. It is intuitively obvious that there was some stability within that society, a cohesiveness that maintained the society in that place for that length of time. Clearly and obviously a society will not persist for long periods without the necessary resources and supports required to maintain the society in a stable state. But mere persistence is of itself insufficient. Simply hanging around in the same spot for a long time is not advancement. There is no growth. There is no furtherance of the human experience. There are no solutions to the problem of existence, and the slightest knock can be fatal to the entire edifice. Any external unforeseen event sufficiently damaging can wreck hundreds of years of patient building. Jared Diamond in his book, *Collapse*, details the collapse of many of the societies of the ancient, and not so ancient world, for the most part

Diamond argues as a general rule that societies have a tendency to self-limit when the culture exceed what is termed the sustainable carrying capacity of their environment, which is the capacity of the environment available to the society to sustain the members in the manner to which they have become accustomed. The carrying capacity is related to the consumption by the culture of strategic resources, which in the case of most societies prior to our technological age were, timber, soils or water, and so on, they had no real in-depth knowledge of water, or of carbon cycles. This overuse of the available resources may result in the creation of a kind of positive feedback loop leading to the inevitable collapse of the society and a loss of that culture, and its associated technological achievements, if any. Put simply, when everything is consumed it's gone forever, and collapse ensues, if no alternative is found.

Purple theory, if it can be so presumptuously described, argues that the sustainable carrying capacity of the modern human environment is infinite, and that the human animal exists in an infinite context with no limits imposed upon us from the environment, or by shocks to the system, whether delivered by nature of ourselves. We have reached a point, or perhaps it is more reasonable to say, we will very soon reach a point, or even that we have already past the point, where it is possible to say that our culture, our society, our empire of technology, is simply too big to fail.

But the question persists, what exactly did they know, these ancient societies, scattered about the globe, from Asia, to Africa and the Middle East, Europe and up and down the Americas? Compared to what we reliably know today in 2016, they knew precisely nothing, is the simple answer, but as in many things to give the simple answer is more or less to give no answer. There is a complexity that may be expected from the facts of the situation. But as we are finding out with each year that passes, human prehistory is, if nothing else, complex and interesting.

Primitive Stone Age civilisations that persisted for long periods of time in a certain landscape, Mayans, Aztecs, etc., are often imbued with certain esoteric knowledge. The knowledge of the ancients, Easter Island, the Egyptians, it is supposed by certain observers, that they knew things or were the custodians of secrets we could never hope to know. They had some connection with the world which we have lost, through our embrace of modernity. They were more in tune with the Cosmos than we, more at ease with the world, a better fit to the environment, and so on. But the number of such civilisations that really discovered anything of truly universal importance or application is very few and far between, the Mayans, for example, for all their precision in stone building never discovered the stone arch. The number Pi for example, the square root of two, the theorem of Pythagoras and other universal truths, eluded them, the spectrum of light, the fact that light is a substance, that the world is round, and of a certain size, that there may be other peoples on other parts of the world, and that great ships might be built in order to explore or go visit. And on and on, the list is a very long one indeed. Again it was Bronowski who pointed out, it was the old world discovered the new. The voluminous nonsense and thousands of books written about the Mayan calendar and the presumed calamities of 2012, and so on, speak for themselves, like Van Dannigan before them the facts simply do not bear the interpretation ascribed to them by the adherents of the various theories. The Chinese, *I Ching*, and the ancient Indian texts, or Tibetan Prayer wheels, hold no secret knowledge or ancient wisdom, no great set of answers to anything. They represent nothing more than attempts at explanations. First approximations at understanding the world, and like all first attempts, they lack the clarity and refinement that comes with further work. That is not to say these cultural items have no value, of course they do, they form an important part of the long human plod out of darkness.

It is nice to think they might have had some secret knowledge, some better way of doing things. Perhaps they had some knowledge lost to us that we now need and that we must rediscover, before it's too late. Perhaps there is something they had, and we lack, and that re discovering it will be the key to saving us from the inevitable doom we are, according to the doom mongers, inexorably heading towards. We might even consider there may have been some connection they had with other beings, ancient astronauts, or trans-dimensional, consubstantial spirit beings from some Astral plane, beings that can only be contacted via the shaman with the special knowledge, received through long years meditating in the sacred places of the world, or in ingestion of the sacred herbs, or whatever.

Clearly, in these assertions, as with many others, the burden of proof will always lie with those making the claim. The evidence is simply not there to support any such assertions. The simple truth is that the ancients knew very little in real terms. They figured out how to move large stones. They could use and direct water, irrigate crops, a form of agriculture, sophisticated in some cases. They discovered and learned to utilise certain easily discovered metals, copper, tin, gold, they practised rule of thumb chemistry, had some experiential inkling about physics, knew a little mathematics, and could do a bit of engineering, they had a certain body of knowledge built up over the period of time, and they had something of a technology as we might define it today. But the ancients had no real method of storing and disseminating their knowledge, they used only what they needed to use, knew only what they needed to know. There was no systematic methodological gathering and storing of knowledge for its own sake as we have learned to do. The best attempt from the Ancient World is certainly, Plini the Elder's 37 volume encyclopaedia, *Naturalis Historia*, from the first century, and while a magnificent work of reference on many aspects of the time, it contains little real understanding of the world as it was, that we would consider acceptable by the standards of our time.

The artisans and craftsmen, of the ancient world who built the pyramids, or who built Stonehenge, or the ancient astrolabe found at the bottom of the Mediterranean Sea would have understood the steam engine, they would have understood Newton's laws of motion. Pythagoras and Aristotle would have understood Kepler and Copernicus, had they ever been able to meet, Archimedes certainly would have been able to assist Newton in formulating the calculus. Any of them could have worked for NASA at Ames or JPL, and their kids could have made it to Stanford.

The Romans needed the decimal number system, it would have simplified things greatly, and if they had made the leap they might have solved the major problem caused by their cumbersome methods of calculation, and devised a universal method of writing quantities. If they had captured Archimedes, instead of killing him, they might have put him to work and discovered, *Pi, i, root two, e, 0,* and *-1.* The Greeks were fine mathematicians, but could never uncouple it from superstition, they knew that, *root two,* was irrational and proved it, but the Pythagoreans kept it secret because it confused them, and was in conflict with their superstitions.

Later on, the Arabs did great work in mathematics during their golden age, including the solution of cubic equations. It was all snuffed out by theocratic pronouncements and the imposition of dogmas, effectively declaring mathematics to be the work of the devil. The Catholic Church had similar difficulties with the number zero, or more properly the concept of zero and its implication of nothingness, this too was branded the work of the devil. It must also be considered that infinity too, known to the Greeks and to Archimedes, was a great cause of concern, and remains so to this day, to many of a religious bent. To mathematicians and physicists, infinity is a convenient tool, and is handled with consummate ease in both disciplines, summation of an infinite series is something the Greeks couldn't do.

What was known to primitive peoples, and the very real limitations of their knowledge can be seen in the following description given by a traveller from Portugal in 1586, Lopez Vaz, who gave this account of Peru and its people at that time. '*Peru is full of people well apparelled and of civil behaviour. It hath many mines of gold and more of silver, has also a great store of copper and tin mines, with abundance of saltpetre and brimstone to make gunpowder.*' The saltpetre referred to came from the droppings of sea birds who lived along the coast of Peru, birds such as the grey pelican, and the white-headed gannet, and a host of other clever birds who made their home along the coast and made use of the Peruvian Current flowing north from the Antarctic. The current carries with it a vast density of microorganisms, which in turn provide food for enormous shoals of fish such as herring and anchovies. The birds feed on the fish and deposit their droppings on rocks and cliffs along the coast, and because of the particular climatic conditions pertaining in the region, dry air and warm temperatures, the droppings accumulate and over the centuries come to form vast deposits of what was later called by "guano" in English. The Moche people, local to the area had used the guano as a fertiliser for many centuries, as far back as 400 CE. Guano was also used during the time of the Inca Empire in 1200 CE to provide nourishment to potato crops. But nobody saw its use as a component of gunpowder, not the Moche, not the Quechua, or the Inca. The Europeans saw it as both a fertiliser and as a component in gunpowder and wars were fought over it, and over who controlled the supply.

Gunpowder was invented in China, the Chinese used it mostly to entertain the Emperor, and the like, and to drive away evil spirits. Very little military use was made of it. It was weaponised in Europe. In European hands that was its prime use, and research and development, and invention quickly followed, and cannon, mortar, grenades, and redundant castle defences were the result. As well as an even more impressive method of moving large stones.

The same can also be said for all of the other key inventions that began our era of technology. The very significant problem solved by the introduction of the printing press, and its double edged result, the development of paper, of proper type, typefaces, book binding, the automation of the various processes, Between the printing press and the telescope the technological way forward was now visible to all, for now you could observe what you saw, and report the facts to many. And no one could stop you. Schools and universities had tools, books and libraries.

Common to all primitive societies, the fundamental lack of understanding and desire for explanation led inexorably to the obvious, universal and immemorial invocation of outside agency. The desire to explain the world in which they lived, and the need to find reasons for everything they experienced, from rain to green leaves, to red sunsets, and everything in between led them to the concept of gods. Good gods and bad gods alike evolved or were invented or conjectured from the human imagination to provide crude, first approximation, explanations. Whether the belief system involved one single god or multiplicities of gods is of no matter; the result was the same, the same end point was reached. Gods became the burden, the retardation, the prevention of progress. The weighty yoke upon the shoulders. Once the god hypostasis rears its ugly head all progress in that society is now deferential to that system of belief, dependent on being able to peacefully coexist with the theocracy. The rule is a simple one, autocracy kills innovation and retards progress.

Societies and cultures can endure in a sustainable way for incredibly long period of time, the Neanderthals hunted and lived in Europe for hundreds of thousands of years, simply doing the same thing, hunting, surviving and living day to day. We may never know exactly how they lived, but we learn more every year. Our own prehistory is a testament to the fact that persistence in a state of primitive pre-technology is possible, and can be sustained for as long as environmental conditions allow. The population of modern humans on earth was sustained at about ten million for well over fifty thousand years. One hundred million at the height of the Roman Empire, and has put on an exponential spurt since then, to be now pushing seven billion, and probably heading for ten billion by the middle of the century. We have done well.

The simple truth is that not a single one of these empires of the ancient world had the economic output of a modern mid-sized corporation or a small modern country. They were fragile in the extreme, despite enduring for centuries in some cases, they were vulnerable to all of the hostility of earth's environment. Most importantly, there was no power source in the ancient world, Rome was effectively solar powered, horse and slave, eating grass and grain respectively, bread for the masses, nothing more. There was some harnessing of wind, very small, and not very well understood by the Romans. They also had some facility with water power, the mills at Barbegal, close to Arles in southern France, are a good example, and tell us that much. They also had an implicit understanding of gravity as in the construction of aqueducts to carry water into towns. Their understanding was rudimentary but practical, that of hands-on, engineers, like the later blacksmiths of the Middle Ages or the millwrights of the pre-industrial age, they had a qualitative grasp of how things were put together, and they operated a form of trial and error, and rule of thumb, they were able to get by. Getting by, by getting the thing done, was how they advanced, when it worked it was used, if it didn't it was discarded. There was no explicit understanding, little or no written record that survives of attempts to master the underlying concepts. There was no fundamental grasp of the theoretical, they were poor mathematicians hampered by their overly cumbersome number system.

We might be tempted to employ the old parlour game of counterfactual history where the historical perspective is revised or reviewed, wherein we consider what might have been, if only, and so on, such as what if the dinosaurs had not died out, and had invented technology. What if the Romans had invented gunpowder, or the steam engine, the decimal number system, the printing press? What if they had evolved the concept of the private or the public corporation? It is all too easy to speculate with hindsight about what could been, and what might have happened if the Romans had made the intellectual strides necessary to invent the above, and a few other items we commonly associate with our industrial civilisation. The 'what if' brigade will always want to generate argument and create controversy with the invention of a million different timelines, but it's a bit like the old Irish saying of playing handball against a haystack. Other than historical speculation for academic sake, there seems little point to the game; though one can imagine such games providing mass entertainment value, or sophisticated educational tools as virtual simulations in the far future. The point is that all we have achieved since the building by Thomas Newcomen of the first steam engine in 1712, could have been discovered and invented at any time by any group of people over the previous 150,000 years. Where would we be now, if they had? But they didn't, and we did and, we are where we are, to use that worn out canned expression. But in our far and distant future we will be the era remembered for finally getting it right.

Counterfactual history arguments are probably one of the most disreputable forms of academic work. We need to remind ourselves that one thousand years after the height of the Roman Empire the dark ages raged across Europe, with all their misery and lack of everything, while America languished in the Stone Age. Three hundred years later and half of the entire population of Europe were wiped out by a plague delivered by a flea brought to Europe by a rat who jumped ship. Adding a further three hundred years only gets us to the Renaissance, a vast improvement but still technologically not much better. In fact, the perspective, and keenly observed figures of a Pompeii wall painting were, in their own way every bit as accomplished as anything produced by Leonardo, Michelangelo, or Caravaggio. We may be certain the Romans at the Height of the most successful of the old world political systems, never thought that after fifteen hundred years they would have advanced barely at all. The frighteningly simple fact is, and needs to be said again, we are only about two hundred years ahead of the Romans in real technological terms. They certainly would have understood the steam engine, the decimal number system, the printing press, gunpowder, and metal ships would have made perfect sense to the Roman admirals. The skill the Romans had with glass is absolutely amazing, their ability to manufacture and to decorate, carve and manipulate it as a material, yet never a lens, never a curved optical mirror, never a prism, never a test tube, or a bell jar. Never an investigation of light. That we know of, yet the spectrum was well known, the edge of every glass bowel is a prism, diffraction occurs all the time, even as simple as a stick in a glass of water. No one ever seemed to ask, that we know of, there seemed to be little or no interest. The machinery used in Rome to build the aqueducts and the Pantheon, all Newtonian mechanics without Newton or his laws, they understood the leaver, the wedge and the inclined plane. It is very reasonable to ask why the Romans did not invent the steam engine, they clearly could have used it, a rail network in the administration of the empire, and so on, and it seems simply to be the case that their focus was elsewhere.

Many people, for a variety of reasons seem to find any discussion of the Middle Ages to be depressing, the architecture, the great cathedrals, the castles, the great works of art, the poetry, Dante et al, all very admirably described by Kenneth Clark in his masterpiece of the 1970s, *Civilisation*, but they are always overshadowed by a sense of terrible loss. Rather than the joy of celebration of what it was or what these accomplishments undoubtedly are, there is an overwhelming feeling that it was all so much of an unnecessary diversion of almost two thousand years. There is nothing there that the Romans could not have produced and indeed did produce and better. And even the Romans could be seen as an unnecessary diversion, and so we could continue, all the way back to the people who lived on the planes of Africa and decided one day to move north.

All of these societies and civilisations ultimately failed because they were solar powered in the old way, in the least efficient manner. Everything that required power, prior to the steam engine, derived its power from the recent sunshine of the current or very recent past. Leaves and grains fed to people and animals, turned into work. The steam engine is the beginning of real technology. It introduced the concept of real power. Mathew Bolton told James Boswell in or around 1790, *we sell here, Sir, what all the world desires to have: - Power.* Anyone who has ever broke wind or cleared a stubborn blockage from a nasal passage with a push of air from behind, would instinctively be aware of the power of compressed gas, and the action of a piston in a cylinder. That is, they would know that if gas or vapour is trapped or confined then released it has a certain force attached to it. Much research effort has gone into attempting to find some origin to the steam engine, and in particular why it took so long to arrive among us. After all, the Romans had problems that would have benefited from the steam engine. The type of machinery needed was common in Mills, operated by the Romans, and had not changed even up to the time they were described and drawn by Leonardo Da Vinci. But it was the problem of a flooded mine, and an aristocratic family bound in fee tail and male primogenitor that gave rise to the steam engine. The Earl of Dudley contracted Thomas Newcomen and John Calley to construct an engine with the express purpose of pumping water from his Lordship's mines. The atmospheric engine built by Newcomen from earlier plans of a pumping machine that appeared in a book, *The Miner's Friend,* published by Thomas Savery in 1702, was so hopelessly inefficient as an engine, that it practically required the output of the coal mine to keep it running. But it worked. The principle was proved. By the end of the century, Mathew Bolton and his partner James Watt had improved the efficiency, and were shipping enormous engines all over the empire from their company in Birmingham. Richard Trevithick was experimenting with high pressure steam and had built the first locomotive by 1801. The steam age and the developments that surrounded the effective beginning of the age of technology are very ably and fondly discussed in William Rosen's book, *The Most Powerful Idea in the World.* And those developments were not confined to steam, the floodgates were opened, and all boats were lifted.

All usable power, even to the present day, is effectively generated by steam engine, or by water wheels, or by wind mill, there is no other way. Water is boiled to steam, and the steam is used to drive a turbine. The heat source is oil, coal, wood, garbage, or nuclear. We are still living in the 1790's. We seem to have a solid state, no moving parts alternative, for everything, but not for power generation, the solid state alternative of solar cells, the digital method of producing power, has not yet proved itself to be as effective as the older methods, either on cost or efficiency. But it will, because it must.

The steam engine was the beginning of technology, proper technology, power driven technology. Power was applied to the world, any amount required, mills were suddenly unleashed by the addition of power, no longer confined to the banks of undependable rivers or invariant wind. A thousand mills listed in the doomsday book, and the millers and blacksmiths, who together kept thee mills running were every bit the equivalent of our technologists and engineers. Overnight, it must have seemed, steam gave power to these mills, and power to pump water, and to move large stones. The mills of the doomsday book morphed into the Mills of the new moneyed industrial masters, like Titus Salt who built an enormous mill in Yorkshire to work Alpaca. Steam gave us the industrial revolution, mills to work wool, cotton, and silk worked on the Jacard loom, the precursor of the digital computer. Iron could be worked as never before, steam hammers beat it into metal ships and locomotives. Coal also gave us new materials, and a new understanding of the underlying structure of the world, suddenly vast chemical works were built, sulphuric acid was deployed, and the chemical revolution, which gave us dye stuffs, and a million other useful chemical products was a part of the landscape. The power of technology and the products of industry produced enormous wealth, which became its own motivation, perpetuating the industrialisation of the world. The external combustion steam engine gave way to the internal combustion engine to produce even more power to apply to the problems of the world, and generate even more wealth, and even more immutable scientific principles. And all because we humans liberated ourselves, and dragged ourselves, inch by laborious inch, out from under the weight of theocracy and autocratic injunctions.

Hindsight, yields great clarity, and perhaps looking back the boundaries needed to be tested, the perimeters of the envelope needed to be pushed against. It is probably true to say that it was necessary for, western democracy, and humanity, to pass through the First World War, the Second World War, and the Cold War in order to be where we are today. And we are not finished yet, there remain the ongoing conflicts between faith and reason, between the pressing need for technological advance, and the narrow parameters of sustainability. The struggle between those who want a future, and those who see no future, because no matter what we do, Jesus is coming back soon, and that's it. The common outcome of all of these monumental conflicts has been a hands down triumph of the western enlightenment based democratic model. In effect the model has taken on all comers and prevailed. It would appear, looking at the rest of the world today, any system stifling individual freedom is ultimately doomed to vacillate between one form of disruption and another, and to generally stagnate. The Second World War was a battle of ideologies, a battle of moral positions, of despotism and lunacy over moderation, pluralism, and the rule of law. The cold war was a battle of the surpluses, to see which system could produce the biggest surplus and therefore have excess to waste on extravagant technological endeavours. America and Europe went to the Moon, mapped Venus, photographed all of the planets out to Neptune, and landed probe after probe on Mars. The USSR and China created Sputnik, put the first man in space, after that, they effectively stopped. Almost everything else they did ended in failure. Soyuz, for example, was their biggest success, and with it they set endurance records for humans in near zero gravity, though in reality it was little more than a tin can locked in a low earth orbit. Every Russian trip to Mars failed, and they didn't dare try for the Moon. Finally, *Star Wars*, the race to build a missile shield bankrupted them, and the western capitalist model won the war. The authoritarian regimes of the East were left with no alternative but to do, as Ronald Reagan said, and tear down the wall; and we laughed at their funny, dirty cars, and their shoddy consumer goods.

Long before confused Islamists, or Tamil Tigers, Japanese Kamikaze pilots engaged in the disturbing act of suicide bombing, and in the process struck terror into the minds of Allied sailors during the Second World War. On the eve of their fatal missions it was the custom that pilots wrote letters home, suicide notes, as it were. Many of the young suicide pilots were trained in the city of Minamikyushu, which recently sent 300 suicide notes to Unesco's Memory of the World Register programme in Paris; a project begun in 1992 to preserve the memory of the world for future generations. Takeichi Kawatoko, a spokesman for the city remarked, "We want to remind people of the horrors of war," they hoped the notes could have been listed in 2015 ahead of the 70th anniversary of the war. And as might be expected there is consternation in China where the proposal was derided as being the equivalent of elevating Adolf Hitler's *Mein Kampf* to the status of a significant work of political philosophy. They failed however, in the home leg of the selection procedure.

The simple fact of the matter is that these kids, for that is what most of them were, school kids, and college kids, had little or no choice in the matter, 4000 of them were for the most part drafted into the military from schools and colleges, and were effectively strapped to a bomb, and ordered to die for the Emperor. It could, of course, be argued that the cult of the Kamikaze was every bit as insane as any other cult of worship, which results in human beings attaching themselves to explosive devices in order to cause destruction, thereby proving some or other point. But then the entire cult of Japanese militarism, and emperor worship as practiced in early 20th century Japan was, by modern 2016 standards, pure bonkers. However, some of the insights in these letters home get to the humanity of the people involved, one letter in particular contains a phrase which cuts right to the heart of the problem of all despotic societies, complaining of the system that was sending him out to die, a young pilot wrote, *Authoritarianism is like building with broken stones.* And for that reason alone perhaps, these documents, and the human beings who wrote them deserve to be remembered for more than just destruction. They are, perhaps, a not inconsequential portion of the human struggle against the autocracy that created them in the first place.

We have a duty to make progress, to achieve our true potential, whatever that might be. And that duty is to our forbearers, to our predecessors, to all those who stand behind every one of us, right back to the inhabitants of the caves. And back to the hunters of the African Planes, the squatters round the camp fire frightened by the dark, the squalid disease ridden vermin infested poor souls of our prehistory who stumbled through time to allow us to be here today. What we do, we do for all of them. We also have a duty to leave something behind, something for those who follow, not to wreck everything, or destroy their inheritance, not to leave ourselves open to liability. Because no one can say for sure how long anyone of us may live. Biology is forever busy, the molecules of endeavour impacting against the resistant barriers of ignorance, slowly but inexorably exerting a pressure, pushing the barrier back into the darkness. We who are alive today may well, if biology succeeds, be present in, and part of, the deep far future, and we may well be asked by our fellow humans to give a detailed account of our stewardship.

Why know well why and how civilisations and cultures of the past collapsed, there are many reasons, but mostly in the past it was because of ignorance and inability. Insufficient size of population, inability to produce a surplus of production. Insufficient knowledge of the world, and our place in it, inability to capture knowledge, or retain any knowledge they did have, brain storage rather than on paper. Inadequate social structure, simply the wrong shape of society, ever changing versions of the pyramid, nothing circular, or cloud-like. Inadequate group structures, a failure to stretch out beyond the narrow bonds of family, kin and clan. Inability to forecast, to look into the future, and reliably predict what might occur, and to record the predictions for testing. Reliance on magic, as an explanation for phenomena and as a solution to problems. Inadequate understanding, or misunderstanding, or misinterpretation of what was observed, with no reliable way to verify what they did observe. It is very difficult to explain something to someone without the aid of pen and paper, or without the ability to draw or render the complex notion for them to see, thirty thousand years ago people were trying to do this on cave walls, perhaps on tree bark, in the dirt on the ground, or the ash of the fireside. We have no idea other than what they left us. But it is now clear to us that they were on the right road, at least some of them were trying.

We are here, and we exist. Now. But we were here and we existed 15,000 years ago, 35,000 years ago, and 70,000 years ago. And for all those long millennia of our prehistory people were neither happy nor content, they couldn't have been, they were too busy with the struggle to exist; for most of our prehistory humanity was simply in a holding pattern. Every bit of technological understanding and ability we possess as a species has been discovered in the last 700 years. What took us so long to get here if technology and advancement born of human curiosity are so inevitable? Perhaps our technological adventure is merely a fluke, perhaps it is an accident, and perhaps our natural state lies in what we did, and how we lived during our interminable prehistory? We only have the one example of this level of applied intelligence, this is as far along the road as we have ever been, or any other species has ever been. One thing is for sure, if the entire world and everything we humans have built was to come to an end tomorrow, we could say with certainty and conviction, that we lived in the best time there ever was. Humans are evolution's greatest survivors, we are the prosperous many, and we live in the golden age of everything.

If we compare ourselves in 2016 to those hunter gatherers who crossed the land bridge into Alaska around 15,000 years ago, they had no idea as to the bounty of their new land, lying about under their feet. They had no knowledge, no information, no ability to unleash the infinite wealth that surrounded them, instead they simply did what they knew how to do. They made stone spears, and set about wiping out the mammoths and other assorted mega fauna; at least the evidence suggests they did.

We live in the age of homogeneity, of cultural homogeneity, of musical, homogeneity, of technological and artistic, homogeneity, social cohesion and alignment, and so on. It is as if the pot, having been stirred vigorously has been allowed to settle itself. This settling out is a process, still underway today, but happening and ongoing, giving us a final contented mix. Tinkering will, of course, continue, and is part of what we are, we tinker and play with our creation. We cannot keep our hands off of it.

Our system is an unashamedly capitalist model of market forces and corporate takeovers, property bubbles and banking collapses. But it's the best we have, by a long way, no other system requires or rewards freedom in the same way. A capitalist system requires that the goods produced be consumed, it requires the broadest base of consumers, slaves do not consume other than what they are given by their masters. Thomas Jefferson wrestled with the complex issues arising out of the concept of a slave owning democracy, he likened it to having a wolf by the ears.

At this present time, 95% of all the scientists who have ever lived on earth are alive and working today, solving problems, collecting specimens, or sitting in offices with feet up, contemplating the mysteries of the universe. It should also be noted that right now, 80% of all the engineers who ever lived are alive today, plying their trade making things work, solving problems as they go, and in command of every little bit of engineering knowledge that has ever been derived through previous hard work.

The history of the human race for all its blood and gore is nevertheless a record of our undoubted progress from skins to suits. Without our advanced technology, humans are little more than tool using animals, progress can only come through dedicated and continuous technological advance. Progress, of necessity, must be allowed to be made questions must be allowed to be asked. The concepts that new is necessary, that different is acceptable, that change is to be embraced, all need to be part of the fabric of our world. Most practicing human beings find the concept of enlightenment values to be a correct, meaningful and pleasing prospect, they find them to be altogether reasonable and agreeable. Most people on objective reflection allow themselves to be governed within such a society, that is, their consent is fully informed and freely given. It is this simple set of facts that allow us to make progress, as opposed to religion, or magic, or superstition, all of which lead only to retardation and stagnation, and ultimately to collapse. That is why we will succeed in what we are doing, that is why it will work this time, and getting out from under that burden must be, and is, what took us so long.

Chapter 2   Why Humans? Why Purple?

*There is no such thing as consensus science. If it's consensus, it isn't
science. If it's science it isn't consensus. Period.*

Michael Crichton 2003

Why humans? What makes us so special that we are at the head of affairs on
this planet? This is one of those non questions that makes linguistic sense
but has no validity other than in a philosophical context. Philosophers and
religious practitioners of all shades have troubled themselves with this and
other non-questions for thousands of years. They need not have bothered.
The question is answered by the self-evident facts of our situation. We
humans were lucky enough to emerge in an unlikely setting from a
sequence of improbable accidents, and because in our case intelligence, the
only asset and advantage we possess, has been better than anything our
predators or competitors could bring against us. We are here because we
have won the prize, the ultimate prize of mastery over everything that has
ever existed. The dinosaurs ruled earth for 160 million years, and prior to
them were hundreds of millions of years of nature spitting out new and
diverse forms for all possible environments. But there has never been
anything like humans. We are the universal constructor, and Nature's best
attempt so far at understanding itself. And our reward is the vast future
wealth of a Purple world which we are in the process of building through
our understanding of science and technology.

We owe a tremendous debt to the past, not just as a species or a
society, but individually, each of us is here because of all those who stand
behind us in time, all the way back to the trees. It's a common truism to say
there's no free lunch, there's rarely something for nothing, and the world
doesn't owe you a living, and so on. But we are all the beneficiaries of the
hard work and effort of the millions who went to their graves before us. The
millions who suffered and lived in squalor, or slaved in early mines and
factories, or were burned at the stake for having a good idea. To be born into
a world where all the heavy lifting has almost all been done, is a great and
largely unappreciated relief. The roads are all built, the maps are all made,
and we can go wherever we choose. We live in a world where the
expectation of long life and extravagant luxury have become ordinary

things, we can feast each and every day, like the kings of old, we know everything that happens as it happens no matter where it happens. Each of us has the wealth of a nobleman to command, it is the very definition of a free lunch, something for nothing, a living from the world.

Neither philosophy nor religion could ever, singly or together, have brought about the modern world. Theocracy in all its forms contains, at its very soul, the primary commandment, *Thou shall not ask why,* and this has burdened us since the fall of Rome. We humans have, in the interim evolved a scientific civilisation, a knowledge and information-based civilisation, in spite of the retarding effects of the narrow-minded theocrats. Following 1,500 years of theocracy, there is no longer a prohibition on asking the eternal question. Why? We no longer live in a world of acceptance. We are no longer supine and compliant. We are demanding, assertive, and above all questioning. We no longer fear sanction for dissent or punishment for demanding assertions be substantiated by the provision of evidence. The stopper is well and truly out of the bottle. This liberation above all else, is what separates us from what has gone before. No more Galileo under house arrest, or Giordano Bruno, burned at the stake, or Montaigne's self-confinement in his tower, or Priestly burned out of his home, and forced to flee the mob. Never again will a scholar the magnitude of Charles Darwin be left sitting on his *magnum opus* terrified of what might befall his family if he dared to publish. Christian apologists often argue that Newton, and many other famous scientists were in fact strong bible believing Christians to a man, and outwardly they were, but the assertion is well rebutted by the fact the last witch was burned in England in 1682, and witchcraft as an offence punishable by death only dropped in the Witchcraft Act of 1735. The thing speaks for itself. We human citizens of the world in 2016, live in a world without chains. The fear is gone. The weight is lifted. Nothing can stop us now.

The difference between us and those who went before is simply that we have technology. We have technology and it works. Not only does it work, it works wonderfully well. We also have a well-earned and well-deserved, bordering on arrogant, attitude towards those who would tell us that we cannot do this, or we cannot do that. We have a nose thumbing, go to hell, kind of approach to those who try to sell us a rancid dead dog. We don't know everything, but we know enough to be able to smell a rat. We have an attitude that comes with, and directly derives from, accomplishment, we won't be talked down to like that, we won't be fobbed off with nonsense, and we don't believe it just because someone says it's so.

That is the difference between those in the past, ruled by kings, and those in the present who consent to be governed in a republican democracy. Even those modern dynamic countries with constitutional monarchies, one thinks of Holland, Britain, Spain, Denmark, Japan, and one or two others, are republics in everything that matters. The royalty aspect, it seems is merely kept as an historical curiosity, the retention of tradition, the vindication of current state structures, and is not meant to be taken seriously. Royalty are retained as a public relations extension, to be rolled out when a visiting head of state comes to call, and as a tourist attraction. In a sense royalty in western democracies are kept as public pets, pampered and paraded, to be viewed or used as required, the exercise of function only, dispensing honours to worthy citizens, but devoid of any real power. The most objectionable intrusion of their function is the notion that somehow they were appointed by an external entity, some god or other. This 'born to reign over us', predestination nonsense is the very definition of magic, and should reasonably therefore, be repugnant to any right thinking objective human. It also has the very real effect of dividing a culture into those who swear loyalty to a monarch, and those who do not, mostly this is nothing of concern in modern states, but occasionally, Northern Ireland being one of the best examples; it causes immense and lasting harm. In other places around the world, there is still the vestige of real power embedded in a royal personage, Saudi Arabia is a good example of this, but frankly, it occurs nowhere of any great cultural, strategic or military importance. Of course it can be argued that all of our preceding examples of growth, progress and stability were monarchies of one kind or another, Ancient Rome, Greece, Cartage, Persia, were all ruled by kings or emperors, one man knows best, the big man, the strong man, a woman sometimes but rare, the pyramid structure, with everything flowing from the top down. They all had one thing in common, they were all inherently unstable, inherently unworkable, they were weak, and nearly always fractured on the death of a monarch. They were greatly oppressive to the base of the pyramid, and incapable of withstanding severe external shocks. They had no powered technology, they never got as far as Newcomen's engine. They were incapable of ever getting to a state of Purple. In short autocracy and benevolent dictatorship, simply don't work.

Why Purple is something that needs to be answered, and why use this colour to describe the final state of human society? Because this final human state will be a state where imperial levels of luxury and extravagance pervade. Not just as we may claim today that even the poorest among us lives a better life than a king of old, Purple is the promise that at some point in the future we will be able to look back and say that even the poorest lives like a billionaire of the twenty first century, that's how much wealth each human citizen of a Purple world will command. We will have it because there will be no limits on our ability to manipulate matter, and no shortage of energy to make our dreams happen. The information deficit that holds us back at this point in time will be closed making the impossible possible, and the wildest dream commonplace. Purple is not wishful thinking, neither is it an unsupported opinion, further it is not a branch of the scientism complained of by Raymond Tallis and others, the mistaken belief that science solves every problem; we know that to be a fallacy. Purple is an extrapolation based on what has gone before, with all due cognisance to the principle of extrapolation, and its many false promises, and previous errors. However, the facts from the recent human past are indisputable, as well as the facts uncovered about the deep past by our science. The Sun has been around for five billion years, earth for four, and life as we define it, about 3.5 billion, humans for about 2 million, technology for about 10,000 years, and our modern technological age for exactly 300 years. Since the blacksmith, Thomas Newcomen, first used fire to drive an engine in 1712. Since then we have advanced rapidly, any point chosen in the last 300 years as a base line is seen to produce progress in any interval of years following the selection. Sometimes technology leads science, mostly the technology emerges from the scientific work. Along with advances in technology has come understanding; Newcomen had no idea of the principles underlying his engine, no one at that time understood the law of conservation of energy, or the efficiency of engines. In exactly the same way, taking today as a base and adding our understanding, not only of science and technology, but also of the problem set remaining it is clear that we will either progress quickly into an era of benevolence and plenty for all humans, or march backward, and possibly even enter some new dark age.

If the work of Leonardo and the later work of Newton, Hooke, and Franklin, et al was our technological childhood, then we may now be in the tumults of our teenage years. We certainly have not yet reached any kind of maturity, there are growing pains, and various phases need to be endured. The question arises as to whether our technological civilisation is yet another false start as pessimist writers such as John Gray, and others would have us believe. It is a very real question and of course it cannot be ruled out as a possibility. We might be heading down the road to perdition, we might be knocking on the door. However, it is not at all clear what could possibly cause the edifice to collapse, or why we would allow that to happen, and not be able to mount a respectable salvage operation. There is certainly nothing and no one on the planet capable of mounting a serious challenge to the western powerbase. Russia can be ruled out, Putin ought not to be there for very much longer, and the Russian people are as western as we are. Not China, although the communist party runs things now, they are instinctively western and non-aggressive, non-imperial capitalists at heart. In fact, both Russia and China are now so firmly in the western camp, they are arguably making a meaningful contribution to the project. When they look forward, they see increasing prosperity, when they look backward they see the abject failure of the communist inspired missed opportunity. Like all reasonable people on the planet they know where the better future lies.

We can guarantee that we will continue to advance and make progress, what is certain is that stagnation in our present state is simply not an option, and is not going to happen. If the Greens were ever right about anything it is that we cannot consume at present levels with the technology base we possess. We need a great deal more technology, and an increased rate of progress. If there's not enough pie then we make a bigger pie. We automate the wealth creation process. We have no real choice, we either make the progress we need to make, or give up. And, if we make that progress, given the current base of knowledge there is only one place we can find ourselves; in a Purple world of infinite wealth.

Anyone who considers the matter with any kind of objective rational sincerity would be bound to arrive at the conclusion that science has imposed upon us a set of irrefutable facts, a list of unarguable certainties, and an unprecedented collection of final answers. Taken together these represent an indisputable body of factual evidence, which are both persistent and permanent aspects of the human record. And because of this understanding, there exists, among rational thinking humans of this world, a vast and growing group of optimists. People who look around the world and consider with great clarity of thought that on the whole the glass is indeed half full. These people use terms such as optimism, hope, abundance, and infinity when they refer to the future prospects of both humans and the planet. Such optimism is grounded in the facts of our four dimensional reality, and in the facts and achievements documented in our human history, and in the progress we have made in understanding our reality, and our place within it. This diverse collection of likeminded individuals have many things in common, they believe, for instance, the future will be better than the past, because we know how to solve problems. They believe hope should be more prominent in our thinking than despair, they believe our best days as a species are in front of us, not in the past. These beliefs are not the irrational beliefs of religious subscribers who believe simply because, the book says so, or because the strong man says so, or because they are told what to believe, and are dissuaded from the practice of questioning or from independent inquiry. No. The beliefs of these rational optimistic thinkers are rooted firmly in the three physical dimensions, and in the fourth dimension of time, in the facts of our reality, and in the proven results of our technological expertise and mastery, through understanding of the world we inhabit. The belief of general betterment is based on real physical evidence collected and tested over the past five centuries, and formed into a body of indisputable fact that points in only one direction, that of an optimistic future for the world, and for people. It is the proposition of this book, or one of them at least, that people who think as described above and use terms such as abundance, hope, optimism, extravagance, luxury, opulence, and so on, when discussing the future are in fact the foundation of a Purple movement. It is the very essence of the meaning this book attempts to convey through the colour Purple. A Purple future will always be taken to mean a future of luxury, plenty, wealth, and abundance. A Purple world will always be taken to mean a world where extreme wealth, abundance, and all that goes with it are the common base expectations of all citizens. We are essentially a finite number of creatures surrounded by the infinities of all we desire, the future could not be otherwise than one of abundant extravagance. Purple is therefore an apt label to attach to a movement of optimistic and hopeful people who have arrived at their optimistic and hopeful outlook by the process of rational thought, and by understanding the world as it has evolved thus far. Technology is the mechanism that allows us to become optimists in the first place, it is technology, and technology alone, that allows for a population of six, going on eight, and possibly ten, billion people to occupy a planet of this size, and to do so with more than enough of everything to go round. And, while we are not there yet, we are very firmly on the road; we are well on the way to unlocking all of the remaining secrets of nature and applying them to create our Purple future of infinite wealth and extravagance. We know that it can be done, because nothing proposed in order to achieve this infinite future is beyond the laws of physics, or the laws of nature, And if it is within the laws of physics than it can be achieved. It is simply beyond our present capacity, outside of the bounds of our current stash of information. We are deficient, we have an information deficit, we have lots of questions to which the answer is, 'we don't know', or 'we have no idea how to do it'. Not as many as we had last year, and we will have fewer again next year, it's still a work in progress.

Purple is an answer to many questions; such as can our technology lead us into a new era of hope, and plenty? Are humans inexorably heading towards disaster? Does it really matter? Purple is a nutshell encapsulating the idea that knowledge and information, allows humans to access and manipulate infinities of energy and matter our forbearers could only dream of and consequently generate all the wealth we could ever need. Purple is a licence to prosper, to consume as never before to squander and party because the future pays for everything. The cheque is in the mail. Purple is the colour of luxury and excess, the colour of unashamed consumption, and guilt free extravagance. Purple is a book of the future, the colour of the future, the colour of success. Purple is a destination; a shiny hilltop in a promised land, but accessible, because humans can solve problems, and create a glorious, wealth-dripping, technological future, the ultimate humanist wet dream. Purple is a story everyone wants to hear, a story of long life, wealth, health, and endless prosperity in perpetuity; who could refuse? Purple is a celebration of the fact, that we are the only technology wielding species in the known universe. Purple is about why we should bend a knee to the memory of those who suffered before us, and give thanks for toilet paper, cotton underwear, running water, transport, processed food, and all the other goodies we ungratefully take for granted. Purple is why Malthus was, and continues to be, wrong, and why pessimism answers nothing. Purple is the ultimate crapshoot; the ultimate set of hopes our dreams, consumption without consequence, greed without guilt, abundance and plenty, without poverty. Purple is the new Green. Green belongs in the same empty sack as religion and Marxist ideology. Purple is the future version of Green. Purple is the new better Green. Purple is not self-indulgent or sentimental. Prediction of future technology is easy, the principle at least is easy, if nature can do it; we can too. Or at least we won't be far behind. Purple is not a romantic notion that technology represents a panacea to solve all problems, it is neither whimsical nor a technological dream, the approach is not ideological but practical. Purple is a tag, a label, and a carefully argued case in favour of the human species and its glorious technological future. Purple is a rallying point, a movement of *Purple People* all loving technology. Purple is the reason why technology should be embraced. Loved and admired, not feared, hated and despised. Purple explains why we need more technology and not less, why we need more scientists not less, why we need to stoke the fire, and fan the flames. Purple invites an inflation of ideas, into a new technological enlightenment. Purple is a promise, it's a best-case scenario, of a technological trajectory pointing towards the ultimate form of our human world, of effective individual human immortality, and infinite wealth. Purple is the new sustainability, sustainability taken to the extremes of the envelope. Purple is a declaration, that we can have everything because everything exists. Purple is the inevitable destiny of the human species, our final resting place. Purple is a world saturated with everything humans could ever want, and getting better every year, a world where we know everything, and can do anything. Purple is about how technology changes people, and the future, not the technology of the future, but how technology *is* the future, the only future. The human project has reached the point that without technology there is no future; there are so many individual humans that technology is the only way to offer them a reasonable, agreeable existence. Purple is also about human consent, and how far our consent should go, and what we must demand in return for our consent. Purple is about addressing any lack of confidence we might see, because we have no right to lack confidence either in ourselves, or in the future, or in our technology. Not only do we not have anything to worry about, we have no business worrying about anything. Purple is not the promise of a new age, nor a new version of some technological utopia rather it is an inevitable and foreseeable result of technological improvement, and the laying down of technological layers, the evolution of human knowledge by the achievement of technological increments. It is only now that we can see so very much further than before. Our problems seem easier to solve, or at least can be described and quantified in such detail as to make their solution only a matter of time and effort. Purple is about human entitlement. Purple is about where we are now as a species, how we got here, where we're going and what might happen if and when we finally get there. It is an optimistic thesis on the human future. Purple is about the pursuit of happiness, which is after all a right, or at least the aspirational birth right of all Americans. Purple is a different way. The way forward. The embracing of technology. The recognition that as the top of the pyramid we have in some important respects placed ourselves beyond the reach of nature. We have no predators, except for the occasional accident; parasites are kept in check, disease is understood. Life is the only game in town. As far as we know there is no other life anywhere beyond earth. Purple is about not accepting things as they are now, it's also about applying pressure to those in power to get the big stuff done.

Why Humans? Why Purple?

The Human race is, above all, a going concern. People must take precedence, human progress cannot be stopped unless we do it ourselves. Modern western society has effectively side-lined all regressive and restrictive forces; science and technology has liberated humans from the ternary of theocratic and despotic rule, and allowed us to come into our own. To stand ourselves up to our full height. All technology has consequences, new technology will always have new and different consequences, unforeseen consequences. But so what? We humans are adaptable, *"the shapers of the landscape"*, as Jacob Bronowski put it. We have nothing to fear. If this sounds arrogant, it's because it is arrogant, because we have a right to be arrogant, Nature made us arrogant, we are Nature's best effort, the result of billions of years of work. Of all the other species of early humans, and proto humans, and other tool using apes we alone made it this far. We alone passed Nature's sternest test. We deserve to be where we are today. And we deserve to continue on towards the fulfilment of Nature's ultimate goal of understanding itself. Purple points to a new human era when the words of Protagoras, *man is the measure of all things,* come into their own. Macmillian's avowal of, *never having it so good,* is the everyday standard. The ultimate goal of technology is that of liberating humans from the drudgery of work, that has always been its primary function; it's why we invented it. We solve problems, to make those problems go away, or to lessen the sting. Hence the ongoing replacement of humans by machines is a natural and desired consequence of our technology; one of the primary reasons for its continued development. Clearly as technology continues to evolve, AI becomes incorporated, and we liberate more and more layers of human society from the drudgery of repetitive or unsatisfying work, leaving them free to be artists, or whatever they wish to be, sport, art science, a society of crafts people, or simply doing nothing at all. Perhaps very poor artists, but at least we'll be available to be poor artists, and if we simply spend our time sitting in front of a TV, wading through endless box-sets, waiting for inspiration or drinking ourselves to death, that's okay too, all facets of human existence are acceptable, everything allowed.

A central thesis of Purple is that we humans deserve everything that comes with technological success because we have won it, we are the very pinnacle of human evolution, the ultimate generation. We are the winners. This is the furthest we have come. Nature's gift to the universe, the best there has ever been. It is also very important to note that we only make progress because we are reasonable people, in the legal objective sense,

69

being reasonable is what has built the world we live in. The reasonable have always outnumbered the unreasonable; this aspect of human existence supports the conjecture that human society is far from doomed; being reasonable is the price of our entitlement. Hence Purple infers a new humanist age to rival the renaissance, and the enlightenment, Voltaire would be proud of us, humanity will be our primary business as *Marley's Ghost* opined, we will revel in the species, and we will have the time and the wealth to do so. And our luxury will be enjoyed without guilt, there will be no beggar at the door.

Humans evolved and survived because we are caring, nurturing and gregarious, we seek the company of others, and share our good fortune; we are naturally altruistic. These facts provide a basis for the assertion that all humans, because of the fact of their birth are entitled to a share of everything we own as a species, and everything we are likely to create in the future. Part of creating a Purple world is addressing the obvious inequalities across the world. The principle of entitlement implies preference for those without. There is no point in the Western World cutting back on our consumption, rather our task is to busy ourselves developing technology to unlock the infinities, and make the cake so big that the self-evident entitlement of all is assured; everyone else comes up to our level, and we push on from there. Technology is the only proven route to making the natural right of human entitlement a reality. Of course this opposes the Christian work ethic, for example, St Paul in the second letter to the Thessalonians, *if anyone will not work, let him not eat.* Which presupposes the availability of work, machines will take away the need to work by automating the wealth creation process. Part of becoming a Purple world means we ensure this action does not create a mob outside the door. A mob becomes a mob for a reason, something started it some spark caused the townsfolk to assemble for the march on Barron Frankenstein's castle, with flaming torches, sticks and pockets of stones. A mob today might have hand guns, or automatic weapons.

There must be an ultimate celebration of humanness with a statement that humans will always take precedence over machines, and over corporate entities. We affirm the value of our human achievement by making sure that no matter what we invent, no matter how smart the machines may become, humanity will always be supreme. Humans must remain the superior entity on this planet; any smart machine or artificial entity created either directly or indirectly by us must be subservient to humanity. All final and irrevocable decisions must be taken by humans, whether we understand the decision or not. We are the different apes. The apes that woke up. Apes with capacity. The ultimate problem solvers, equipped with the quality of empathy. The ability to see the other side. To imagine the other person, to align oneself with others, and form groups. Alone we humans are mere food for predators, together we are mighty, we clearly learned that at an early stage, and practice it to a high degree. Humanity, is it the biggest group of them all. The sum total of all of the rest. The umbrella that covers everything it is the entire unfinished human project. We are all too well aware of our animal like behaviours, they emerge with clockwork regularity, can be predicted, provoked, and encouraged to the surface with ease. It's a face we all recognise in ourselves, and in others, from eating, to rutting, to war and murder, we show it. But the social exceeds the animal, the refined outweighs the darker tendencies. The philosopher and writer, John Gray asserts in his book, *Straw Dogs: Thoughts on Humans and Other Animals*, quoting Lovelock et al, that a human being is nothing more than a colony of bacteria. We must reject the concept behind the idea, even if it is true in a strict biological sense, in the same way that we reject the notion that a cathedral is simply a pile of stones, or that a microprocessor is merely a jumble of atoms. Clearly there is something more, clearly, means it is self-evident that such is the case. There are self-evident emergent properties of the human that are obvious to other humans, we are connected, we are all part of a greater whole, and different. Flamingos will resume dancing, feeding and mating, while a fish eagle dismembers one of the population on the shore of their lake in Africa. Wildebeest will resume grazing while a pride off lions feeds on one of their number nearby. By contrast, if it was one of our own, humans would stand together, we would arm ourselves, and look to settle the score with the lion, or turn the fish eagle into a headdress.

And our Purple world will be a meritocracy we still put a value on being the best, the best human chess player is still valued even though machines have robbed the title of being the best in the universe. Only the best golfers can play in the US Open or the Masters. Only the best players will compete on the world stage, and those fractions of endurance and perfection, and timing that mark out the truly great from the average player in whatever discipline will always remain. We will always strive to be better than we are, to go further, higher, faster, as the Olympic ideal instructs. No matter how many technological enhancements we might acquire or employ we will always have the few who rise above the heads of the many. The meritocracy of sport, of art, of music, of good looks or screen presence will always be a part of human endeavour. The rest of us have no choice but to watch.

We are today, in a pre-Purple state, and it is self-evidently dysfunctional on many levels, while we in the affluent West marvel at our technological mastery; other populations have never seen a smartphone, or cast a vote. The Middle East remains condemned to test and retest the absurd notion that one mythology is superior to others. Technology is the only great liberator; its advance and global spread, coupled with the defeat of theocracy will deliver the modern human world times ten, and what we call problems today will fade to Purple. We can't help ourselves, we can't stop speculating about this future, we will probably never see how it all turns out, and so speculative scribblings are as close as we can ever get to being there. So we set to the project with the zeal of a convert, developing an obsession only bettered by Gollum for his precious. The long path we have travelled in the formation of our modern western society, requires that we look after it very carefully. If it fails we humans may never get a chance like it again. It is argued that the functional portion of the human project, is not a pyramid, like the structures of old, nor is it a square, or a circle, or any other kind of a shape, rather it is an amorphous blob with no discernible form, and ever changing blurred edges. It operates on numerous levels and in countless dimensions, and constantly rearranges itself as it evolves, grows and makes progress. It has no direction or purpose, no collective consciousness, it simply exists, and it is largely a meritocracy with peaks and troughs, branches of this and that activity, and outposts of endeavour. It has no end point that can be discerned from within its structure. It has no objective function in absolute terms. It has no external forces guiding, watching, or exerting influence for good or ill. It's like a gigantic human brain. It is a microcosm of the universe. It is random in its nature, and chaotic in its complexity. It provides a home and protection for even the weakest and most insignificant of its members under the self-evident naturally evolved structures of its mass. It is a project worthy of our intense interest. It is too big to fail. It is too smart to fail. It is too precious to hand over to machines. It is the only game in town, because it is all we have.

Chapter 3   Our Empire Of Technology

I firmly believe, that before many centuries more, science will be the master of man. The engines he will have invented will be beyond his strength to control. Some day, science shall have the existence of mankind in its power and the human race commit suicide by blowing up the world.

Henry Adams, 1862.

What we've built today, in the face of all of the retarding forces trying to stop its construction, is so improbable an event, and so complex an undertaking that it took us 250,000 years to get it done. Western technological civilisation is a worldwide consensus of almost universally accepted systems, rules, principles and methods that have provided the highest possible standard of living to the greatest number of people in the history of life on this planet; a realisation of the dreams of Jeremy Bentham. It is nothing short of an empire of technology, it is our empire of solved problems. It is a world populated by nearly seven billion people all beneficiaries of technological advance. There is an apt line in the Monty Python movie, *Life Of Brian*; when one of Michael Palin's characters declares, *as empires go, this is the big one*, it was certainly true at the height of the Roman Empire, and applies here too. Empires come and go, but there has never been anything like our empire of technology. Technology will continue to advance, because it has the almost unanimous permission and encouragement of the entire human population, if we agree on nothing else, we all agree that we love our technology, and we want more. Nature shows us the way, provides our list of problems not yet solved. We are in a fearless time; we have no fear of the future, no fear of magic, or of superstition, or mythology or religion, nor do we fear the consequences of asking why. There have never been so many scientists and technologists chipping away at our remaining ignorance, drilling into the depths of their subject, exploring every hidden niche. Human technology is incapable of making anything as interesting as a fly, or a spider, an ant, or a flower, a bacterium or a virus, and hence we clearly see how very much more we still have to do. For that reason alone, stagnation is simply not a possible option. Humans and our technology are fundamentally part of nature. Nature provides the impetus. But we have never been as far along the path as we are now, and clearly that fact alone unnerves some people. We are not travelling in circles like mice on a wheel, we have direction, and something of a purpose. Speculation about the final form of our technological endeavour has become a staple of modern writing, with more than a few scribes wondering if we have perhaps built something beyond our capacity to control.

There are so many things we can do now that were unthinkable even a very few years ago. Almost every virus has a vaccine. The human genome has been sequenced to the point where a new field of epigenetics is now thought as part of biology. Many cancers are now curable, and yes, the word *"cure"* is used, that was never the case before, there is a whole field of immune system stimulation which promises to cure the remainder, and for the first time in history more people survive than die from the disease. Although the sceptics among us point to the facts of the situation, and the long list of previously hailed breakthroughs which turned out to be clinical dead ends. Cynics also point to the fact that a great many very much simpler conditions, baldness being one, remain untreatable or uncured. In physics, the standard model of particle physics has been vindicated by the Large Hadron Collider. The Kepler telescope has discovered thousands of exoplanets, so many that they can be catalogued as earth-like, super-earths, and so on. There may well be ten earth-like planets for every human person on earth, in theory at least therefore, we are already rich beyond our dreams. There are realistic plans to get to Mars within the next twenty years, the long overdue return to the Moon promises to be made with a vengeance, by creating a permanent, and ever expanding human outpost. Smartphones and tablet devices connected to the internet, itself the marvel of the age, potentially provide the entire information store of the species to any member of the species, the ubiquitous 'Chinese rice farmer', for example, has the contents of every library at his fingertips, depending of course, on whether his government allows him to access such information. The ever watchful party machine is deeply suspicious of the educationally aspiring farmer class. The Internet expands by the hour, the super fibre links that span the globe mean that potentially, and soon, every home on the planet will be able to stream all the live content they can handle. The science of Big Data finds patterns no one knew existed, crowd sourcing can answer your most obscure question, or fund your business idea. The crowd knows everything. There are mumblings of viable quantum computers, and AI seems to be rolling itself out faster than the early pioneers in the field could ever have imagined. Basic science continues to surprise and astound us with discoveries of things we had no idea existed, and Nature continues to reveal secrets and surprises we never suspected. Those who have written in the past that we had reached the end of science or the end of knowledge are proved wrong with repetitive regularity, and silenced by terms such as dark matter, or quantum gravity.

Technology has also retained and continues to enhance its subversive reputation. Anything you can do can be done better with a 3D Printer connected to the internet. You send it a file, press a button, go off to lunch, and when you return you have a new tool kit. They've been printed while you ate. Make anything you like, as many times as you like. It's happening now, and the limitations are being squeezed out, the process is being refined. Already there are 3D plans for various types of guns online, they can be found at Pirate Bay, and all the best torrent sites; with the obvious limitations of plastic or acrylic, but there are obvious security implications. It is a kind of subversive technology, make anything anywhere, and to hell with the factory, or transport, or middlemen, or royalties. Synthetic biology, and new open source methods of biological assembly have put biological tools and technology into the public domain that were expensive and time consuming work for skilled practitioners only a few years ago.

Technology can also happen fast, sometimes so fast we are left behind and bewildered. The first steam engine in the world was built by Thomas Newcomen in 1712 to pump water from a mine, it was an awkward atmospheric beam engine with a zero point five percent efficiency. It was a further one hundred years before James Watt invented the condenser, and improved the efficiency from 0.5% to 2.5%. A further one hundred years later the Titanic made her maiden voyage powered by huge steam engines, the technology had hardly changed with two hundred years of development of steam, and water power, and wind power, and the industrial revolution. On the other hand. The wright brothers were the first men to strap an engine to a pair of wings, and get the both man and wings off the ground in 1904, and return them in safety. Ten years later Barron Von Livendorfen was clocking up kills in his Blue Max. Charles Lindbergh crossed the Atlantic in his Spirit of St Louis in 1927, and shrunk the world to a manageable size for the first time ever. Flappers were being taken for a spin by rich playboys in their metal fixed wing machines in the late 1920s, and in 1937 Blitzkrieg hit the headlines as German planes dive bombed over Guernica in Spain, thirty years later a man stood on the Moon.

In 1974 the first microprocessor, 1981 the first IBM PC. Sometimes technological growth from emergence to widespread take-up can be slow and plodding, sometimes blisteringly fast with near reckless abandon. Nature can fly, so can humans, Nature can travel at speed, so too can humans, if Nature can, we can too. If it took Nature four billion years to evolve a cockroach, it doesn't really matter if it takes humans a few hundred years to perfect the synthetic version, and when it's done it stays done, because we write it all down, and put it somewhere everyone can find it. But unlike the blind watchmaker approach of Nature, through trial and error, and survival of the fittest over millions of years, humans are keen sighted watchmakers of extraordinary skill, and we can figure things out, we can see what we are doing. We can see which roads to take and which to avoid, and we can see or imagine the goal at the end of the road, and we operate in parallel. We know it can be done when we see it has been done, and we can learn from the master how to do it better.

Technology is just like a shopping list. If science is a list of things we know about the universe, then technology is a list of things we can do. It's a pile of books. It's a collection of files, a list of methods, of techniques, and of processes. It's the art of humanity, the collection of solutions to problems we have encountered. How to get the next meal. How to stay warm at night. How to stop the big guy in the next cave from taking the fish you just caught. How to stay alive in a hostile and uncaring environment. It's only the sum total of human know how. We also have a list of things we might be able to do someday, because doing them is not against the laws of physics, and there probably are millions of examples. All the cool Si-Fi stuff would be on that list, like getting to other stars, or making a hole through the earth, or flying to Mars and back in a day. But we cannot do them because we don't know how. Not yet. We also have a list of many things we want desperately to do, but can't at the moment, because they are just beyond what we know, not so far beyond that we can't imagine them being done, but far enough beyond us to be tantalisingly out of reach. The list includes, curing cancer, living as long as we wish by curing old age, making whatever we want out of whatever we have by manipulating individual atoms and molecules, like plants do, and making so much of it that no one ever finds themselves short. And all the other stuff that we would love to do but can't, growing new limbs, repairing tissue damage, baldness, obesity, and so on, it's a very long list.

Technology is neither a tendency nor a an entity, the term is imposed on an aspect of the human project by humans, it is merely a collective term for certain human activity, there is no direction, no goal, no purpose, and its progress in time is random. Giving it a personality is something people do, a peculiarly human trait, to personify inanimate things or strange concepts, personification is nothing new, the Greeks did it, the Romans too, and the Romantic Poets. Technology is nothing new, nor is the human love affair with technology only a recent phenomenon. Many writers from the ancient world recognised the many imperfections in the technology they had inherited or evolved, they knew it was possible to improve on it and do better, and they looked forward to a future improved by better technology and methods. For example, Archimedes. *For I assume that through the method I have demonstrated some persons who now exist or who will exist in the future will arrive at other theorems that have not yet occurred to me.* And the first century Roman writer Seneca,   *As a result this knowledge will be revealed through long, successive ages. A time will come when our descendants will be astonished that we were ignorant of things that are so apparent.* Even later, St. Augustine, writing in his *City of God*, in 426 AD *...has not the human intellect discovered and put to use so many and such great technologies... ...What skill in measuring and counting! With what sharp minds humans grasp the paths and arrangements of the heavenly bodies! How great the knowledge of this world humans have filled themselves with! Who could describe it?* However, 426 AD was a long time ago, and somehow the learned people of the world went from loving and admiring technology to fearing and loathing technology. Science and technology fell from grace, and we got the Dark Ages, and a theocratic nightmare every bit as evil as any in the Muslim World of today.

What a difference technology makes in our world and in our lives. *We've never had it so good,* is the standard reported paraphrase of Macmillan's statement of 1959, it was true at the time, and is doubly true now, and increasingly true with each and every passing day. Matt Ridley in his book *The Rational Optimist,* paints a pleasing and agreeable picture of the modern world from the perspective of the average 2010 consumer living in the modern western democratic state. The example of an average 35-year-old French woman is taken, and she and her life requirements are compared with Louis the Sun King prior to the revolution just over two hundred years ago. There is no comparison, between the two lives, and it matters not which metric is chosen, modern life wins hands down. And if we go back further to Henry VIII and the luxury and excess of Richmond palace or Hampton Court, or back even further to the time of the Emperors of Rome. Modern life wins hands down in every department. Whether food, personal comfort, education, medicine, transport, communications, entertainment, or whatever else we have, our modern life is vastly superior. Even the poorest of modern western city dwellers lives better than a Roman Emperor or a King of France.

Everything we do is on a grander scale. We eat when we want, talk to whomever we please, travel wherever we wish, at speeds the ancients would never even dream of, and at a price affordable to all. We are surrounded by technology and experiences unknown to our predecessors, we control vast wealth, comparatively speaking, each of us probably has a wardrobe or two full of the finest garments money can buy, fit for a king. Shoes of leather, and other exotic materials, the comfort of which would have been unheard of in previous days, probably several pairs, none of which leak; except for golf shoes, they all leak. We sleep in comfortable beds free of parasites, our food comes prepared to the highest standard, from the very best menus, prepared in the most meticulous manner by professional catering people who serve us unquestioningly, we command the night, bathe in warm water, whenever we wish. We are not confined to travel at the speed of a horse on a road unfit to be called a garden path. The streets we walk on or travel round are free of animal excreta, and not just the odd dog mess. Prior to the age of the internal combustion engine, city streets were not unknown to cattle, pigs, horses of course, dogs, and a veritable menagerie of goats geese, ducks, and every other creature we have a use for. Any food we eat today is free from bacteria, worm eggs, and any and every parasite imaginable. We work, probably as and when we wish, even though most people need to work and do, if that work disappeared in the morning most of us would not die for the want of it, no one would starve in the street, you would more than likely not lose your home, and while you might find yourself in reduced circumstances for a period of time, the chances are that you would recover and get back on your feet. But even modern life on welfare is a blessed existence compared to the agonies, terrors and fears suffered by the aforementioned anointed ones. In short our human lives are an improving work in progress. We are not there yet by a long way, but we will get there, and soon.

We humans have reached a consensus. We have a system of government, of politics, and of economics, and administration that works. That is, it all seems to work. Despite various perturbations. We have arrived at that system by trial and error; the process of evolution. It has grown out of the ruins and ashes of everything that has gone before, there has been a baptism of fire, and a long and arduous struggle for establishment and acceptance. But we have arrived here none the less. A common set of languages, English, Mathematics, Music, a common number system, and a way of collecting facts and verifying each other's work that we call the scientific method. We have chosen the political system we have, because all of the alternatives have been tried and found wanting, our chosen one is, *the best except for all the rest*, as Winston Churchill put it. The only system shown to work is the technologically based liberal democracy. The western European, American system comprising the consent to be governed, freely given to representatives, trusted by the people to leave office if and when asked. Watched by a free press, and a technologically armed, ever vigilant populous; this system is, our empire of technology.

There is a very wide spectrum of opinion to both left and right, and both extremes have been tried before, and imposed on various sections of the global population. We have probed and tested the darkest recesses of both left and right, our bloody recent history is a testament to that fact. It was done in the full glare of publicity, and fully documented. And both extremes were found wanting, neither extreme is capable of producing a working, system, neither produces an amenable, agreeable, habitable space for normal human beings to live and grow. Neither fascism nor communism, nor any variation is capable of creating the conditions under which any branch of humanity can flourish. Nor does military dictatorship, nor theocracy, nor any form of corrupt democracy provide a safe dwelling place. In fact the so-called political spectrum is not a spectrum at all. The extremes of politics actually stretch away from the centre along a curved path, much like a leather belt stretches away to either side of the buckle. And where the two extremes of left and right meet, right above the crack of your ass, is where you find a type of perverted amalgam of political extremes, a form of fascist-communist bastard condition that results in states such as North Korea under the Kims, or Cambodia under Pol Pot. The theocracies of Iran and Saudi Arabia would also fall into this broad stretch of dark, ass-adjoining leather. And almost without exception these places are populated by people covetous of what we have built in our empire of technology, and as soon as they get the chance they will cut the head from the snake and sign up for a prosperous new future.

Empires go through phases, they rise and fall, and historians have great fun-filled careers pondering the reasons. Our empire of technology is the latest great empire in a long succession of empires, all of which have failed. It is an empire unlike all those which have preceded it, and this one looks like it will survive. It's not an idle boast of techno-utopianism, or irrational-techno-exuberance, to pervert Alan Greenspan's phrase. It's what we might choose to call techno-optimism, and it's about the facts. The fact that we have come this far means that we are capable of coming this far, and it means the doom mongers and the naysayers of ten or twenty years ago have been proved wrong. In fact, all of the techno-moaners, all the way back to the dawn of technology have been proved wrong. Empire has become a dirty word in some quarters, our empire of technology is broadly welcomed by everyone whom it embraces, everyone, even its detractors, is a willing participant. Those subjugated by the empire are willing accomplices, its conquered peoples want nothing less, its spread is vast and its rule benign. It is an empire that promises imperial levels of wealth for all its citizens without the uncomfortable baggage of imperialism. It has formed not because of any political will, but because of a tendency in nature for complex systems to sometimes reach a collection of stable states which are symbiotic, and remain stable for a prolonged period; it is as natural as a rainforest. The empire of technology is therefore probably the last ever human empire, the end of our imperial questing, the beginning of the final form of human society, an agreed solution to the human problem. The forces raged against it are so few and so weak as to be non-existent, because our emperor is the human spirit.

There does however, seem to be some confusion about what technology really is, its extent and its parameters, and a great deal is written trying to define the meaning of the term. On the whole it is probably true to say that we know it when we see it. Technology is our hand outstretched, to mould, to shape, to help, to protect, to guide, to solve the problems of human existence, and advance the human project. To coordinate and extend the human eye, and the human brain, and the collective human consciousness. Technology comes to us from the human compulsion to solve problems and understand our surroundings. Technology is a promise, above all technology promises to provide us with a future abundance undreamt of in any previous era.

Technology is a metaphor for what humans have done. It is a description of a list of achievements and accomplishments born in science, and assimilated into everyday use. It is a list of stuff that works, a collection of what we know that allows us to live the way we choose. It's the reason why we do not need to forage, or scavenge, or chase wild animals around the Savannah or through the Jungle. It is the reason we are the creators of the landscape and not merely a figure within it as, Jacob Bronowski put it. But it is not a complete work, rather it is a work in progress. Technology is an incomplete record of human endeavour, an unfinished record of our achievements, the partially concluded history of an ability to solve problems, it is therefore, also a long list of problems yet to be solved.

Some people think there's no difference between humans and animals, they believe we are part of nature, like dolphins, or dogs, just another species, nothing special, and probably destined to go the same way as almost every other species ever evolved on this planet. Others think that humans are special because, the magic man in the sky did it, he made us in his image, created and not evolved, and he protects us to this day, takes sides in our wars, and listens attentively to our prayers. Both of these opinions are nonsense, they are little better than unsupported garbage, based on pure speculation or at the very worst wishful thinking, being as they quite clearly are, contrary to the observed laws of nature. Or to be charitable in the extreme, such notions are unlikely, given what we now know. The truth of the matter is that we are the apes who one day took hold of a stick and used it to lift the rock, or get to the higher fruit, or defend our food from the other fellow's advances, or chase off the lion. The difference is that we invented the stick, when we picked the stick up, we invented technology, and we did it first. Humans alone are just another species, probably on the same path as Homo Heidelbergensis or the Neanderthals. With technology we are off to the stars, we live forever, remember everything, and make the earth glow in the dark. That's the difference. Technology is the great leveller of species, it gives humans the edge over all other animals, it derives from our evolved intelligence, and allows the smaller weaker human to master the stronger and more vicious species; not by hiding or by running faster but by standing our ground, and facing them down with tools, sticks, stones, and fire. Our technology started when we were still monkeys. One of us stood on his hind legs and reached the higher fruit. He found he could reach further with a stick. The same stick could be used to poke at a nest, to prod at a beehive, to collect ants, and so on. Stones, sticks, bits of bone and old teeth were the earliest tools humans used, before we were even humans. Technology is as natural to us as is breathing or standing up. It has come with us as we evolved, and evolved with us, it is not separate from us, it is our thicker skin, our sharper teeth, our set of claws. Humans can no more do without technology than other species can do without their skins, sharp teeth, claws, feathers or whatever else gives them the evolutionary edge. Brawn has shown itself to be no match for brains.

The most important thing we may say about technology is that technology is not an entity, nor a thing, nor a person in its own right as a body corporate or a government or an organisation containing a steering committee. Technology is a history of human endeavour, and a never ending interplay between problems and solutions, the never ending search for a solution, and the creation of further problems in the process. Those who invented the cannon also invented the problem of how to make the cannon ball fly further or higher, or smash through a thicker wall. And at the same time created problems for the builders of castles against which the cannons were massed, problems of the, 'what the hell do we do now', kind. The creator of the plastic bag while solving the problem of carriage at that level, also and unwittingly created the problem of disposal. Technology is therefore also a measure of the ingenuity of a species of the creative extent of the human mind. Just as the law will not see a right without a remedy, technology is our great unwillingness to leave a problem without a solution.

The notion that technology is a living breathing thing, and an entity in its own right with an independent existence seems absurd. Equally absurd seems the idea that technology is a creature of ours, fashioned by humans in antiquity, and growing in an exponential manner towards some conclusion. With the expected natural consequence of alienating the individual human and causing the demise, destruction or absorption of the human species into its greedy malevolent form. Such a view seems both childish and naive, it is very much in the same vein as the old Ptolemaic view of the heavens of spears within spheres musical in their rotation, and everything moving in Devine circles with earth at the centre, and Man at the centre of the earth.

Yet this is how some influential people on the West Coast see technology, Kevin Kelly for example, in his work entitled, *What Technology Wants*, seeks to personify technology as though it had a mind of its own. As though there exists some emergent property of technology as a whole, some property of its existence more complex than the technology itself. But this seems to be nothing more than the old trap of the pattern seeking mammal. People have been personifying inanimate objects and concepts for as long as people had the capacity to think in abstract terms. The representation of the Gods as people, by all civilisations from the Greeks backwards, to the first cave dweller to carve a stick or a bone, is nothing new. The romantic poets personified everything, especially natural phenomena the Wild West Wind, daylight, night-time, and so on. Assuming technology is a living, breathing, thinking entity with an agenda of its own, and a plan, is as natural a thing to do as smoking Pot.

Real technology, that is technology with real power behind it, derives out of necessity, and has its origins in the year 1712, when Newcomen first set fire to drive an engine. Everything prior to that was leading up to it, or was a false start. Prior to the steam engine the only use of fire was to heat things, to make them hot so that they could be changed, such as cooking, or smelting of metals, or working of metals by a blacksmith. Fire was not used as a power source; to do useful work like drive an engine, to replace people or horses. But there was a necessity that required a power source that simply did not exist, the wholescale pumping of water from mines. In a climate where rain is a dominant factor, it is simply impossible to dig a hole of any depth without it filling with water, and that water needed to be pumped out, winter and summer, round the clock. Wind couldn't do it nor could water power. Something new was needed, something big and powerful, and world changing. And once it was made the world would never be the same again. The steam engine was as radical an invention as the telescope, or the printing press, as potent as the fire used to drive the thing, it was nothing short of a revelation. And once the problem of a flooded mine was solved, its application to other fields was as inevitable as was its improvement. The wealth generated as a result was as nothing that ever been seen before, and left the kings of old and the emperors of Rome looking like paupers.

And with the vast wealth came the obvious and inevitable conclusion that further and better improvements could and would lead to even more wealth, suddenly money became a driving force behind technology. It became clear too that these further and better improvements could only be made if the world was better understood, and that meant turning to science, the real originator of technology. Science is a body of knowledge which consists of a collection of facts derived from empirical observations followed by questions, followed by the formation of theories, formed out of conjecture, and derived from imagination and the *what if* approach. Science is not truth nor is it ever held out to be truth, though it does have a nasty habit of uncovering the truth, and revealing some very uncomfortable facts, such as the age of the earth. By the process of anthropomorphic protectionism, science, like technology, is often personified as an entity and endowed by the projectors with qualities it can never possess.

To a first approximation people don't care about the origins of the universe or much else for that matter. Most people are preoccupied with the vagaries of life, *chewing upon life's gristle* as the Monty Python song goes.

Most of us are self-obsessed, bordering on narcissism, or otherwise occupied with our kids, with the future, and the day to day maintenance required to remain alive. We fuss over family politics, our jobs, our golf handicaps, sports results, and how we spend our money. Hence there is little or no room for the head-hurting stuff, such as the contortions of quantum gravity or space-time, and the like. We have others to do that for us. We have in effect willingly outsourced our thinking. And therefore we need to trust those who are thinking on these matters for us.

Our Empire of technology is built upon science, and therefore science needs to be understood, not the specifics of any particular branch in any great detail, but simply as something that exists, and occupies the time and effort of a great many people, and absorbs a very great deal of public money. In one, if not all of his many public lectures on science, the great physicist Richard Feynman explained the scientific method in simple layman terms. First we start with a guess, then we test the guess against observation, then we experiment to see if the guess holds up to our scrutiny. If it holds up then it's a good theory and we can continue to work on the basis that it is a good theory. If, however, the experiment shows the guess to be wrong then it's wrong, and no matter who thought of it, whether or not he is an important man or a Pope or a King or whatever, or what weight of belief is attached to the guess, or how many people think it's the right guess. If the experiment designed to test the guess is wrong then the guess is wrong, and we throw it out and guess again. It's that simple. It's the falsifiable nature of science. For example, peer review and the weight of scientific data suggests climate change is happening, and that if we don't act we might suffer consequences. Or the laws of physics may be employed to state as a matter of fact that the study of astrology is nonsense. It is a waste of creative effort, and pointless for intelligent people to engage in such a deceitful and fraudulent racket. However, Astrology is a business, and a pretty big business at that, and the business of astrology is to provide its customers with such products and services as they are prepared to pay for, and so on. So from that point of view, astrology has a perfectly valid place in the modern world. The fact that it is clear nonsense, and dishonest bunkum does not lessen the rights of the customers to have those products and services provided. In exactly the same way science can state as a matter of fact that there is no credible evidence to support any claim of the supernatural. From which we may reasonably conclude that the Catholic Church is nothing more than a vast multinational real estate investment corporation, dispensing products and services to those willing to pay, *salvation by the shilling,* as Robert Bolt put it in, *A Man For All Seasons.* Science cannot answer questions of a philosophical nature but it can speak with authority when something is manifest nonsense; such as supernatural claims.

Of course while almost everyone loves technology, not everyone wants to do science, or even cares to do or to know anything about science. It is reasonable to say that many, if not all people are content to leave science as a mystery they would rather not know about. This tendency among most people goes to the heart of what C. P. Snow was so worked up about in his influential 1959 lecture, *The Two Cultures.* The simple fact of life is that people opt for either art or science, rarely are they able for, or interested in both. Not since the days of the gifted artisan have we seen both. Polymaths are a rare occurrence these days. It's hard to be very good at more than one discipline, science degrees are harder to get, you might need to learn two things, physics also requires mathematics, to be good at biology you need chemistry, and chemistry requires both biology, physics and mathematics; engineers need to know everything. Most people find learning mathematics to be a chore, it is dull and boring, it's the old adage that the most boring thing to do with money is to count it. Most people are proud to say they don't know or care if they never came across group theory or differential equations, regardless of what attempt is made, most people have a natural tendency to find science and mathematics dull and incomprehensible.

Not everyone knows everything. The smartest theoretical physicist or biologist on the planet may be completely ignorant of the works of Goethe or Shakespeare, may never have read the works of Kafka, Proust, or Beckett. May have no clue as to French Cinema, or the paintings of Claude Monnet, or Van Gogh. And so what, if that were the case? She might get round to it someday. That does not mean he would fail to pass the Turing test or be unable to articulate her way round modern culture. The mathematician Paul Erdos was such a person. Known as the hobo mathematician, he spent most of his time in airports, travelling the world collaborating with his fellow mathematicians, he had no home, no wife, no family, and his life's possessions could fill a single shopping bag. Of course he is the extreme case, and single minded artists have also been well documented, Vincent van Gogh, and Paul Gauguin, for example. The well rounded individual of any great depth is a rarity. We don't have time. We don't live long enough. And it's true of all professions, the requirements of the profession ensure all else takes a back seat. The astronomer Carl Sagan wrote novels. Very few novelists would be comfortable reading his papers on planetary atmospherics, or be prepared to pay the entry fee to do so. Richard Feynman played bongos, and was competent in rendering the human form, especially those he observed in local strip clubs, it's doubtful if any competent painter would tackle quantum electrodynamics, there are very few if any examples of crossover with a real contribution being made on both sides of the art-science divide.

The two cultures are not the same, and science is not the poor relation. It is probably self-evident to any casual observer that it is so very much more important to have a knowledge of Shakespeare than to understand relativity. A very superficial knowledge of relativity is of no use whatsoever to anyone, but a very superficial knowledge of Shakespeare opens up the world. It is perfectly acceptable for people to walk away from science, and declare that they neither understand it nor care to. It is however entirely unacceptable to do a faculty, or field of knowledge, reversal, and be a scientist who presents himself in public as someone ignorant of everything outside of his narrow scientific field. Not only is it unacceptable, it's impossible. There is an ignorance about science within society in general. But that's okay, it is socially acceptable not to do science, the entry fee is significant, and why bother paying it. The entry fee that must be paid, to get anywhere is significant in terms of personal commitment and intellectual ability, time and effort. Hence science and its principal tool, mathematics, have always been minority pursuits carried on by those perceived to be distant and overly cerebral. There's an old saying, if you wish to study physics and mathematics, you had better be good, there's no room for the average.

The arts, on the other hand, are absorbed by osmosis, it is impossible not to be exposed, no special tools are required to be interested, or to understand. Most scientists do a bit of art. On the other hand, name a single prominent artist who dabbles in science; the odd pharmacology dabbler for sure but nothing serious. It's one of those ramps that only operate one way, it's easy to lose money, not easy to make it, easy to gain weight, not easy to lose it. Any fool can be an artist simply by saying so. Anything can be classified as art just by saying so. Sometimes they are right. *I pose, therefore I am*. Two of the seminal, *ought to read*, books referred to throughout these chapters are concerned, with the history of science and technology, on the one hand, Jacob Bronowski, *The Ascent of Man*, and with the history of the arts, on the other, Kenneth Clark, *Civilisation*. It is instructive to see how each deals with the art-science divide, Bronowski spends a great deal of his time discussing art objects, and great works of art. Clark by contrast, declares that he knows nothing about science, and has nothing to say on the subject, other than as a route to technological and industrial development, or a potential destroyer of civilisation. Yet by and large, it is the science, through the fruits of technology that pays for the art, and creates the surplus so necessary for art to be able to be valued.

Leonardo Da Vinci was no doubt an accomplished individual, in art and technology a true renaissance man, he studied and observed nature in great detail to discover how things worked. He imagined a flying machine, but it bears not the slightest resemblance to a Jumbo Jet or an F-16 fighter. Not only is our technology stranger than he imagined, it is stranger than he could ever have imagined. Today we have many Leonardo's, many great scientific and technological thinkers, with unlimited imagination. And it is not an unreasonable prediction to say that what they will produce, and what will evolve from their work in the future, will be as far removed from our imaginings as the Jumbo Jet is from Da Vinci's drawings. That is the principle of our technological advancement. Because we are now free from the chains of theocratic interference, and dogmatic belief in magic, we are capable of making progress. If it can be imagined, and is not contrary to the immutable laws of nature then we may well be able to achieve it. Like Leonardo, we are in the process of plundering the natural world for its secrets, nothing is left unstudied; we demand to know everything.

There are nevertheless those who fear technology and see the empire of technology we are building as being something to fear, and something to rage against. There are a wide and assorted collection of individuals and groups, variously described as being opposed to one or more aspects of technological advance. At the extreme end of this spectrum would be the anti-civilisation*ists*, anarcho-primitives, and Luddite sympathisers, who collectively spurn modernity, and would like to see a return to a world of very few humans living a tribal existence without recourse to technology. Even back to the Stone Age to a time when all we had was stone, stick and tooth. Unfortunately many of the articulate points to be raised against technology by such people, were those in Theodore Kaczynski's, so called, *Unabomber Manifesto,* a wholly discredited piece of dystopian nonsense written by a deranged murderer, and unworthy of further comment.

The more moderate objectors are those who fear that technology is advancing into a future of unintended consequences and indeterminate outcomes. Many fear the consequences of having the ownership of technology concentrated in very few, very powerful hands. And if technology is actually a living, breathing, plotting, malevolent entity then who owns it and who controls it? The promised future technologies are more potent, with more potential for damage to the species, and to the environment, many fear we are incapable of controlling the genie once released into the world.

But the response must be the same response that has been issued to all techno-objectors over the history of technology, those who complained that steam power and travel on rails flew in the face of god, or was unnatural, and inviting disaster, and the many early rail accidents involving train crashes and boiler explosions seemed to echo their concerns. But we survived, as we survived automobiles, air travel, and the weapons of war. Technology can kill, of course it can. But the Holocaust was not caused by technology, or facilitated by technology, nor was it carried out because technology made it possible. It was carried out by men wholly motivated by hate, had technology not been at their disposal, they would have employed sticks and rocks. Bronowski said it in his famous speech from the pond outside Auschwitz, and while he was speaking in defence of science, the same defence may be mounted for technology, and the long list of solved problems amassed by humans over the last few hundred years. Technology is not a living-breathing thing it is not an entity independent of humans. It is not something we need to get emotional about.

Do we need a global minister for technology or someone responsible for regulating it, and making the important decisions? Someone to decide if a piece of technology is, as the character Deckard says in the movie *Blade Runner*, referring to intelligent humanoid machines, *either a benefit or hazard, if they're a benefit they're not my problem*. Deckard had the simple binary, policeman's, approach to technology, of course it's never going to be that simple. There's really no way to tell until the work is done, and the technology is invented. We could ban certain fields of research, but that has its own problems. It's never that simple. Everything we have ever done in technology is a double edged sword.

Technology also liberates, as can be seen by the number of countries which either ban or restrict certain technologies. The liberating effect of technology can be potent and long lasting. Anyone who does not believe technology liberates ought to ask feminist historians to quantify the effects technology has had on the lives of women over the centuries. A female finger can pull a trigger every bit as easy as a male finger, or push a button or press a *return* key. Technology empowers, it liberates it elucidates and defines the lives of people, not just nerds or technocrats, but ordinary people, in dark places. And for those reasons alone, and for cotton underwear, and lavatory paper, we should be thankful for technology. But we should do more, we should embrace it and love it, and throw ourselves into it, and understand it. Its importance cannot be overstated, we should celebrate and affirm it. We should play with it, and grow with it; it's as human as we are.

Technology is also a process. Something we are going through. And defining it as a process implies there may be an end to it because there was a beginning. The end of technology has two clear meanings, the first being the end as in the goal, or the end point following thousands of years of building upon the foundation we now have. The other end is the, crash and burn end, the demise of our empire of technology, when we finally realise that it's all over. What if anything will be the end point of human technology is not something anyone can call. Being a work in progress we will continue refining our work, caressing it, improving it, tweaking it, and tinkering with it. It is in our nature to do that, so we might say there can therefore never be an end point to technology. But what happens when we have achieved complete mastery over everything in this universe, and no problem can be imagined that has not been solved or cannot be solved? We have no idea, no doubt these questions will keep futurists employed for centuries to come. But will technology, as we know it, come to an end? Will our empire of technology collapse and fade to grey dust? It is very hard to see how it might, the source of the collapse would have to be internal there being nothing outside powerful enough to cause alarm. The question therefore becomes what internal forces could arise or converge, to bring to an end the most fruitful period of time the human species has ever known? *"It is by luxuries and the vanity of women that empires fall,"* according to Benjamin Franklin, but we need more than that, the effects of technology upon people are not as bad, or as pervasive as certain authors imagine. People are plastic in mind and attitude, and in outlook. We adapt, we improvise and we overcome as the Marine Corps are fond of saying. While people are plastic and adaptable they are also sceptical, doubting, mistrusting, and eminently capable of making up their own minds, take it or leave it. People are not as stupid as they ought to be, or as some portions of the media and government might like them to be. If technology did come to an end there would be chaos, and an end of civilisation. Each and every one of its willing participants has therefore a vested interest in maintaining and perpetuating it. Because we love our technology almost as much as we love ourselves, we embrace it, we will not let it go without a fight.

Technology is here to stay, perhaps it will eventually tell us what it's all about. One thing is for sure, the human project has never been in a better position from which to make progress. We have no idea where it's all going, we are a bit like the fellow standing in his garden, he has a pile of lumber just delivered, a set of tools, nails, screws, bolts, and a very sturdy oak tree. He doesn't know the details, but he knows he will eventually finish a treehouse, and his kids will be able to play in it all summer long. Surrounded by computer screens and smart phones, we still buy books. Bogged down in the technological quagmire we still have a use for the humble pencil. We couldn't stop making progress even if we tried. We are in a very good place.

Chapter 4   Negative Solves Nothing

> *Ignorance more frequently begets confidence than does knowledge: it is those who know little, not those who know much, who so positively assert that this or that problem will never be solved by science.*
>
> Charles Darwin, *The Descent of Man*

There is a puzzling question why so many otherwise intelligent people seem so pessimistic on the future of humanity, why so many seem to have lost hope, and seem to spend their time decrying the entire human project. Purple is a reason why Malthus was and continues to be wrong, why philosophers like John Gray, and pessimists in general are wrong, why Islamic, Jewish and Christian extremists are wrong, and need to drastically rethink the meanness they show to their fellow human creatures. Purple provides a reason why the absenters and withdrawers, and those who opt out and give up are wrong, why the globalisation protesters are protesting about the wrong things. It is also why Green politics and those who advocate sustainability are mistaken in believing there is some finite carrying capacity of this planet which we dare not exceed. Purple is an assertion that progress is not such a dirty word, and that leaving the world in a better condition than you found it is a noble ideal.

The fundamental self-evident fact is that there is not a single problem faced by human society, now or likely to be encountered by our species in the future that is soluble by negative or pessimistic pronouncement. Purple is therefore about the hope humanity must have in facing the future if we are to avoid the adverse consequences predicted by those, Raymond Tallis, in his book, *Aping Mankind: Neuromania, Darwinitis and the Misrepresentation of Humanity*, calls the enemies of mankind. Purple is a celebration of the work done by the human animal, and the human project to date, and working off of the solid ground we have prepared, there is nowhere to go other than bigger and better.

Of course the right to free expression is paramount, and must be cherished, everyone has a right to their opinion, and we know what opinions are like, everybody has one, as we know, and so on. However, much of the negative opinion expressed seems to have no basis in the facts of the situation, or is wholly erroneous, or derives from a belief set that is in direct contradiction with the laws of nature and the facts of the universe as understood by modern science. Because while Tallis raves against what he calls Scientism, defined by him as, *"the mistaken belief that the natural sciences (physics, chemistry, biology and their derivatives) can or will give a complete description and even explanation of everything including human life"*, he would nevertheless have to admit that while it may not be complete, it is perhaps adequate. And adequate may well be sufficient to our purpose. Science is the only probe we have to feel into the dark corners of the world, and seek to understand what makes it work. And having understood how it works, we then apply what we have learned to solving the problems we face. We are nothing is not problem solvers, and technology is the list of those problems we have so far solved. Of course, we're nowhere near finished yet, but then that's the fun part, so he must also be forced to concede that technology is the only game in town. While technology doesn't solve every problem, and is not the cure-all, it is the only edge we have over the other animals, it's what has brought us to where we are now, and made us too big to fail.

We in the West have built a wall to enclose our empire of technology, we maintain that wall, and man it with stout fellows armed to the teeth with the finest weapons our great wealth can provide. We have also created a situation that no one outside the wall will ever become powerful enough to cause a significant breech, their best efforts and their biggest armies would not be enough to cause a dent. Our wall is also transparent, it allows those without to see exactly what we have built, and challenges them to join us by overthrowing bad governments, refusing to obey bad laws, and ending the reign of religious bigotry so prevalent across vast tracts of the globe. We show those out in the cold that there is hope, and help is waiting. But if they see what we have built, and make the decision to have nothing to do with it, that's okay too. The stark reality is that it matters not a damn what goes on outside of the wall. Nothing outside of our empire of technology is strong enough or powerful enough to topple it. No manmade power that is, no nation state, no movement, no religion or system of thought is radical enough to leaver it from its foundations, because anything outside of it that is armed was effectively produced within, or is a variation of what lies within. And the overwhelming evidence points to the fact that every human living outside wishes to live within. Very few, outside of ascetics would sample both lives then consciously choose the one on offer beyond the wall.

Like Marley's ghost we declare that mankind is our business, and it's our business because we make it our business, and we do so because of our innate properties of compassion and empathy which evolved with us and brought us to where we are now, and because our great, and ever increasing wealth, allows us the luxury to care. It's the well-known mathematics of games theory, we help ourselves by helping the other fellow too. We want more because we always want more, and because we can see that there is so much more to be had. We've barely scratched the surface of the possible, and we have an insatiable desire to see what wonders next year will bring. Bringing everyone in the world up to our standard of fairway homes, and two car driveways, and year round consumption is our primary task, we then have a world where no one has any business complaining. And technology is the route to that end, technology is the only way to make it happen, by closing the information deficit and unleashing the infinities of energy and matter. We need to progress only a decade or two more at the pace we are going to be able to automate the entire wealth creation process, thereafter we can increase the clock-speed, as it were, by a factor of ten, or a thousand. Thereafter, we are all richer than god, every one of us, and without exception. And, as we continue to build upon it, we will end up very long lived, and very wealthy, because we also have an infinity of time.

On the face of it we would appear to have achieved a great deal in a relatively short time, but we are not there yet, and there have been mistakes along the way, we have gone down roads we ought not to have travelled, and we have a great deal to undo and put right. Human suffering continues apace, great swaths of the world go to bed hungry and exhausted every night, and have no access to the basics of a decent life, their life expectancy is not much better than it was in Roman times, there is no educational expectancy, no expectancy of pleasure, or the great comfort we in the West have come to know and love. Not that all is perfect in the West, no one in a modern western city need look too far to see someone worse off than themselves. But as far as the material wellbeing of members of the human species, we would appear to have achieved a great deal of progress, and all of it worth celebrating. Of course we can criticise the defects, and the wasted opportunities, and the long list of things to do is long indeed, but progress there has been. Progress can be seen very clearly when we take longer time intervals. Would any of us wish a return to pre industrial revolution days, of itchy woollen undergarments and no mass production, cotton underwear alone, that great gift of the mechanised loom, is cause for celebration. Would anyone wish to return to the days before aesthetic dentistry? De Quincy's observation that a good deal of human suffering was toothache was all too true at the time, and is one problem we have thankfully left behind. Would anyone wish to return to a pre antibiotic world when every hospital had a sepsis ward from which there was little chance of recovery, and you might end up there simply by scratching your leg getting over a fence? No one can argue that the world is not better than it was in the 1970s, or the 80s, and so on, it clearly and obviously is. Hence the self-evident conclusion that progress is a process, sometimes gradual, sometimes rapid, sometimes discernible, sometimes not, but is it a continuing process, and shows no signs of slowing or coming to a halt any time soon. How could it? We're hardly begun.

Reasonable criticism about what we have built and where it seems to be going is to be welcomed, debate is always healthy, but winging and moaning about what we have built, gets us nowhere. Naturally there is the obvious question about where the one ends and the other begins, and who's to say, and so on. It is probably true to say that most reasonable people understand the difference. For example, claims that our inevitable evolution of artificial intelligence will lead to mass displacement of working people, with potential adverse consequences is not an unreasonable proposition. However, claims that our inevitable evolution of artificial intelligence will lead to the human species becoming extinct within one hundred years require a little more evidence. The one position would be objectively seen as a reasonable debating point, the other is fanciful alarmism, if not arrant nonsense, and whinging for the sake of it. Other negative pronouncements of religious origin may also be taken in the same manner. Claims that there is no point in achieving anything because we are in the end times, and the world is about to be destroyed must be taken with a pinch of salt, and a request for some evidence. Or Islamic claims that the entire world will become a glorified Caliphate with all infidels either killed or converted, are equally bonkers.

The Biggest moaners are by far the religious fanatics who simply never give up. They have a book, and the book tells them everything is wrong, and from that set of initial conditions, let there be moaning, and let it continue regardless, and let not the facts or the evidence get in the way of a passionately held belief. Nor is it only a single religion, or just a single faction of a specific creed, it seems to be all of them, the one thing that unites otherwise incongruent warring factions across the world is their malcontent with the way things have turned out for human society, and their opposition to the western way of life in general. The fact that life for the human animal has improved over the recent past, and continues to improve is a great source of disappointment and sadness to them all. Most annoying by far is the interminable interference in matters scientific and technological, on the entirely erroneous pretext that somehow religion, and no one else, has the right to speak, and pronounce, indeed dictate under threat of violence, on moral issues. Attempts to set limits or restrict the type of question that may be asked, such as Pope John Paul II telling Stephen Hawking that it is inappropriate to ponder what might have happened before the big bang are also as dangerous as they are laughable. Our dismissal of such suggestions as risible is of itself a measure of progress. We might also tag onto the religious fanatics those of our fellow humans who occupy themselves with endless tales of conspiracy, and plots hatched by governments and clandestine organisations. Whether it's the moon landings, or 911, or the military industrial complex, they believe fervently in the theory that there exists a conspiracy, a grand plan that involves others and excludes them, and the entire structure is about to collapse. Just like religious belief their convictions bear all the hallmarks of a faith, including rituals, meetings with others, holy texts and messianic prophets. It would be nice, for example if the entire Western World was one huge American conspiracy involving the military, and the corporations. It would be so much easier to deal with, and understand.

Next come philosophers and economists, too numerous to name, with little or no knowledge of science or technology, who between them have never solved a real problem, or made an accurate prediction, yet feel free to pontificate in the negative about the future of the human project. Otherwise intelligent academics who ought to know better, relying on outdated or discredited research or methods, such as Malthusian notions that we humans are engaged in a never ending binge, consuming all before us, and causing the inevitable progress trap to spring at any moment.

Next come globalisation protestors of all denominations who decry the continuously expanding western democratic model, and see nothing but danger in the ever more global approach of governments and corporations. And we can include here in this group those largely right of centre governments and parties who play the nationalist card at any opportunity, insular types, who wish to absent themselves from the world for purely selfish reasons, good examples being Putin's Russia, and the right wing parties and politicians in Britain, France and USA.

Next on the list are the Green Party poopers, and climate change dictators. These are fanatical autocratic groups and individuals who insist and demand that everyone in the world curtail their energy consumption habits on the off chance that the global temperature might rise by two degrees. It might very well rise by two degrees by the time we get to the other side of the transition period in which we find ourselves right now. But long before we turn the planet into a desert populated by a few wanderers reminiscing about the time there were trees, we will have closed the information deficit to enable us to suck carbon dioxide out of the atmosphere, and use it as the main input of a million industrial processes. Rather than sobbing about the rise in temperature these characters would be better employed promoting new technology and solid state methods of doing what plants do, and pressuring governments to throw billions at the problem. Eventually we will figure it out, and make it cheap, it will probably come out of an oil company then talk of global warming will fade in the same way talk of the impending nuclear winter faded after the Berlin Wall was carted off in trucks. There is also a cohort who like Lovelock believe the earth is suffering, from an infestation of humans, far too many of us behaving like an infecting organism, in his own words; *"Gaia is suffering from Disseminated Primatemaia, a plague of people"*. Nothing could be further from the truth, given the right information, and unfettered access to the infinite energy of our own personal fusion plant in the sky, the entire population of the planet could comfortably exist in extreme luxury at suburban densities on a plot of land no bigger than Australia; we could allow the rest of the planet to return to growing wild. The truth of the matter is that we have no need to conserve energy or any other natural resource, we are surrounded by infinite amounts of energy and matter, and require only the necessary information to make whatever we wish from the combination. We can continue to consume everything at our current levels, or at increased levels, in the sure and certain knowledge that by the time we run out we will have solved the problems holding up progress. So many people are working on these problems, and the rewards are so huge that we will reach a solution. If plants can do it, so can we.

Further along the list are all those who oppose science, largely out of ignorance or out of unfounded fear that science is inherently bad, and that sooner or later something is going to escape from the lab and end it all. Not everyone wants science, not everyone can handle science, and not everyone can do science. Some people are simply attracted away from science, repelled by it, they hear mention of the word, and shut it out. Such people are on the whole harmless, they are probably artistic, and simply lean away from science, it holds no appeal for them. Others simply fail to see the logic of the argument or simply dismiss any arguments that contradict their worldview. There are also the wilfully ignorant, like the ardent UFO believers who know that every light in the sky must be alien in origin. What else could it be? The conspiracy theorists, who denounce the honest labours of millions of early Egyptians, and instead give the credit to extra-terrestrials. The religious nuts who lump every scientist and technologist into the same conspiracy to teach evolution to their children. The *'don't know, don't want to know'*, people can be hard work. It's of no matter. It all adds to the mix that is the modern world. Such people are required, and are to be tolerated, loved, cared for, and kept around, because we are human, and there's room for us all. Humanness, the essence of being human, or what we humans have achieved to date is not a gift from anywhere or from anything. It is simply a fact of our being here, nothing more. We are here because we have survived to be here. Our survival is its own testament to our ability to survive. Like the mountain climber who labours to climb the difficult peak then stands to admire the well-earned view, and wonders why. The answer is simply the self-evident fact of standing there, the climber stands at the summit because the climber is capable of standing there, nothing more.

Progress is possible because nothing we meet in the future is likely to be as harsh as the trials of the past, the chances of our empire of technology coming to an abrupt end via some cataclysm or other are remote. We are too big to fail. We exist therefore we deal with it, we get on with it. We endure, having no alternative. We do what we have evolved to do, to be better than what came before, and to continue moving forward. There is nothing else to do, nothing else we can do. Therefore those who deny our ability to fashion better tools, and make further progress know little of humans or the ability of humans to survive, and are therefore simply wrong. Of course if you are prone to believe that we were created and not evolved regardless of all of the evidence, and if you are convinced that it all happened by magic in a garden 6,000 years ago, there is little hope, if you cannot be convinced by reasoned argument, then don't be. At the end of it all we are animals, apes to be precise, living in an uncaring and hostile biosphere, we have no one to turn to, there is no helping hand, no guiding light, there is no big brother or sister who had been along the path before us and can show us the way. We are in uncharted waters, every day we wake we are the furthest along the road we have ever been. In the final analysis, you can't argue with success.

Chapter 5   Purple Is The New Green

Mao Tse-tung

On the face of it, Green is a good idea. Green commands that each of us takes no more from the world than we need, and puts back whatever we take. We live a carbon neutral lifestyle, with a low carbon footprint in an operationally carbon neutral world. It sounds great until you do the math. What it actually means is that we halt all further development, we all cut back to what the Green people refer to as, sustainable levels. We give up our large homes, and move to small insulated properties that can be kept warm with body heat, we drive tiny cars, with tiny engines running on synthetic fuel made from organic manure, or old newspapers, or whatever. It is possible to go on painting this dull picture of the Green idyll. But what's the point? Nobody in the Developed World wants to live this unpleasant dream. We've come too far and endured too much as a species to end up living like that, if that's the best we can do then we ought to pack it in right now. One of the central themes of this work is that sustainability at present levels of development is effectively stagnation, and is therefore unacceptable, stagnation is simply not an option.

What we want is a Purple way of living, we like big homes and large cars, and luxury and plenty, and warm sunny vacations, above all, we want all the technology we can get our hands on. The only way to maintain that lifestyle, as well as bringing everyone on the planet up to our affluent western level, is to close the information deficit, and let loose the infinities of matter and energy that surround us. In other words to learn how to do everything we wish to do, or that can be done by applying energy to work on matter, via information. Purple is the new Green, Purple is a better green, Purple is a green that will get stuff done, Purple is a practical Green, it is Green version II, the only Green we will ever need. Green thinking and Green politics could be seen as a precursor to Purple, a first draft of Purple, or a first attempt at Purple, and like all first attempts it's not the one you stick with, it is not the final version.

According to the Green movement, generally taken to mean those individuals who believe passionately that the earth is sacrosanct, and is all there is in the universe, and must at all costs preserved and protected for future generations. If we each live the life of one of our pre-civilisation hunter-gatherers, frugal in our living, light in our footprint, conservative in our outlook, taking only what we need from Mother Earth, then perhaps the earth could support 40 billion people, or something. If on the other hand we live like kings, like western billionaires squandering as we go, energy wasters, resource consumers, the throwaway society, like devouring tigers, then the planet can only sustain perhaps 100 million people. The Green movement like any other broad based collection of like-minded individuals is a spectrum, from the sandal-wearing loonies at one end, to very reasonable and serious people at the other. But what they all seem to have in common is a desire to restrict human progress. What they collectively fail to recognise is the fact that we cannot and will not cut back, or reduce our footprints, and hence another way forward must be found. Purple, is the way forward. At the end of the day, it doesn't really matter how many billion tonnes of carbon dioxide we pump into the atmosphere if we promptly suck it all back out of the atmosphere and use it as a raw material, to make the stuff we consume, like food, fuel, large homes and big cars. We can't do it today, but we're working on it.

There is no point in coming as far as we have clearly come, and enduring all we have clearly endured as a species just so we may live like our primitive ancestors. And the poor of the world are right to demand the lifestyle they see the western elite enjoying today. Why shouldn't we all have the affluent western suburban lifestyle, why should we not indulge ourselves, and bask in the glow of our substantial well-earned wealth? Of course we should, and we ought to do so without shame, we ought to demand the best, and as much of it as we want, we should demand abundance and lots of it. We deserve it, and we are entitled to it.

That's the difference between being Green and being Purple. Purple demands that we live as we are without cutting back, without worrying too much about the consequences. We have the ability to solve problems, we evolve technology to take care of any mess we make, we pay vast amounts in taxes, and we therefore have a right to expect that any problems caused by the fact that we exist will be solved. This is why the Green agenda is fundamentally flawed, because we are at a point in our technological history, and our understanding, where we can clearly see that future technology has the capacity to deliver a bountiful new world. Green folks have been barking up the wrong tree. Beating about the bush with the wrong end of the stick. Cutting back will never provide an answer, it is at most a stopgap, a delaying of the inevitable collapse.

The reasons why Green politics and the Green movement, and their sustainability arguments are wrong, comes down to the list of wants and needs of individual humans, and the sense of entitlement we all share from the simple fact of our existence. Who among us does not think we deserve everything we have and more, if we are rich, we damn well deserve it. If we are not so rich then we damn well deserve a damn sight more than we have. When asked to cut back, or take less, or pay more, our first inclination is to protest, or reach for a gun, or bang on the table. We know what we are worth, and we want to be paid accordingly.

And if that sounds greedy and ignorant, it is not supposed to, it is intended to be provocative. Our survival as a species is an explicit goal to which each and every one of us is committed, each one of us has a substantial investment in the future, and we therefore cannot behave towards our planet as though we are not going to be here. But we are here, and there are probably going to be ten billion of us before the population peaks. Hence policies that are workable if one assumes a carrying capacity for the planet of only two billion simply do not work, and can never work. Because of their extreme views and discredited arguments about sustainability, and in particular the economic costs involved, the Greens have been pushed to the margins, they have been left preaching to the choir, the congregation being immune and uninterested have left the building.

Our current phase of burning fossil fuels, industrialisation, environmental destruction, and other ills complained of by Greens is simply an undesirable but necessary phase. It is of course regrettable that damage should be caused to the planet, and to other species during this phase, but it is a necessary evil in order to reach the ultimate goals of the human project. It simply cannot be helped, because there are too many of us and we all want, the not unreasonable, comforts of modern life. We are at a crucial point in our project, and while the last three hundred years have been somewhat destructive, they have also been beneficial, and not more destructive than any of a number of natural calamities that have befallen the earth since humans evolved, and throughout the history of the planet. Think of ice ages and super volcanoes, to mention only those. The Green view is that we must follow a sustainable path, with low levels of consumption, and low carbon footprints, that we make do and recycle, and drive small cars, and use public transport, and an assortment of other small things that have no real impact whatsoever. Tim Flannery in his book, *Here on earth*, urges us to create and propagate memes that decry excessive consumption in the same way as the anti-smoking meme has spread throughout the developed world. The Purple view is that everyone on earth is entitled to live as we do in the West and can be accommodated if the appropriate technology is evolved. Along with Green ideology there is also a confused and irrational Christian view, which is even worse, and far more objectionable that we should get back to the pastoral scene, become sheep herders in simple costume, with old faithful dog and pan pipes; no doubt. Because that's how they lived in biblical times, and if you want to be close to their philosophy then you ought to live like them, because ultimately the end is so close it's not going to matter anyway. Or even worse again, the Hippy, New-Age Commune approach, that we should all live in the woods, and learn to forage and hug trees, and commune with, and be at one with, Mother earth. Such solutions are simply unworkable, because they cannot be rolled out across the planet, the only people who can live at such low-carbon sustainability levels are wealthy people from the West who own sufficient land to conduct such temporary experiments. And while the Greens, and the Hippies, and the proto-Christians are all happily communing, or seeking pastures new, the rest of us can starve or kill each other for scraps of food in the failed cityscapes, ruined buildings and burned out cars of a dystopian techno-wreck.

There is frankly a great deal of alarmist nonsense published about the state of human society, the pressure of numbers, the decline of this, and the rise of that. Much of it is clearly designed to be controversial and provocative. Purple opposes this narrow thinking and seeks the opposite approach, to optimistically enable people to do whatever they like in a guilt free environment in the sure and certain knowledge that scientists and technologists will find a way to solve the problem. In the movie, *Barbarians at the Gate*, which ought to be compulsory viewing for every Green fanatic, the character of F. Ross. Johnson, refers to lawyers and bankers, saying that all he wants from a banker *is a new calendar every year*, and the only thing he wants from a lawyer is to be sure *he's back in his coffin before Sun-up*. Some people today apply similar language to refer to scientists and technologists, which is shorthand for their proposition, or at least is the perception in certain quarters, that technology cannot solve our problems. But we also know that perception is not reality and while scientists and technologists have something of an image problem, it's not quite to the same degree of public odium as bankers and lawyers. Science and technology have delivered the world we now inhabit, good and bad, and the simple fact is that there is no one else to solve the problems and deliver the future we want. Much of what has been espoused by the Green movement over the last few decades is simply misguided at best, and plain wrong at worst. In discussing the earth, the atmosphere, the oceans, and other complex systems, perturbation results in a change of state, which can never be quantified, based on the initial conditions alone. In a world governed by chaos and complexity, perturbation of these systems leads to new, different and unpredictable states of the same system. These perturbations are on the whole not capable of being quantified with any acceptable degree of accuracy. Clearly the measurements tell us that there is substantially more carbon dioxide in the atmosphere than ought to be, and it is probably causing the climate to change and the planet to heat up. But we need not necessarily be too concerned about this, we are in a transition period from burning fossil fuels to direct harnessing of the Sun, it's going to take some time, it's a global problem requiring global money and real leadership. We probably have one hundred years to actually get it right.

There is a prevailing view among many authors, Tim Flannery et al, that if everyone on earth lived, as we in the West do today, we would bankrupt the planet, or that we in the West must cut back. Resources are scarce, and so on, water, oil, and everything else must be conserved and used sparingly, in case we upset some delicate balance. Purple as a philosophy profoundly disagrees with such poorly conceived and simplistic notions, the contention is that we may continue as before consuming all before us. Sustainability is a myth, everyone in the world today living in poverty has a self-evident right to live as we do. That is, at the highest standard possible. And we all have the right to live as billionaires, if that is possible, and it is. We all have the right to live to the level of the late Gunther Sachs, for instance. Our rights cannot be translated into reality right now, and remain only aspirational, but what we can see from the technology we have unleashed on the world to date is that these aspirations are possible, and more than that, they are very likely to be made real by the time the transition period is over. It needs to be said again, the only difference between what comes out of the tailpipe of a car and what goes into the gas tank, is information, and it is information that at this moment in time we lack.

How very patronising and colonial of us to preach to the poor the doctrine of waiting to see how things pan out. To hang on as best they can for some indeterminate length of time, after which, we may get round to helping, so long as it's not inconvenient to us, or impacting on our affluence. The poor can get their water from a communal well with bucket and rope, or out of a dirty puddle, and live in a mud hut, walk ten miles in bare feet to school. We may someday get around to helping. We send them our pennies along with our toxic waste, and obsolete technology. The old notion of charitable giving has always been, from Victorian times to date, derived from biblical sources, those who give should be celebrated and applauded, and those who receive should be grateful, two repugnant notions wrapped up in one detestable concept, as the late Christopher Hitchens oft opined. Simple charitable giving will not solve the problems of people outside our wall, it serves only to ease the odd individual conscience. Big problems required big ideas and big solutions, the only way forward is technology, big technology. Are they not human too? If they are human then they have an entitlement, deriving from that simple fact, to live as we do. The West has won the debate on every level, and our values and technology prevail. There must be a massive transfer of wealth from us to them, and a massive creation of wealth by us and by them. The continuing and seemingly never ending portrait of despair and human misery that is the continent of Africa is just as bad and as sorry a sight today as it was when Gordon was holding out in Khartoum, and Victorian notions of Christian charity were being formed. Yet the continent remains the richest place on the planet in terms of natural resources, and by comparison to other parts of the world, underpopulated.

The most compelling reason why Purple is the new Green, and a better Green is the assertion that stagnation goes nowhere. Sustainability at today's levels is the same as stagnation, which is the same as giving up, which is the same as condemning most of our fellow humans to remain as they are, living miserable lives devoid of technology or comfort, with little or no future. Our present position in time is transitory and cannot be the end point of any trajectory of the human project. The billions before us did not die just so we could stop here. If we stop our adventure at this point our species may as well end here. We have not made sufficient progress to call our project complete. The consequences of stagnation are unthinkable, in particular the denial of hope to the billions beyond the wall of our cocooned western model. If this is all there is, and this is as far as the human project goes then we have simply failed. And if that is all that Green can deliver then Green has failed, and Green is the wrong way, and was always the wrong way.

Purple by contrast is an appealing get rich quick scheme for the entire world. People want to spend money without feeling guilty. We all suffer the blunt force trauma of modern existence, we need to learn not to feel guilty every time we flush a toilet or buy a chocolate bar, current human problems cannot be solved by cutting back on anything, or restricting technology. The only solution is to spend more, and develop more, more and better technology, everyone in the world needs to be brought up to our western level. There is no middle ground. We live on a planet that glows in the dark, because we made it glow, and we understand why it glows, and we can travel far enough away from it to see that it does indeed glow. We should express no intention of cutting back, because those of our fellow humans outside the wall don't want to be poor. We need to unlock the infinities of energy and matter to make the rest of the world glow just as brightly; it's the only way forward. Purple calls for an inflation effect, the inflation of technology and progress. The need for greed. We are altruistic and greedy at the same time, we don't need to go the Ayn Rand route; both need and greed can be accommodated, and are not mutually exclusive.

Stagnation was clearly the norm in prehistory, and in recorded history until the advent of the industrial revolution. A pre industrial revolution blacksmith would have found work to do in any period in the preceding twelve thousand years, there was no change, horse, wheel, copper, bronze, and iron, wood, thatch and weave, silk and cloth that was as far as it went. We cannot afford to stagnate; we have reached beyond the tipping point where stagnation in the progress of our species is a viable option. We stand on the brow of a hill, like Sisyphus we have laboured long to get the stone to the top, but unlike Sisyphus the stone stays put. We have always been at a crossroads always on the brink, close to the edge, there's nothing new in that, offering sustainability and making do solves nothing. Accessing the infinities of matter, energy and time is everything, mastering the information deficit is the only answer. What we have done already must and does count for a great deal, it is a foundation of unsurpassed solidity, and will not fail us, it will not be washed away by anything we have seen before. The foundation built by the Romans was unstable and soft, it was horse powered, and slave powered, and was washed away by a tide of theocracy and infighting. They fell into a stagnation trap, they grew to strain the limited technological base they had built, and could ultimately go no further. Their entire world was powered by grain, they did not understand the power of fire, and how to control it. We do. We will not go the way of Rome.

In his book, *Where Good Ideas Come From*, Steven Johnson, the author of several US bestsellers, and a contributing editor of *Wired*, tells us good ideas have historically come out of cities, and other innovation hubs. He traces the history of good ideas and concludes that with our modern technological environment in place, the conditions are perfect, there has never been a better time to have an idea, and that many more of our ideas will necessarily become good ideas. But we've always known that good ideas and innovation can emerge anywhere, brilliant minds and genius will always find a way. No one knows where the next spark will come from, what we do know, however, is that ideas come more easily in a society where free thinking is encouraged and rewarded, an environment where people are not constrained in their thinking or hindered in what they may think about. The western democratic tradition is such a set of conditions, we have the JFK spirit of going to the Moon. *We choose to go to the Moon,[ ]because it's hard*, we choose to go to the Moon because, *it's there*, because *we can* go, because *we're not afraid* to go, and because we're not afraid, we can choose to go.

It is therefore not possible for any such society to stagnate, it will always have its collective mind focused on the next big thing, the next problem, and as a result be vibrant and forward looking. You might think! Yet there was no further Moon work, no one has been back there since Eugene Cernan stepped back on board his flimsy Apollo spacecraft in December 1972. Technology has improved in leaps and bounds yet there is no Moon Base Alpha, as predicted in the 1970s TV series *Space 1999*, instead we got forty odd years of low earth orbit space programs, and what many consider to be a stagnation. Tyler Cowen, in his book *The Great Stagnation* points out that low hanging fruit was the cause of America's rise to world leadership, free land, cheap labour, abundant resources, and so on. He points out that America and the West are now in a time of stagnation, low to moderate growth, lack of opportunity, low wage growth and more, and all due to a failure to recognise that the low hanging fruit has all been collected and consumed.

The low hanging fruit argument has also been applied to technology and future developments by many authors, prophesising the end of science, or the end of technology, or the end of chip manufacture, or agriculture, or whatever we have. The proposition is that we find ourselves on a technological plateau with further progress difficult because the easy things have all been done. But there is an obvious cause for optimism which arises out of a recognition and understanding that the low hanging fruit is well and truly gone, and there is therefore, a pressing need to seek alternatives. Clearly the alternatives are all around us, and we are left with the self-evident fact that better technology is the only way to go. Anyone who advocates an end of technology and a return to a pastoral existence could easily be equated to a genocidal lunatic on a par with Adolf Hitler, Pol Pot, or Stalin.   Make no mistake, a world without technology condemns billions of nice pleasant fellow humans to death. Marching backwards towards un-invention of all the goodies we love and cherish leaves only a pile of ashes. We cannot all live in the woods on berries and worms; there are not enough worms, berries or woods.

By the same token anyone who advocates that we remain as we are now, or try to share what we have among everyone who lives on earth is every bit a genocidal lunatic. The reasons are very simple indeed. There are too many humans on earth for everyone to live in some ideal pre technology, or frozen 2016 technology setting. It simply cannot be done. Ending technology and walking away from the technological present would simply cause billions to starve. Everything we eat is technology. Freezing the clock at where we are simply condemns all those outside of the wall we have built to be frozen out in the cold, we deny them hope, condemn them to a future no better than their present, which is a present none of us, within the wall, would care to adopt for ourselves or our children.

Another claim by Green people is that modern humans are in some way disconnected, there are those who claim modern city-dwelling humans are disconnected from nature in a way our ancestors were not, or that our experience of nature is a sanitised one, removed from reality are is therefore wrong. Clearly such people don't keep pets, do not engage in gardening, nor do they play golf, or hunt, or attend winter sports meetings or visit their local zoo or park. Nature is all around us we are deeply connected if we only look around. We are every bit, a part of nature. We are nature. Our technology is a part of nature. Tim Flannery again, he criticises both technology and what humans seem to have become as a result of using our technology, he describes people as, *"helpless self-domesticated livestock"*, a term most people would soundly reject as being as offensive as it is untrue. As an aside it is also interesting to note that such authors seem to discuss the human project absent any understanding of, or discussion of the corporate nature of our technological civilisation, and the fact that the next stage of human evolution is going to depend very critically on what corporations do with the many and various forms of artificial intelligence they are busy developing.

One of the core principles of Purple thinking is that better technology is possible and that solutions to even the most challenging problems are possible with better technology. Better technology is inevitable if the problem admits to a solution. Sometimes waiting on the better technology is by far the better option. Hence sometimes when we seem to be doing nothing about a particular problem we are in fact waiting for better technology, and therefore we are doing something. It's called the Proxima Centauri argument. Consider for example the technology needed to get to our nearest stellar neighbourhood. Suppose we wished to fund a mission to Proxima Centauri, a red dwarf star just over four lightyears from earth. Perhaps we discover an earth-like planet in close orbit, and we want to launch the mission right now. Suppose a movement is formed to pressure the international community to build a probe to go there, and further suppose that ten percent of the world's resources are consumed in the process. The probe leaves our planet travelling at the fastest possible speed our technology can manage, the journey is going to take one hundred years. Ten years later, technology advances in leaps and bounds, and we can build the same probe for one tenth of the price, it will travel at three times the speed. So this probe is dispatched. It overtakes the first but will still take thirty five years to complete its journey. Now further suppose, twenty years after launching the second probe, and both probes are largely forgotten, the wormhole technology developed by an artificial superintelligence can possibly send a very small probe to the nearest star in a millisecond. At no cost whatsoever.

You can see where this is going. Was it a good idea to send probes to the nearest star? Was it a good use of the world's resources, and so on? We are beset with problems of this nature for which there is a solution right now, but unfortunately insufficient technology. Getting a group of humans to land on Mars is one such problem, three years to get there, three years to get back, and a year on the surface. We could do it, but it's at the very limit of what we can do in space, so why not wait until the technology catches up. Perhaps, we could go back to the Moon first, and build everything we need to fully idiot proof a Mars trip while on the Moon. And then perhaps send massive resources to construct a base on one of the moons of Mars. From there it must be possible to construct using intelligent robots, a very solid base on the surface of Mars, ready and waiting for the first humans, so they may set down on a site fully prepared with all the comforts of home, and be in no doubt that we are on Mars as a long term project. Our infinity of time allows for this approach. There is no point in trying it next week, we don't have the security of technology necessary to do it right. Energy is another such problem. We know what the ultimate solution is going to look like, everything must become solar, eventually. There is currently no other option. And it's probably going to have to be based on the Moon, and be a massive international effort, costing trillions, and with the stated goal of providing energy to the entire planet. But until we get the international structures in place, and certainty of technology, countries like The Philippians will continue to construct large numbers of coal fired power stations; they currently have no alternative. We in the West could be accused of applying Proxima Centauri type logic in constructing vast wind farms all of which may be redundant by the time they break even. The list of problems is enormous, from population growth, to pollution, to global water distribution, mineral resources, and feeding the potential ten billion people expected to inhabit this planet within fifty years.

We know at least two things, the first thing we know is that all of these problems are soluble, they are not contrary to the laws of nature, and are therefore problems we can solve, the second thing we know is that these problems cannot be solved by Green proposals, sustainability is simply not strong enough. The argument in this book is that we need new Purple solutions to these problems, heavy duty solutions that preserve the western way of life, and offer the same or better to every other human on the planet. These solutions will only come with a new technological impetus. Purple solutions are different. They are technology solutions, not Band-Aids to stick over something to help us limp on to the next election, or the next climate change conference.

Ideas derived from nature, modified by humans and applied to the problem are what we need. For example, grass species used on golf courses that can be watered with seawater would constitute a Green solution. The Purple solution is to use the original grass, evolve the technology to desalinate the seawater using solar power, and build the plant big enough so that water is never again a problem. Solutions to environmental problems can be Green and Purple, the same solution to the same problem, but the core principle of Purple is that sustainability is not an option, and we in the West do not need to cut back on our consumption levels. The solution is more technology, bigger and bolder, we need to grow our technology, and close the information deficit until we are in a place where the problems go away. On the face of it, many of the methods might appear to be Green solutions, but in fact they are not Green they are Purple solutions. Purple is a better Green.

Another part of this Purple thinking is the idea that things do not have to be as efficient as possible, wasting energy and resources is something we humans are good at doing. The first steam engine, Newcomen's atmospheric engine was installed at the head of a coalmine with the express job of pumping water from the mine. Its efficiency was little better than zero, but that didn't matter, because it was working within a coal mine and therefore the cost of transporting the coal was zero, the cost of the coal was also zero, otherwise the proposition would have made no sense. Were the engine installed at a zinc mine ten miles away, and had the coal to be transported to the engine, the costs might have outweighed the usefulness of the engine, it would probably have been more economic to employ men and boys with buckets. But other people saw the potential of this first steam engine, and for them the cost of transporting coal was a huge cost, and a big issue, they needed the power, and knew it could be done better, and eventually James Watt made it happen, and gave the world real power that could both be relied upon and reproduced.

Now suppose we discover a method whereby solar power is converted in the usual manner to electricity, and the electrical power is then used to drive a solid state converter whose job it is to extract carbon dioxide from the atmosphere along with water vapour, and convert them into alcohol, or any number of a series of programmable stable organic molecules capable of being used in an engine directly or capable of being burned to drive a steam turbine in the conventional manner, or deployed to make food or plastic, as required. This is the goal of a great deal of research around the world, it is difficult work because as every chemist knows, carbon dioxide is a very stable chemical waste product and does not reduce easily to something that can be used. The best we seem to be able to do at the moment is to reduce carbon dioxide to carbon monoxide using UV light and a metallic photo catalyst. Long term we are searching for a complete solution, of the solid state black box variety, a plug and forget system, perhaps a long way off, perhaps just around the corner. The company that evolves this technology will grow to be bigger than either Apple or Facebook, so the incentive is already in place. Of course we would like the process to be as efficient as possible. But at each step there is an efficiency issue, conversion of solar energy to electricity in the solar cell has a certain efficiency, conversion of the electricity into fuel has another efficiency, and finally conversion of the fuel into electricity has yet another efficiency. But finally the efficiency is of little concern if the process works and is clean, we have an infinite amount of solar energy to play with, and if the entire process is only one percent efficient, we simply build the damn thing on an appropriate scale. It's exactly like having an inefficient steam engine working atop a coal mine, get it working in principle then we can worry about improvements. The Purple solution is developing cheap and robust technology to use carbon dioxide and water vapour as the input raw materials, and makes a damn sight more sense than mere conservation of what we have.

Another reason why Purple is the new and better Green is visible in examining that ultimate solution in search of a problem; the electric car. It could be argued, Electric cars are silly, pointless and going nowhere, and more than that, they are completely unnecessary. They are a good example of an immature technology being foisted on an uninformed consumer base for questionable philosophical reasons, and at an extortionate price. Electric vehicles are an example of the environmental lobby, hell bent on human stagnation. They would have the compulsory introduction of small electric vehicles for everyone, vehicles barely large enough to accommodate a single driver and a bag of groceries, for a short trip, vehicles that require to be charged up overnight to accomplish this modest task.

Of course we have Tesla, and of course they have big plans to roll out stations all across America and they have plans for more efficient batteries and you can get *up to* 300 miles per charge, but they are premium electric vehicles, and really only the tip of the spear, as it were. Real world electric cars must by necessity be small, light, and require to be recharged over night or over extended periods. They are a very good example of an aspirational technology, a kind of, 'wouldn't it be nice to have', technology. But we are not there yet. Not by a long way. There is an old saying in electronics, you can't shrink the PSU, what they mean is, the PSU, Power Supply Unit can only be reduced to a certain size. Power supplies are not efficient, they lose power as heat, and heat needs dissipation, the more power you require, the more heat will be generated, the laws of physics cannot be changed. Hence average electric vehicles of today have, the very distinct, golf cart feel, they require to be as light as possible, with everything minimised and slimmed down to the smallest amount, they have a very limited range, which is fine for a trip to the store, but not for a long cross country vacation with a family of four and all their many and various items of personal comfort. And more significant is the road haulage aspect of electric vehicle technology. Most freight is delivered by road, big trucks, small trucks all laden with heavy goods from polystyrene, to lead blocks, it all needs to be moved, and there is at the present time no more efficient engine than the internal combustion engine to accomplish that task.

Consider what a proper electric vehicle ought to be, in fact must be, to be acceptable technology. Consider a standard family SUV, like a Jeep, or a large Mercedes S Class saloon, our fictitious automobile contains a battery that can be charged, from empty to full, in the same amount of time that it takes to fill a conventional gas tank, and can travel the same number of miles as you can get from the full tank of gas. When such a vehicle can be put on the road at the same price, or better, as the gas version, then there may probably be wide consumer take up. But even at that, there will always remain those who wish to use the old fashioned internal combustion version, so called 'petrol heads', people who wish to feel the roar, or the purr, of a real engine on the open road, and we have, and will always have an enormous legacy of older vintage vehicles. Even in 2016 there are approximately four thousand steam engines licenced and approved to drive on the roads of Britain, over one hundred years after the last one drove off the production line. And if at some far future time such massive advances in battery, and power delivery, engineering, via bioengineering, or high-temperature superconductor, increased motor efficiency, or magic crystal technology, or all of these technologies, we will all be effectively driving green vehicles anyway. The same technology capable of delivering a, *proper*, electric vehicle will also find application in the home manufacture of the fuel required to drive a conventional vehicle, by efficiently absorbing solar energy, storing it, using it to manufacture the fuel via micro manufacturing technology utilising carbon dioxide, methane, water vapour and nitrogen, from the air. In other words, it is entirely feasible that internal combustion engines of the future will be entirely carbon neutral, or as green as they could possibly be. These would be *Purple* internal combustion engines. The substantial investment of time and effort, research and engineering that has gone into developing the internal combustion engine, since the first gas powered engines replaced steam engines in the pumping of flood water in Cambridge, in England in the mid nineteenth century will not be lightly set to one side. Steam engines still generate almost all of the electric power we consume to this day, as well as driving ships and other interesting applications. Just because a technology is old does not mean it needs to be abandoned. If it works, don't fix it. Tinkering is a different thing, but dumping a proven technology simply because of philosophical opposition is hardly justified, and almost seems spiteful.

We have no idea what might be possible down the road, perhaps even electric aircraft will be possible with advanced mega power generation, motor and battery technology, imagine a large passenger jet of today powered by an on-board battery. Whatever we do in our future vehicle choices, depends on our ability to manipulate energy, and convert it from one form to another, and do it quickly, and at sufficient power levels to make sense. It also depends on how we approach the massive number of problems still to be solved in this area, from increased efficiency of solar collection, to storage, to the manufacture of petroleum alternatives; it's all interconnected. The research currently underway in electric vehicles is not going to be wasted, the applications will be many and probably very significant all across the human project.

The Green movement are also quick to point out potential progress traps, the classic progress trap example given by many authors of early humans hunting mammoths, one mammoth caught in a trap, or a spiked pit, or killed by a rock dropped from a height is considered progress. The application of human ingenuity to hunt and subdue the larger mammoth implies humans hunting in cooperation; all good progress. But causing a herd to charge over the edge of a cliff is progress too far. A progress trap. It may or may not have happened, we don't know. This analogy extended to modern society, and the burning of fossil fuels for example, burning some is okay, but burning the lot is a progress trap because it will do irreparable damage to the planet. Again, perhaps it will, we really don't know, the climate is a chaotic system with non-traceable inputs, we don't really know, but everyone agrees that we really would rather not find out. However, even if humans did wipe out all the mammoths in an organised campaign of wanton destruction, no consideration is given, by the Greens to the fact that the very regrettable wiping out of mammoths, and very probably all other very large land animals, the prehistoric megafauna, did not lead to the extinction of the hunter. This predation by humans also led to the extinction of all of the large predators, in the process claiming the land for human settlement. It also had the effect of forcing humans through necessity to farm or hunt other food, and innovate their way out of the absence of mammoth meat, another version of overcoming the low hanging fruit theory. Clearly what appears at first sight to be a progress trap may in fact be construed as the opening of other paths. There is only one hope. More and better technology. The current set of problems faced by humans on earth are far too big and far too many for us not to evolve the technology we need. The alternative is to allow the problems win the day, but that is not how humans operate, it's against our nature.

Purple is not a different way of looking at the same thing, Green people are concerned about the environment and the planet, but what makes Purple the better way forward is that Purple is more than just concerned, Purple does not see any problem without a solution. We are not entirely independent of nature. The world we share with the plants and animals is a series of interconnected systems of various species in many different degrees of balance. But it is nowhere near as delicate as the Green movement have led us to believe, much the work of early environmental campaigners has been superseded, by an understanding that the biosphere is far more complex and far more robust than they were ever able to compute in early models. The recuperative powers and adaptation ability of the planet, and the biosphere are much more hardened and developed than we ever believed. The more disasters we see, both natural and manmade the more we marvel at the power of the planet to recover, and of species of plant and animal to bounce back. Purple opposes Green when it comes to Green assertions that there is such a thing as dangerous over-consumption, or that we can ever do irreparable damage to the planet or to its biosphere. But that does not grant a licence for wonton destruction. The atmosphere of the planet is twenty-five miles thick. The diameter of the earth is approximately eight thousand miles. The atmosphere is therefore the thinnest little bubble of protection, delicate and precious, so it makes no sense at all to poison it, there's nothing else between us and the radiation of space. We therefore need to look after it, and this attitude is not Green, it's Purple, the difference being that the Purple solution will be to develop the future technology necessary to repair any perturbation caused by previous technology. Not wanting the river where you like to walk your dog, or do some recreational fishing to smell of sewage or have the PH of battery acid is not Green, it's Purple, it makes good economic sense. We want it that way because that's the how rivers should be, not polluted or used as means of disposing of our waste. After all, what use would a Purple world be without clean rivers and old-growth forests? But if we have been guilty of such behaviour in the past, and we have. Then we need to devote some portion of our resources to putting the matter right. In a Purple world we will have accessed the infinities of matter and energy by eliminating the information deficit, our disposable wealth will be enormous by comparison with the resources of today, and the problems of today or yesterday, up to and including species extinction will become the success stories of future generations of caring and enormously wealthy humans.

Another compelling reason why Purple is the new Green is the recognition that unleashing the infinities also means understanding that we have an infinity of time stretching before us. We can therefore accept that leaving problems for our children to work out is a partial solution to that problem we may face. We can therefore come to a new acceptance that we humans as a species, or as a snapshot of the species in time have nothing to worry about. All the problems we create, in fact every possible problem we are likely to be capable of creating can be solved by technology, and in particular, if not by our technology, then by the technology developed by those who come after us. In effect our successors will clean up the mess, if it might be put in those explicit terms. It's a bit like inheriting an old house from your grandmother, it may be run down, with no modern conveniences, and the wallpaper is, '*oh damn*'! So What? You inherited the place, and with a little work, a bit of money, and a few weekends it's good as new. Our successors in title will inherit a property, which will perhaps not be in ideal living condition, but they will also have the advantage of the technological equivalent of economic inflation, and the technological equivalent of compound interest. No matter what the condition, the technological advances between now and then ought to provide them with the tools, the money and the time to effect repairs to their exacting specifications. What inflation does to money, technological progress will do to the problems we pass down to our future selves. Some problems simply dissolve with time, and there are precedents that we should note with interest, the prediction that London and New York would be under two feet of horse shit by 1910 is one such problem that morphed into oblivion. The Victorians had no choice in some of the problems they faced, they did not have the technology, or the capacity to solve them, and so they were left for future generations. Poverty, for example, along with many aspects of public health. Nor did the Victorians ever worry about how to get a robotic probe to land on Mars; the problem didn't exist. In fact all of the problems faced by the Victorians, from pollution, to disposing of garbage and sewage, to urban planning and disease epidemics were all dissipated by the technology of their successors, who are now our predecessors. We are in a much better position to provide for the poor of the world now than the Victorians ever were. Progress has been made and the poor have benefited, the rising technological tide has indeed lifted all the boats. We therefore should not feel in the least bit guilty that we pass on to our successors some of our more intractable problems, they will ultimately thank us for it. It seems only reasonable that we should place the very minor burden of saving the planet onto the broad shoulders of future generations. This approach makes Purple a much better version of Green, by allowing us to continue in our conspicuous consumption, or our dangerous over-consumption, with an easy conscience.

Another reason why Purple is the new and better Green is that for years the Greens have been telling us we are running out of stuff, peak oil being a prime example, Purple does not accept this proposition. So far noting has run out, nor does it look like anything ever will, there is no evidence to support the Green position. There are no shortages in a Purple world. In dealing with perceived shortages, it is necessary to understand that at the back of much of the current nay saying is the Malthusian negativity of Lovelock and other stalwarts of the Green movement, and their presumption that resources are finite and dwindling, that time is running out, or that the tipping point is almost reached. The tipping point has always been in sight for as long humans have been able to stand upright. For example, we have been at peak-oil since the 1860s, yet the world is awash in oil as better technology allows us to extract it from places never considered practical. The activities of the horizontal drillers and the *'frackers'* in the USA is a classic example of technology solving a pressing problem.

The contention of Purple is that we exist in an environment of infinities, energy, and matter, the only deficit is in the information required to match the one with the other to furnish all of our needs. The Green view on the other hand is that everything is in short supply and must therefore be conserved. The basic raw materials of the future will be carbon dioxide and water, the only difference between these two ingredients, on the one hand, and gasoline or plastic, on the other, is the addition of energy and information. Other resources such as water, minerals and land are not in short supply they simply need to be better managed and recycled.

There exist a standard set of presumptions pertaining to the everyday things people need and use, hoard and store, keep for later, or squander, in other words, our consumption, and most of the presumptions are either erroneous or unsupported by the evidence. These presumptions were derived or evolved at a time when it was believed that everything had its season, that, *waste not want not*, should be a guiding philosophy that conservation and stagnation were to be the watchwords of the human future. The presumption that resources are scarce, or that things in general are running out, or are even capable of running out, or that once a substance is used it cannot be used again is a nonsense. The proposition that we are in some way consuming, all before us, like a veracious plague of rats, even to the very elements of the periodic table being consumed faster than they can be replaced, is one of the biggest fallacies of the last two centuries, and is simply the wrong way to look at the world and the stuff of which it is made.

No one has ever gone to war because air was in short supply, there has never been a shortage of oxygen nor is there likely to be. It seems absurd to us that there would ever be a complaint that air is in short supply and we must therefore cut back. Water on the other hand seems to be forever in short supply, from California to the deserts of North Africa to the lush green of Asia, everyone seems to have a problem storing the abundant rainfall that pours onto the earth each and every year without fail, all powered by the sun. The global water problem can be solved by solar power, by a combination of large desalination plants resembling oil refineries, and local small stand-alone facilities. Proper storage and collection facilities are the answer, the problem of water is a solved problem that has not been implemented globally because there is no global agreement on how it must be done. Technologically the problem is solved, politically water remains a serious problem; but it's neither scarce nor running out.

Metals too are a similar story to air and water, the earth is full of metal ores, no matter where we dig a hole we find something. Economically viable is a different matter, and it is also true to say that many of the more productive mines are in remote places. But that is merely a cost of business and does not change the fact that there is no shortage of metals on earth. The other reason that metals are not a problem is that on the Moon, and in space generally, metals are in infinite supply, even a modest near earth asteroid probably contains more metals than all of humanity has mined since the invention of the shovel. The fact that we cannot get to them is just an issue of technology and information. One day we will get into space as easily as we dig holes on earth, and the infinite wealth of space rocks will be let loose.

The age of oil will soon come to an end, or is close to a conclusion. But not for the reasons usually advanced. We love oil, we are addicted to it and we need it, it is the most efficient method of doing so many things that simply cannot be done any other way. There is nothing humans have invented or can utilise that has the convenient energy density of oil. We will probably continue to use oil for as long as humans inhabit Earth. But as long as our activities involving oil continue, and are carbon neutral there is no problem, and we can plod on in our consumption of oil to our hearts content. The problem we face is that of finding a new source of future oil, and the answer is that we will make it ourselves. Again it needs to be said, all we are missing is the know-how, the wherewithal to convert carbon dioxide, and water, and nitrogen, into the complex organic components that comprise oil. The deficit is knowledge, but our ever-decreasing knowledge deficit implies that we will get there someday. It is an inevitable fact that technology will produce the solution, the solar powered black box that uses air as its input, and produces usable organic compounds as its output. Such black boxes will be manufactured in the billions and become a common sight across the world. They will probably make the extraction of oil from the ground a prohibitively expensive undertaking. Everyone in the oil business knows this to be true, and there's not a thing they can do about it. It is effectively a solved problem. The only issue is the cost. It will be one of those global events that will have deep and lasting effects. Just think of which way the politics of the Middle East might go when we in the West have no further need of their oil.

Our entire fuel and energy future is, of course, solar, and we are now in a transition to that future. It might take us twenty years, or one hundred years, but we have the time, and we know we will be able to make it work. Make it from the sun is where the money will be. That's the infinite source of energy that will ultimately power the human project for the next million years. The almost the worst kind of cliché to say the Sun puts out more energy in a fraction of a second than humans have used in our existence to date, but nevertheless true. It is entirely our fault we have not engineered the required solutions necessary to harvest this infinity. The facts concerning the Sun's output are readily available on the internet, and in numerous books on the subject. The only statistic needed here are the fact that only one billionth of the Sun's output reaches earth, and of that most is radiated back into space. This is the real Purple solution.

Food, or the provision of food via agriculture is another area of human endeavour where the advances of technology have provided abundance on a scale our forbearers could only dream of. Ploughing a field with a horse can only get you so far, our modern methods mean that there is more than enough food on the planet to feed ten billion people, if we ever need to feed that many. Further efficiencies are also on the drawing board, the concept of vertical farming and the technology of producing synthetic meat are in their infancy. When mature these technologies will cause almost all of agriculture to effectively move indoors and off the land. The days of the farm and the farmer as a small production unit in the rural setting are likely a thing of the past, the corporations controlling our food production want things to be more controllable.

People too, as a resource required for the human project, are also something in plentiful supply, nor is there a population crisis in the world. If the population of the planet peaks at ten billion or even twelve billion it's of no matter. Eventually it will decrease to a level of about two or three billion people. It might take one or two hundred years but it is likely to happen. Educated people on the whole do not have enormous families, they tend to be small with one or two children. This at least is the western experience. Hence the solution to the population crisis, if there is a crisis, and if it is a problem we want to solve, is to educate people. In particular the education and the empowering of women. This is the real solution to poverty and everyone involved in the poverty business is well aware that this is the case.

Land is also something everyone will say is in short supply, those of a religious bent proclaim that god who made the land only made so much of it. Yet the simple truth of the matter is that we live on a planet of more than sufficient land to go round, in fact it could probably be argued that we have infinite amounts of land. Up until the 1780s no one in the developed world even knew that there was a place in the great Southern Ocean that we now call Australia. A large continent sized piece of real-estate missing and unaccounted for, since it was stumbled upon about thirty thousand years earlier by a group of hunter gatherers, who continued to hunt and gather until their way of life was irrevocably disturbed by European settlers. There is more than enough land, water, energy and everything else humans need in Australia to support a human population of ten billion people. The land mass of the Australian continent is some eight million square kilometres. If we all moved down under, ten billion people could live a very agreeable life at western suburban densities. While an interesting exercise to show the theoretical possibility, in practice it is a clear demonstration of limitations imposed on human society by the current information deficit, in that we do not yet have the necessary information needed to implement these requirements given the entire land mass of the planet. The final simple truth of the matter is that there is no critical resource that has even begun to start to show signs of running out. We're not running out of anything, there is no commodity on earth that can ever be considered a finite resource. Infinite energy, and matter are manifestly available. Not having the technology to make use of the stuff that lies all about us is not the same as not having any stuff. Given time people and energy anything can be accomplished. With unlimited energy, unlimited resources, and unlimited time the only limitations will be those imposed upon us by ourselves.

Underlying a great deal of Green thinking is a latent mistrust of technology and modernity, and a wish not to be associated with it. The belief seems to be that we don't really deserve technology, we don't really deserve advancement, we don't really deserve to be the masters of the planet, there's something higher, something out there looking our way, watching over us, ready to thwart, ready to scold and punish, wagging a paternal finger. We need to feel guilty. We need to feel undeserving, like Alfred Doolittle in *Pygmalion*; the undeserving poor. But that's human too, and it's okay. Purple represents a guilt free future existence, a well-founded belief that we can have it all, and that we can do it in a way that is not disrespectful to either people or the planet, by harnessing the infinities of matter and energy, and closing the information deficit. If Greens are akin to extreme Marxists in their zeal and dogmatic idealism, Purple is more like moderate social democracy with a zero budget deficit. The Purple approach does not make demands on people it makes demands on leaders, it demands they get their collective finger out, so to speak, and do whatever is necessary to deliver us a Purple world.

Green doesn't work. Green politics. Green policies. Green taxes. The Green agenda won't work, can't work, what works are people. People work and the world moves forward. Science works. Technology works. Corporations work. Governments work. Their efficiency is another matter, another day's work if you will. But essentially the world we humans have built works, it functions and makes progress in a forward direction. We might not see this, or appreciate this day to day, but taken decade by decade or generation by generation, we have made and are making progress.

Purple was the old Roman colour of luxury and plenty, the colour of importance, the colour of the patrician class, and the colour of excess. It should now become the colour of progress and abundance, the colour of the future. It's not a romantic notion that technology will solve every problem. But that better technology or the right technology has the capacity to solve every problem. Many commentators maintain that technology alone cannot do the job, the Purple approach says that the right technology can do the job. Opposing Green does not imply an incompatibility with certain principals espoused by the Green movement, Purple is a love of technology and a love of the world, along with a love of the progress we make year on year. Much of the traditional Green agenda is not only wrong but is also unscientific, which is the more unforgivable conceit. Much of it is derived from an admixture of aspiration and nonsense, with its origins in the dropout culture of the 1960s and 70s and the *Heit Ashbury*, 'over thirty sucks', attitude. We cannot afford Green, Green is an exclusive doctrine for the very few, because a green planet could only sustain a very few. And for that if nothing else, Purple is the new Green.

PURPLE

Chapter 6   Inevitability and Infinity

Things that are invisible and things that don't exist can sometimes have remarkably similar properties.

Old physics truism....

Inevitability and infinity are two words used frequently and without hesitation throughout this book, and therefore require to be clarified as to specific meaning and application. Taking inevitable first. Inevitable in general means, not if but when, it means something is on the way, it's only a matter of time, it also implies something predictable based on current knowledge. If you drop something heavy from a high building, it will inevitably hit the ground, and so on. Infinity too has a simple first approximation meaning, and that is taken to mean a very large, more than will ever be needed. For example, if you lived on ten dollars a day yet had an income of one million dollars per year, you might consider yourself to be infinitely wealthy. Most people live for today with an eye on tomorrow, however, the future has a nasty habit of turning up unannounced and unexpected. Time ticks mindlessly by, and humans continue their problem solving activities; all of a sudden we're in the future; just like the 1985 of the movie, *Back to the Future II,* turned into the 2015 it was predicting. Survival into the future from here will be as a result of humans almost becoming a part of our technology; we therefore have no choice but to contemplate the future.

The Greeks imagined a man flying under his own power, but he ventured too high, incurred the wrath of the gods and fell back to the ground. Flying has always been a human dream, we knew it could be done because we could see the birds and insects, even fish doing it, and if the birds can do it, we can too. Leonardo Da Vinci imagined flying machines, he drew them in great detail, but his drawings bear no resemblance whatsoever to modern flying machines. The principle of inevitability would propose that if something can be done, and is not beyond the laws of physics then it probably will be done, if we continue working the problem, and if it makes sense for whatever reason, to individuals or groups of people, to governments or to companies to do it. Because the Ford Model T solved the fundamental problem of low cost personal transport, its take up was inevitable, despite the fact there were almost no roads at the time with a suitable driving surface. The associated events that followed that success, such as, improved road networks, supporting industries, oil refining, and so on, were also inevitable, and could have been, and indeed were predicted. This logic of inevitability applies across almost all of human technology; certain imagined future outcomes will always be inevitable once a pivotal problem is solved. From our perch on the shoulders of yesterday's giants we see so very much further than was ever possible before, and that implies a greater certainty, and a greater precision in our ability to forecast inevitable outcomes.

We exist in a time of continuing progress and there is no sign that the rate of progress is slowing. Whether it will continue at today's pace or even follow an exponential trajectory is something we can argue over, but progress there is and growth there is also. As a species we love to solve problems, and as individuals too, this innate ability is what brought us to where we are today, and from this obvious fact certain inevitabilities obviously follow; if a problem is perceived to exist, people will get to work on it. We also recognise that on the whole, groups of individuals can achieve more than individuals; and that motivation and incentive usually fuel achievement.

The addition of time to almost any technology results in change. Sometimes the change is improvement; sometimes the change is mere design. For example, the Wright brothers solved the problem of powered flight in 1903, over a century later the shape of aircraft is pretty much as they designed their first powered aircraft, wings and a method of propulsion, were they to visit an airport of today they would fully understand what they were seeing. They would not see cubes and spheres, or pyramids floating inexplicably in the air, for example. Sometimes these changes lead to new foreseeable consequences, and sometimes to unexpected consequences, but always to a changed state. Certain inevitabilities can be inferred from one state in predicting the next. For example in the 1990's cell phones became a mass-market product. Networks were built out, the customer base grew, predictions were made about the future, devices would be smaller, faster, easier to use, more features, on-board processors would become faster, memory more plentiful and cheaper, power requirements would become less, the addition of the internet would lead to more uses for the new networked devices, and so on, were all predicted. Most of it came to pass or was surpassed, the *iPhone* and all that surrounds it, the distinction between the pre *iPhone* and post *iPhone* world is stark, the concept of Apps and the entire ecosystem surrounding them, was probably an unintended consequence; sometimes we get more than we thought we would.

Given the notion of inevitability, our continued creation of better technology, should by extension, eventually and inevitably lead to a day when all human problems are effectively solved, and further innovation is simply not required, John Horgan's 1996 book, *The End Of Science*, discusses this assertion, that our best discoveries in both science and technology were made in the past. It is self-evidently not the case, since by definition, there cannot be an end of something that is of itself incomplete. What this means is that science is a field of human endeavour which is self-evidently incomplete and unfinished, you need only take the time to look in a mirror and ask if you are immortal. Of course you are not. Therefore medicine could not possibly be considered a finished product. And since medicine depends critically on inputs from all other sciences, and from all branches of engineering, it is quite clear that these fields too remain open to further discovery, innovation, improvement and refinement. A field such as religion or astrology, on the other hand, might be considered to be finished or closed by virtue of the fact that the field itself is a proven nonsense, and therefore further work yields nothing only additional nonsense, with no discernible advance other than the inner refinement of that which has already been shown, by science, to be irrelevant apple polishing by the otherwise unemployable.

We might also consider certain societal inevitabilities. Given an autocratic despotic system where people are afraid to think out loud, it is inevitable the society will get only so far than stagnate, this follows from the basic human instinct to be able to follow their own path, and make as much money as they can by deploying their skills and attributes as they see fit. No reasonable human wants to be locked into a state of doing only what they are told. Given a society founded on enlightenment principles, where human rights prevail, and people are given the freedom to create and express ideas, and the encouragement to make money and create wealth, there will inevitably be a growth in culture, leading to the potent combination of technology and the freedom to pursue wealth with the inevitable explosion in growth and prosperity. And everywhere this social structure is deployed the result is the same; vibrancy of culture, creation of wealth, and a largely contented population.

We live in a world where certain facts are obvious, and by virtue of that fact certain other facts may be derived or follow naturally as a consequence to become inevitabilities. The Sun shines is an obvious fact, it produces a billion times more energy than strikes the earth, and of that which strikes the earth, only a very small fraction is used, in fact, if all of the sunlight striking just a very small area of the United States was captured and stored each day we could power the entire planet, and have lots left over. That simple set of facts leads to the very inevitable conclusion that eventually we must develop the technology that enables us to efficiently capture and store this abundant energy, and put it to use. And if our technology is sufficiently improved from today, then it must inevitably at some stage become very much cheaper to utilise abundant solar power rather than going to the expense of pumping oil out of the ground and refining it, or operating coal mines.

Given the starting point at which we find ourselves in 2016, certain specific inevitabilities may be expounded without fear of contradiction, they are not predictions in the sense that a date is suggested or a specific type or format is advanced, rather there is a generally discernible fact that may be stated as an inevitability given the initial conditions.

The fact that intelligence exists in many creatures, from fish through birds and mammals right the way to humans at the top of the pyramid means that intelligence is something we ought to be able to duplicate in an artificial way. The very existence of the human brain implies that the creation of artificial intelligence is inevitable at some point in the future. Already we can cram an awful lot of information into a very small space, it seems therefore only a matter of time before these small spaces may begin to surprise us. Couple that with the fact that every corporation in existence wants AI and is investing heavily to make it happen, and you might reasonably think you arrive at a very unambiguous inevitability. The specific formats or timescales are of course capable of being discussed, but the inevitability of the occurrence seems beyond question; you might think. Of course it is not known whether or not the human brain is a one off, or whether it is an inevitable evolutionary outcome. We have only one example. We simply do not know if intelligent, technology wielding, life is common in our universe or if we are the exception, the one-off freak of nature. Despite the fact that intelligence is reported in many other species, technology has only emerged in one single species, that we know of, to date. The dinosaurs didn't do it, and they had a great deal more time than we ever had, there were dinosaurs with brains as big as modern monkeys, tens of millions of years before the asteroid struck, but they never evolved further. We also have the examples of isolated populations of mammals in places like Australia and Madagascar, and the Americas, as examples of where technology did not evolve, even though they had as much time as the mammals in Africa. So, the usual caution must be advised when claiming certain paths are inevitable, the only thing we can say for certain is that because we have the example of the human brain, we know that such a computational capacity exists. We are therefore determined to do whatever is necessary to translate that capacity into our technology, and repeat natural evolution as far as we may be allowed to go.

Viruses, bacteria and the cellular basis of life are being investigated and understood as never before, because we know from our studies that these tiny factories effectively produce everything we use as a species. The fact that we do not control, and do not understand this area of knowledge in anything like the detail we aspire to, means that the research will continue at full steam. The fact that it holds such promise in terms of the potential bounty of new materials and methods, implies that a vast amount of research funding will continue to be spent. All of which leads to the inevitability that we are entering a new era of nanotechnology and useful artificial life.

Long-lived animals and plants spur our interest in human life extension, the field is of considerable importance to every human alive. There is a certain grim inevitability in it, you could say, but because this problem attracts so many brilliant people and such vast funding, the inevitable progress must be expected.

The general human aversion to work means corporations, by their nature, and because of the nature of the underlying law, are driven to achieve greater efficiency. Companies will continue to automate every process and production path they can, this process is already underway with a vengeance, and must result in inevitable displacement of vast numbers of people.

Because most of our vast future wealth is to be found in space, space exploration on an increased scale is inevitable with consequent predictable technological advance. The problem of overcoming gravity coupled with improvements in all kinds of technology means that what once required a satellite the size of a bus to get the job done may in future be accomplished by something the size of a beer can. And why stop there if the same job can be done by an instrument the size of a cell phone. And if AI is part of the space future, it should be small matter to send intelligent decision making probes to anywhere in the solar system to stake a claim or establish a foothold.

The perception of a problem implies an attempt will be made to solve it, leading inexorably to new technology or to a refining of existing technology and a redefinition of limits. While there is no analogue in nature for some elements of our technology such as, a bicycle, or a space shuttle, or a laser. There is also no analogue in our technology for, an ant, or a spider, or an oak tree. Yet we can see that if the technological analogue were to be invented it might be usefully put to work doing for us the exact same things that ants, spiders and oak trees do in nature. Not just the harnessing of these creatures as we already do with bees, but the independent creation of technological analogues. Armies of synthetic ants capable of building structures in space, grain by grain, as real biological ants do here on earth. Or how about creating the synthetic equivalent of fungus, we see them at work and we envy, they are capable of infiltrating ice and rock, just the job for use on asteroids, or in mining deeper than we ever thought possible. Our inability to do these things coupled with our relentless ability to dream leads to the inevitability of entirely new branches of future technology.

Of course further invention is inevitable, because what is common to all invention is a problem to be solved. An invention solves a problem, important or trivial, global energy needs, or how to stop a martini olive being left in the glass, a problem that gave rise to the cocktail stick, and the entire cocktail stick design, manufacture and distribution sector. The other common thread is the human brain. If a problem exists then it exists for more than one person. If a significant problem, it is also a highly visible problem, with well-known dimensions, many people are working on it, or thinking about it, and if soluble there is likely to be a common solution. Both Newton and Leibnitz worked on problems which individually and separately required the invention of what Leibnitz called calculus and Newton called fluxions. Darwin and Wallace both arriving at the theory of evolution by natural selection over the same time period having both read the essay by Malthus on the reasons why populations of animals exceed food supply. It happens all the time, many scientists on reading the same paper or abstract are moved to think in a certain way or perform an obvious next step.

The point is not that the light bulb was inevitable, because it was inevitable, but that the problem existed to which the light bulb provides a reasonable and workable solution. Problems exist and solutions are found to solve these problems, and if the solution is reliable and practical it is retained, copied, improved, and so on, to become part of our technology. The steam engine was invented to solve a particular problem. Other problems all of a sudden become soluble such as how to run a mill without a water wheel. Once a problem is solved, the solution invariably finds other applications. Other discoveries are not necessarily as inevitable, the World Wide Web, for example or its derivatives, the search engine, and social media platforms.

The term infinity requires some elaboration as to its specific meaning in the context of Purple. It is an easy term to use, but not necessarily easy to define what infinity means, the concept is relative, and depends on the method of measurement, and the specific nature of the subject under discussion. For example, the Sun has been shining for five billion years, and will do so for five billion more, the earth receives only a fraction of its output, and humans use only a very small fraction of that, by any definition therefore the Sun is an infinite energy source. The same argument is made for matter, most of what we use, eat, wear, live in, desire or can imagine is made of a very few of the 92 naturally occurring elements of which there seem to be infinite amounts. For example, carbon, oxygen, hydrogen, nitrogen and iron are in such abundance on earth as to constitute an infinite source. Of the other elements, what we have mined so far is but the scraps from the table compared to what exists on earth and beyond. We might also confidently state that we have an infinity of time, because destruction of our world, via nuclear war, grey goo, virulent pathogen, or gamma ray burst, are such unlikely events this side of us having the technology to deal with such scenarios, that we are more than likely going to be around for a very long time indeed.

There are of course, different embodiments of the term infinity, what it means to a mathematician, physicist, philosopher, priest, economist, politician, or a corporate CEO, will not be the same thing. Philosophers and assorted religious individuals have the silly notion that infinity is, in some way, a number, and can be treated as a number therefore delivering, on command, the hoped for, logical absurdity, for example, if you have an infinite number of apples and you take away five apples, you still have an infinite number of apples, hence there is a logical absurdity. Infinity is not a number, pure and simple. Numbers can converge to infinity, which is a different thing altogether, for example, counting from zero adding one each time gives the sequence of real positive integers, said to, converge to infinity, or become infinitely large. However, if you take an infinitely large number and count backwards, you eventually reach zero, a very definite limit.

Extending matters out as far as the entire Universe, we find that the universe may be unbounded and expanding forever into the infinity of space, it seems to be this way as measured using the cosmic microwave background. But counting back, at least as far as the evidence is concerned, there is a kind of definitive beginning. We call it the big bang, because we don't yet have the physics to describe the proposed quantum singularity.

Sometimes the concept of infinity is surprisingly easy to handle. We are often told, for example, that there are an almost infinite number of stars in our galaxy, and that most stars have planets orbiting around them, and clearly this almost infinite number is so big that no human lifespan could even count them let alone know anything of their contents. Even if we had the most powerful telescope imaginable and could see every aspect of any other solar system in the galaxy, we could never study them all. However the concept of self-similarity comes into play here. There might be a near infinite number of stars in the galaxy. But so what? They are pretty much all the same, in that they are all made of the same stuff, and behave in pretty much the same way, they are subtle versions of the same thing, of course there are infinite variations but again, So What? We see the same thing all around us, there might be an infinite number of oak trees in the world, or clouds, in the sky. But they are all the same, or at least self-similar, once you've seen one cloud you've pretty much seen them all, and one oak tree is pretty much like another. An infinity of variations is not the same thing as an infinity of different types. We can therefore expect to see repeated patterns of self-similarity everywhere we look, and no doubt enormous numbers of variations on the earth-like theme.

It is therefore possible to make a statement that as far as humans are concerned there is an infinite amount of energy available to us from the Sun. There is no logical absurdity arising from the Sun's finite nature. The simple and accepted fact is that our Sun will continue to light this pleasant world of ours, as it does today, for at least the next five billion years. The numbers don't lie, there really is that much free energy reaching this planet each and every day. We simply don't use it. Because we don't know how to use it. Plants do, and have been doing so for billions of years. But we are smarter than plants, and sooner or later we will learn to unlock this very real infinity by closing the information deficit; see for example the, *National Geographic Magazine*, for September 2009.

Along with the infinity of energy from the Sun there is an infinite amount of matter here on earth for us to manipulate with our infinite energy source. None of the 92 elements are in short supply, there is more of any element than we could ever use, including gold, and the other very rare elements. This is the Purple definition of infinity, an abundant supply so large as to never present a problem of short supply. And we might add to our list of infinities the infinity of people, we have so many people alive today that we will never become short of people for whatever purpose could be imagined. And to go along with an infinite number of people there is an infinite food supply, we produce far more food than we can ever consume. And we do it to such an extent that obesity is almost the normal state for a growing number of people in the West. In short we can conclude by saying that it seems a self-evident fact that we are surrounded by infinities of every conceivable quantity required to maintain and advance the human project. The only deficit is in the information available to us to efficiently and properly exploit these infinities so that the finite human population can have all of its wants fully and completely met, now and into the very far future.

There are clearly consequences and implications arising out of the fact that we are effectively surrounded by infinities of all we require. The notion of the Malthusian trap whereby we devour the entire supply of whatever named quantity you might care to choose is revealed as being a nonsense. Human technology renders all manifestations of the Malthus conjecture redundant. Nothing is in short supply, nothing ever will be in short supply. This fact carries the clear implication of infinite wealth.

We have seen something of what it is like to live in an infinite world when we look to certain nation states in the Middle East, namely the oil rich countries where money is no object. Their near infinite oil reserves, by comparison with most other nations, has generated near infinite amounts of wealth, and the ability to create whatever is desired, and create a class of individuals characterised by enormous personal wealth. We may therefore expect that future access to infinite amounts of energy, and infinite amounts of matter, and the advanced technology needed to allow the free interaction of the one with the other, will as a natural consequence, lead to hitherto unimagined levels of wealth across our entire society. Effectively we will create a state of infinite wealth on a global scale, meaning that both individuals and states would have access to the means to carry out their every wish. Of course we can expect the all the usual outcomes of an infinite life of abundance, excess, abuse, decadence, narcissism, perversion, and the like. There would be an effective population of billionaires pampered by semi intelligent servant machines, and wanting for nothing. We may find it hard to fathom the concept of an infinite world where everyone has whatever they wish as a consequence of unlocking the infinities, but wealth and levels of wealth have always been relative. Consider books for example, in the era before the printing press, books were laboriously and individually produced by skilled craftsmen, writing with quills on calf skin. A tenth century nobleman might boast a library of five books, and be considered wealthy. Today anyone could have a thousand books in a small apartment. Or even in a small pocket. With access many millions of other books.

The implications of infinite amounts of time are also interesting. Individual humans are obsessed with time, forever pressed for time; time is money, and so on. The universe is not pushed for time. And time is not weighing heavy on us as a species, we may therefore take as long as is necessary to do what needs to be done. If anything we've wasted a great deal of time hanging around doing relatively nothing much throughout the incredibly long human prehistory. It could almost be argued that only the last few hundred years of our history has any real significance. The implication is that there are no impending deadlines on the species. No looming calamities about to befall us, there is no requirements to get to the stars right now; we have plenty of time.

The prospect of effectively infinite future wealth generated by unlocking the infinities that surround us is tempting in more ways than one. The fact that we can in effect presume we are enormously wealthier than we are now, means we can in fact, borrow from our future selves. The concept of borrowing a finite amount from an infinite future is not an unreasonable concept, the size of the amount is irrelevant, as long as it is a finite amount. The future automation of the wealth creation process implies future wealth on a level we cannot now contemplate. We could and probably should consider this to be a cheque in the mail scenario. If you knew for certain there was a cheque for one hundred grand in the mail, you could and would happily borrow a few hundred, at any rate of interest, to see you through to the lodgement date.

Recognition of the concept of infinity also has clear consequences for the politics of the future. There is clearly no point in going to war over resources that are effectively lying around and easy to find. No one has ever gone to war over, who controls the world supply of sand, or the world supply of nitrogen, or the supply of fresh air. By the same token, what would be the point of going to war over oil when all that is needed to make your own oil is a steady supply of carbon dioxide and water vapour? And who would risk retaliation over a supply of metal when it may be cheaply, if not freely gathered from deep in the earth by an army of autonomous, worm like, machines. So a recognition of the facts of infinity really ought to diffuse tensions that may otherwise build by considering resources to be limited or scarce. They are not.

From these inevitable solutions to perceived and real problems, faced by our modern world will invariably arise some unintended and unpredictable consequences. Some of these outcomes are obvious and some less so. For example, we know nature can build such creations as a flower, an oak tree, or a fly, or a human brain. Human technology cannot. Therefore Human technology cannot be considered a finished enterprise. So it will continue, and progress, and advance until there are no further problems to solve, or at least none that really matter. Inevitable means that it must and will happen at some time, all we need to do is keep working on reducing and eliminating the information deficit, the difference between what we actually know and what we need to know. This action unlocks the infinities that surround us. We can see an oak tree is almost perfection, and a culmination of billions of years of evolutionary endeavour. But humans can neither build an oak tree of our own, nor live as long as one, so we have a very long way to go. But our current levels of science and technology are the most solid platform for further progress that we have ever been able to build. If we consider right now to be the beginning, there are no limits, David Deutsch calls our time *the beginning of infinity*. We stand at the beginning of a very long journey. It seems there is nothing likely to prevent us from making the journey, wherever it might lead.

Chapter 7   We Don't Believe In Magic

> *From Cahill's corner the reverend Hugh C. Love, M.A., made obeisance unperceived, mindful of lords deputies whose hands benignant had held of yore rich advowsons.*
>
> James Joyce, Ulysses

There is no problem any single human being faces, or has ever faced, or humanity as a going concern, faces today, or has ever faced in the past, to which religion in any of its many thousands of formulations could ever hope to provide a solution. This chapter concerns the defeat of theocracy, and how we have won the battle between ignorance and superstition, or between science and religion, or between faith and reason, or between enlightenment values and the top down autocratic approach. The concern here is how science and technology, and the rational, empirical, evidential approach have relegated religion to the same space as astrology, psychic predictions, spoon bending, ancient aliens, UFO sightings, alien abduction stories, and water divining.

Religion in all its forms is a divisive and retarding enterprise wholly unnecessary in our modern empire of technology. Religion makes reasonable people do, say and believe, unreasonable things, it makes otherwise intelligent people do and say things that are not in the least intelligent, and provides immoral people with the moral excuse and justification they need. It has been best said by the Nobel Prize winning physicist Stephen Weinberg, "*...in an ordinary moral universe, good people do what good they can, bad people do the worst they can, but if you want good people to do wicked things, you are going to need religion*".

Religion is nothing more than another form of magic, and we have no need to believe in magic. All that we do is rooted in reality; our guiding presumption is that we inhabit the real world. We must proceed from the assumption that we do not inhabit a dream, such as suggested by the famous nursery rhyme, *Row, Row, Row Your Boat*, or the movie *The Matrix*. Nor do we inhabit some virtual reality, either of our own choosing, as in a simulation we designed ourselves, or as a form of education imposed upon us as a kind of school project of the far future. Nor should it be considered a form of therapy, for example, to illustrate to us just how good or bad, things really were in the past, or as a form of punishment, perhaps a custodial sentence imposed for a minor transgression. Even though simulation theory is a sound mathematically based and valid proposition, and could be a reasonable explanation of our reality, it makes no real difference to human experience. Neither is there any evidence that are we under external control outside of ourselves. We must believe and operate on the presumption that we are here, we exist, it is what it is, what you see is what you get. We inhabit a four dimensional universe, three of space, and one of time, and at the level of our existence that space has a certain set of properties, it is flat, it is unchanging and unchangeable, it is unbreakable, we only know the horizontal and the vertical, there is depth and solidity of matter, and of light, and heat, and gravity, that is the reality we inhabit and everything we do must conform to it, and must be related to it, and rooted within it.

Hence we must believe in the laws of physics that space and time exist, that matter exists, that energy exists and that the relationships between these things exist, and are as science has discovered. In fact, we have no choice but to believe in the laws of physics, they are immutable and universal; they came into existence when the universe formed and are its very structure. We must accept that the passing of time occurs in our lives in one single direction, and cannot be made to speed up or slow down except as allowed for in the theory of relativity, we must accept, for example, that time travel is, among other things, a violation of the first law of thermodynamics. Of course quantum theory is more than a bit strange and photons, particles and even atoms have been observed to behave in what we normally consider odd ways, but we are far removed from quantum effects in the real everyday world of our direct experience, and hence we can only deal with what we experience directly through our senses, or what we are told by reputable scientists and others. We dwell in the intermediate, in the space between the vagaries of the quantum and the extremes of gravity, on a logarithmic scale at least, we are half way between the largest and smallest things in the universe; as we know it, as we have revealed it to be.

So we start from the presumption that reality exists and we inhabit it, a kind of logical positivism, if you like to attach philosophical labels, the kind of philosophical label, Stephen Hawking believes, our understanding of the laws of physics and the structure of the universe have made redundant. Magic does not occur in our reality, superstition has no place, snakes do not hold conversations with sane and rational people, the Sun cannot be stopped in the sky, and dead men do not get up from their bed and continue the next phase of their life, as if they were never dead. Water remains water, it is not a solid surface upon which one may perambulate, nor can it become dark, and exude a grape-like aroma.

Magic in all its forms, has always been, and remains to this day, one of the principle retarding factors preventing us from achieving a Purple world; our deployment of rationality to defeat magic is one of the most important attributes of a maturing society, it is a measure of the strength and solidity of our civilisation. However, magic in its many forms still conspires to retard our attempts at progress, it even occasionally protrudes violently into our society, and continues to act as a brake, we are therefore not quite finished yet, and a great deal of exorcising of magic remains to be done. Our objection to magic must be that it provides no insights, no answers, tells us nothing, and reveals nothing. Invoking magic as an explanation for anything we observe in our reality simply stops the process of questioning, it stops the learning process dead, if magic is the answer than we need look no further. Saying, god did it, is simply a copout, it is tantamount to saying, I don't know, and I don't care, or, I don't know, and I don't want to know. God did it, is not an explanation, it is an abrogation of our responsibility to know, to gather knowledge and find stuff out; it is a lazy and indolent attitude to individual existence. There's an old saying in politics, *if you have to resort to violence then you've already lost your argument*, this is exactly the reason why magic is unacceptable in any discourse in our modern world, *if you have to resort to magic then your explanation is nonsense*. It really is that simple.

Why magic persists is one of the big questions of our time. Magic persists in our world because vast tracts of humanity persist in believing in it, and therefore magic clearly exists and persists. Its continued existence also owes much to the fact that a great many people derive a very reasonable living out of peddling magic solutions; if there was no money in it, it would be gone by next Tuesday. By magic is meant, any and all explanations for any phenomena, experienced by our senses, which are neither rational nor scientific, or those that are based on unproven or impossible to prove, assertions, presumptions, suppositions and inferences. Any explanation, or claim directly contradicting the known laws of physics. In fact it may be plainly stated that if any claim directly contradicts the laws of physics, that claim is nonsense.

The most vicious and pervasive form of magic in our world today is undoubtedly religion, and its invocation of magic must be the primary objection to its continued existence and prominence, in our culture and society. Because religion relies on magic for answers to questions, and to support the outrageous and extraordinary claims it makes, it has the effect of punctuating or terminating the learning process and retarding rational enquiry, it stifles and calls a halt to natural human curiosity.

At last count, according to United Nations records there were over four thousand separate and distinct religions operating on this planet. That means over four thousand different opinions, and belief systems all separately correct, and in possession of the one single and absolute version of the truth behind the universe, and underpinning human existence, where we came from, where we might all end up, and how we ought to live while we are here on earth. Each separate version makes its own set of extraordinary claims, contrary to the established laws of physics. There is variety, if nothing else, four thousand is a lot to choose from and clearly there must, by definition, be differences between them, because they are separate and distinct. Hence, it is by the smallest leap of axiomatic logic that we conclude they cannot all be right, or be the one single absolute truth behind the universe and human existence, and if one version is wrong, then why not the whole lot, why not all four thousand? That is the most reasonable result, given the preponderance of opinion.

There is no intention here to discuss religion in general, its different forms, global distribution, numbers of practitioners or any other aspect of the phenomenon. There is such a wealth of garbage written on the subject of the finer dichotomies of the various sects and factions, there is no point in adding to it further. Our concern should be that religion has been, and remains, a retardation to technological, and therefore human progress, a roadblock on the way to a Purple world. Religion is out of place in a modern technological world, having been disposed of, by science, long ago. Nevertheless, it stubbornly remains with us in the world today, like a foul odour hanging on the air, it continues to assail our nostrils; its persistence at remaining with us is not difficult to understand.

Most scientists abhor religion; they see it as an irrelevant anachronism, and something we would do well to leave in the past, along with stone tools, and witch burning. Although we must be troubled that according to widely reported surveys in the United States, 90% of the general population believe in a personal god who responds to prayers, and involves himself in human affairs generally. The figure reduces to 40% for scientists, and for the elite members of the national Academy of Sciences, the number is 15%, which is both deeply troubling, and the reason why it can only be stated that most scientists abhor religion. Most rational thinking people also agree with most scientists, and in fact, most people capable of honest rationality, agree with most scientists. But stubbornly it remains as part of this world, because clearly a great many people cling to it for a variety of reasons. And while the comments here are directed at the Christian version of this nonsense, because this book is written in and directed at the Western World where the dominant religions are flavours of Christianity, the observations must, necessarily, be taken as applying equally to all four thousand varieties. In fact, trying to distinguish which religion is less nonsensical than another is a bit like trying to resolve degrees of probity among politicians. Why would you bother? We must operate from the presumption, that there is no god, or rather from the self-evident fact, that there is no credible evidence for the existence of a god, or for any other extraordinary claim ever made by any religious sect. There is no god in existence other than as a figment of human construction, we must also proceed with the assumption that there has never been a god, and never will be a god, and that humans have no need for a god. There is no human problem outside of theological nuance and narrow philosophical endeavour that requires the existence of a god or gods.

There is a counterfactual history argument concerning Rome, along the lines of, what might have happened if the Romans had invented the steam engine, the decimal number system, the printing press, gunpowder and any other of a variety of fundamental enabling technologies. The conclusion is always the same; that if these things had been invented in the Rome of the Creasers, we might have been born among the stars. Instead, the Romans faded away, their society and technology were lost, and we got sixteen hundred years of Catholic theocracy from which we are only just now, in the last few hundred years or so, liberating ourselves, in fact our liberation can almost be dated to the first peering of Galileo through his home made telescope. It is logical to argue that we are still trying to wash the blood from our hands, and the stink from our person. The objective rational view on the subject is quite straightforward, there is no problem that humanity or individual humans now face or have ever faced to which religion provides a solution. The interested reader is referred to the book by Richard Dawkins, *The God Delusion,* and to the book by Christopher Hitchens, *God Is Not Great,* for a complete exposition of the malevolent influence religion has had over the years, since grabbing power from the

Romans, and to various books by Sam Harris, and many others on a similar theme, for cogent arguments to counter religious claims. These authors have done the work, as it were, on god and religion, and there is no need to effectively repeat what has already been said. However, there are a few points, which ought to be made on the subject, and one or two aspects that are worthy of restatement or embellishment, because of the way religion continues to permeate our world, and pollute many aspects of rational discourse, and because of its relevance to achieving a Purple world.

Liberation from theocracy is our aim, religion has over the centuries, effectively hijacked a great many aspects of human society. Perhaps, everything that humans have ever done or have ever tried to do, has been tainted by religion in one form or another, and ascribed the status of a gift from some omniscient deity, for which we must endlessly praise. In doing so, religion has interceded in human affairs in a perverting way, altering the manner in which progress could be made, acting as a filter, and a barrier, like a glorified Maxwell's demon allowing certain ideas through and preventing or diverting others. Religion is by its nature derived from the minds of humans, and purports to impose a set of magical explanations onto the physical human world offering what Karl Popper and David Deutsch, refer to as *bad explanations* for ordinary human activity or the natural observations of the human world and the universe of which we are intricate. These bad explanations were then enforced by violence giving rise to what Jacob Bronowski characterised as a culture of *ignorance, arrogance and dogma*. It is possible to make a long list of hijacked properties and observations to include such objects as, thunder and lightning, the Sun, the stars, the movement of the planets, the arrival and departure of the seasons, the growth or plants, childhood, and so forth. These, along with more ephemeral qualities of the human world such as love, charity, human kindness, beauty, law, morality, and many others, were all claimed, by those practising religion, as being god-given and as such requiring no further inquiry or explanation by humans. Acceptance of what was handed out by the almighty was mandatory, and was not only professed as the singular truth, it was positively enforced by the certainty of violence, *whoever is not with us is against us*. And whosoever is against us may be freely burned at the public stake or hung from a gibbet or have his or her head forcibly removed.

Charity is but one example of a hijacked human attribute religions claim as their very own, activated and motivated solely by god carried on by humans in order to please or appease some divine presence. The simple truth of it is that charitable giving has always been a part of human existence, the care of the sick and the young, the formation of bonds between people in a society or community, bonds that held us together in small groups beyond the family, as a protection against the hostile biosphere. Charity is as innate and self-evident a part of human existence as is walking upright. Otherwise we would not be here. In every archaeological dig where human remains have been discovered, it has always been the case that skeletal remains are found which exhibit injuries that had repaired over time, injuries that would have killed the individual had they been left alone, clearly indicating that the injured party had been cared for and looked after, while the injury repaired itself. It is a natural trait of the human species, to care and reach out to others, the simple fact of our evolution is that if we cared for the old, they in turn, would care for the young, we could then hunt and provide food. Wild animals probably ate those early humans who found themselves alone, or who shunned their fellow humans. It really was that simple. Humans are simply not physically fitted to survival as loners in the wilderness. We are far too vulnerable, far too fragile.

Charity, we have always been told, is a Christian virtue. Only Christians give. An atheistic society would be devoid of altruism, atheists would follow the Ayn Rand approach of exclusive self-reliance, and zero altruism. Clearly, as charity and altruism pre date any Christian church, that is not the case, and we may therefore liberate altruism from the clutches of the all-grabbing churches, and the 'preach and screech' brigade. We like to give, of our time, and ourselves, and get a kick out of it, giving generates endorphins. Giving some loose change to a person on the street can often generate a hit more potent than a double espresso. And charity is not confined to the human animal, anyone who has ever watched a pair of garden birds feeding and caring for their young, knows exactly what charity and selfless giving are all about. In fact the practice is widespread across the animal kingdom, it is a survival strategy evolved over the millions of years, and has nothing to do with any perceived divine presence.

Where do objective moral values and duties come from, if not god? What is the basis in our reality for the adoption by human animals of a moral position? What causes a human animal with nothing beyond the human to arrive at a decision about what is right, and what is wrong? This is another facet of the human project that has been hijacked by the clerics over the years. These values are not objective, in any natural sense, they do not come to us naturally; that is, no human is born with a fully functioning moral compass. But certain of these values seem to be inbuilt in most humans or in sufficient numbers of humans so as to ensure their widespread adoption. That is, there seems to be a group consciousness effect, a moral consensus that some things are beneficial, and some things are not, in the same way that we know to stay clear of fire. Moral values come from us; they are an innate part of each of us, which begs the question. Who or what put them there? To which the only answer can be: we did, of course. At some point along the path of our long evolution we learned that to cooperate together and be nice to each other was, not just preferable, but essential to our survival as a species. And since we did survive as a species it is logical to make the assumption that somewhere back along the path of our evolution we must have begun thinking in that manner, again those that did not, were probably eaten by wild animals. As has been mentioned in a prior section of this book, at some time in the past, approximately 80,000 years ago, there was a *'Great Pinch'* in the number of humans down to possibly as few as 3,000 individual members of our species, the entire population of the planet could have occupied a small theatre. We could not have survived had we not cared for each other. The fact of our survival speaks for itself. Innate human compassion is what got us through that crisis and all subsequent calamities; it is the very basis of all morality and law among humans. Religion has nothing to do with it, those who make extravagant claims to the contrary have never provided the requisite evidence in support. As with all other extraordinary claims made by the religious about their practices, no evidence is provided, because faith steps in, and belief in the claim becomes its own reason.

Morality as a concept has been ruthlessly and shamelessly hijacked over the years by the superstitious, and those who profess to know, they claim it can only be derived from some external source, by which of course they mean some god. Human morality does not come from a god any more than wine arises out of water. Human morality, and the standards and mores we impose upon ourselves arise out of a set of self-evident facts, grounded in the reality of our existence. Our morality is firmly based upon rationality, and the acceptance that humans only exist because we care for each other, our innate inbuilt compassion. Good manners arise out of respect for the self, I am alive, and therefore I respect your right to be alive too, and we go on from there, it is a simple self-evident rationality. The biosphere we inhabit is so overwhelmingly hostile to us as individuals that we could never have survived unless we cared for each other, stood together, and cooperated our way through evolution to where we are now. We also however, have evolved an opinion based morality, moral issues have become more complex as our social interactions have evolved, and our technology developed to allow more and more options that did not previously exist. Moral issues become matters of personal individual opinion regarding the quality of what is right and wrong, according to one's personal moral sense. Abortion, for example, the consensual removal of foetal material from a human female body is a simple medical procedure, there is no right or wrong, it is simply the business of the patient seeking the procedure. Those of a religious bent who believe otherwise are, of course, entitled to believe what they like, and when emotive language such as babies, murder and holocaust are deployed, the waters are well and truly muddied. It is probably safe to say that no human person ever went to medical school in order to pursue a career as a murderer of babies. But as with so many religious positions what lies behind it is the arrogant dogma of an intransigent belief, in the case of abortion, the belief that foetal material has a soul, given to it by some god. The notion of a human soul is an unproven nonsense grounded in magic, there is no evidence for the assertion, but those holding the belief defend it by declaring it to be a matter of faith. It therefore cannot be discussed, the conversation is brought to a screeching halt. Like the amplitude of their full width at half maximum, a Christian's faith may not be discussed. The death penalty, is another right or wrong issue with no absolute, as is the legalisation prohibiting drugs, and a host of other moral issues that are irrelevant to or independent of the facts to hand. In all such cases each human being makes up his own mind, and has the right to hold and express that opinion, based on the available facts. In many other cases the law makes a decision, public representatives, with an ear to the public voice make the decision, and legislate for prevailing opinion, and attempt to reach a consensus, which sometimes works and sometimes not; or at least that's what ought to happen. The law also intervenes in certain cases; the old adage that hard cases make bad law is as true today as it was in the time of Cato.

The notion that there is some connection, causal or otherwise, between morality and religion needs to be put to bed for good. There is none. Religion is to morality as economics is to topiary, there is a connection if you want to discover one, but it requires hard work to justify the effort. Religion and religious people, are connected to morality in the same way as business and business people, are connected to an economy. There are many business leaders who fervently believe that because they can run a company, big or small, and because they can, as they say, deliver shareholder value, they can somehow run an economy or make economic pronouncements.

They can't, necessarily. It has been shown over and over again that they can't. Anyone doubting that fact need read no further than the very slim volume produced on the subject by Paul Krugman, entitled, *A Country Is Not A Company*. Economics is not business, and a business does not run like a national economy. In the same way, religious people have no more insight into good and evil, or right and wrong, than nonreligious. They claim that they do. But they don't. Like all religious claims it ultimately reduces to magic, and closer scrutiny usually reveals the Emperor to be in a state of undress.

The hold religion has over humans is that it sets itself up as the chair of moral arbitration, promising the true path, the next life, and damnation if not followed, god is the pinnacle, all derives from the higher source, the ultimate CEO, the top of the pyramid. We now know enough to be able to say that good manners alone negates the need for a god, nothing in human existence, or here on earth or visible in space that we can measure requires a god for either its explanation or its existence.

We are, each of us, the proud owners of a brain that exhibits the phenomenon of conscious thought as an emergent property, we have a duty to be proud of that, and a requirement to show it the respect it deserves, and hence to respect others in turn. We exist individually for only the briefest period of time, relative to our surroundings, like a solitary spectral line in a wide expanse of continuum, *the light gleams an instant*, as Samuel Beckett put it. It would not take much of that precious time to see both sides of the argument as to whether Shakespeare or the Bible gave more to western civilisation; the speech on the quality of mercy by the learned judge in *The Merchant Of Venice* alone is probably enough to sway any reasonable vote.

The Problem of god is that he exists, we made him, when we were young, as an explanation for what we saw in the natural world, and could not explain in any other way, and therefore god exists. We got together, formed a committee, and made a long list of god like attributes, and applied them to our creation, and proceeded in our devotions. But now the problem arises that we have grown up and don't need him anymore, but he refuses to, or won't, go away. Or rather the problem arises, that god is believed to exist by differing groups of people who then behave towards everyone else as though they had the ultimate and only revealed truth, to which the alternative is death and damnation, and this is very definitely *our* problem, because of the implied licence such doctrines convey. There are so many problems with god and religion, and belief and faith that rational people simply give up, and call it a day, nonsense is nonsense after all no matter which way you look at it. It is hard for rational thinkers to accept nonsense as anything other than what it is, they are more likely to find themselves aligned with the many public sentiments of Christopher Hitchens, along the lines of, *I will not be spoken to like that*. What we know now allows us to arrive at a simple fact, if anyone makes a pronouncement that is contrary to the known laws of physics, they are talking nonsense. All religious claims are based in magic and the supernatural, are contrary to the laws of physics, and are therefore nonsense.

Most people are well acquainted with the arguments in favour of there being a god, or perhaps more properly phrased the underlying reasons for the existence of Christian theism. Arguments such as the, *why there is something rather than nothing,* argument, we exist therefore *god* must've *done it*. Or the universe *fine-tuning* argument, which postulates that we can only exist because the universe was specifically designed and tuned for us to exist in, therefore *god done it*. The argument from the *beginning of the universe,* that someone put it there, therefore god must surely have *done it*. The *mortality of the human being* argument, we are here, we die, and there must be a purpose to it, therefore god. The many and various *ontological* arguments, that the universe exists and has certain properties or a certain structure, and this therefore gives rise to the existence of god. The *observed complexity* argument, that because the human eye, or the brain, or the world, or whatever, is so complex, there must be a creator, and that creator is god. The argument that in a *naturalistic universe,* rationality would not be possible, therefore god must have created all rational beings. Another argument relies on the experiences of human beings both believers and non-believers which seem to indicate the existence of god, sudden conversions, miracles and the testimony of believers. There are also the arguments, or rather the claims from history; that Jesus existed here on earth, and could not have been other than the Son of God, therefore god exists. To the rational reasonable mind these and other arguments are as old as the desire to believe, and are not in any way grounded in our self-evident three-dimensional reality. All such arguments are easily refuted, and again the reader is referred to the work or Dawkins and Hitchens for refutations better than this author is able or willing to provide. All such, *god exists* arguments, are quite cosily put to bed by the simple request for those making the assertions to offer some kind of evidence. And let there be no argument as to where the burden of proof lies. It lies with those who assert, that is a maxim of law in most jurisdictions, if you make a claim, or an accusation, or an assertion, you have the burden of proving what you assert. It is not simply a request, it is a maxim of the law;

the burden of proof lies firmly on the shoulders of those making the claim. The religious, of course, can never offer proof, and it is therefore a most reasonable question to ask, what is the problem to which religion seeks to provide a solution? More to the purpose what is the problem to be solved by god? The question may be asked of any religious believer, whether a Hare Krishna selling books on a London Street, or a Bible thumping Baptist in Alabama. The answer will probably come back as the same, the problem to be solved is life itself. When pressed on the matter and specifically why they should see life in those terms they will resort, as all religious believers do, to the circular argument that: *it's true, because the book says so*. Or some king of magic has been witnessed. The same answer is given in *Monty Python's Life Of Brian*, to the question, why can't women attend a stoning? *Because it's written, that's why*. What passes for explanation and proof of wild, and often crazy assertions is little more than the same asinine nonsense contained in the holy scribblings being vigorously defended.

If we were to start again from the beginning, following some calamity, a flood, for example, which destroyed the earth, and we recovered to start again with a much-reduced population of grateful survivors in a much-changed world, would there be a role for religion? It is a matter of grave doubt whether it would be our top priority. Religion, not being the solution to any human problem, being manmade and obsolete, ought reasonably be expected to fade away in the face of better explanations for our reality and our increasingly technological society.

Yet religion persists, and will probably persist long into the future. There are so many problems with the god hypothesis that it beggars belief or stretches credulity, if you prefer, to think that anyone with even the slightest spark of intellect, and any kind of exposure to an enlightened education system would indulge in religion. And yet, some very, otherwise intelligent people indulge in it, and not only take it seriously but also claim to live their lives by its tenets. Most rational thinking people have the greatest problem trying to square that particular circle, and simply have to be content with amused bafflement. Be very clear on this point, practicing a religion, on any level, demands that you dispense with the use of your critical faculties.

Rational thinkers also live in state of complete disbelief and non-understanding as to why many religious adherents seem so immune to hearing any counter argument or of being affected by argument or evidence, especially new and convincing evidence. To the rational thinker such a mind-set seems to be nothing more than wilful ignorance and intellectual fraud, arising out of wishful thinking. Any attempt to engage with, for example, Catholics, or *born again* Christians, or Muslims in any meaningful manner concerning the specifics of their beliefs usually results in irritation at their profound and wilful ignorance, and utter contempt for any argument against dogmatic positions. They quickly become personally insulted, and, like the demented preacher in the movie, *Inherit The Wind,* resort to playing the, *they mocked him too* card. Unyielding dogma, and the closed mind are common traits in all religious believers, across all versions of the nonsense, questions as to why they believe what they do in the face of a body of scientific evidence so manifestly at odds with superstition are usually parried with scripture quoting, or further dogma, or accusations of blasphemy. There is no appreciation of the other side of an argument, no possibility of compromise, or of tolerance. It was true in the time of Galileo, and remains true today, the singular adherence to a code of belief, to such an extent, that nothing to the contrary could be entertained, much less allowed, and any argument to the contrary is dismissed as required by the doctrine. The difference being that in the time of Galileo those raising doubts would have been punished in the harshest manner available, all rational thinkers alive today, would surely have joined Giordano Bruno and others at the stake had they lived in 1600. We may rejoice that those days have passed into history.

Hence religion is defenceless against the onslaught of rationalism, and cannot escape from the facts, which are entirely against religion. There are only a very few facts worthy of mention. First and primary is the fact that there is not a single piece of credible evidence to support any of the claims made by any religious believer, nothing claimed stands the test of rational scrutiny. Second is the fact that because there are so many different variations each opposed to the others, often violently so, the very likely and most probable objective conclusion to be arrived at by the rational thinker is that they are all mistaken. The third fact is that all claims involving magic as a reason are nonsense, being contrary to the laws of physics. The fourth fact is that the burden of proof for claims made lies firmly with those making the claim.

We humans share 99% of our DNA with chimps, our closest animal relatives, if we are made in god's image than so too are chimps. Debating this point with a closed minded religious adherent, Christian or otherwise, will usually return the simple retort of disputing the fact that we share anything with a chimp. In the same way that other Christians will adopt the hard line on evolution, stating categorically that evolution is merely a theory, thus attempting to reduce it to being simply the opinion of certain scientists, use of the term, certain scientists, leaves the gate wide open to the clear inference that there are, certain other scientists, who follow some other theory or adhere to the biblical truth.

Common to all such dogmatists is the so-called gift of faith, which is supposedly a gift from the almighty. Faith works in a peculiar manner, the magnitude of the faith is in direct proportion to the probability of the nonsense to be believed. The more you believe something that is contrary to the laws of physics, the stronger is your faith, and the stronger your faith, the more your god will love and bless you, both while alive, and after death. To rational thinkers it is little more than the gift of ignorance, the gift of dogma, the gift of misrepresentation, of supplication and submission, of unquestioning obedience to theological authority and nonsense. It is unquestionably the gift of closing one's eyes to the beauty and majesty of the real world, and substituting it for apocryphal gibberish.

There must also be the most profound misgivings about claims by almost all adherents to the Christian belief system that they enjoy a personal relationship with the founder of the creed. This claim is almost universal and vehemently defended by all who believe, and in fact, is one of the strongest reasons for the persistence of individual belief, and in common with all other religious claims, is of course, always unsubstantiated. And because it is a matter of personal faith, cannot be questioned. Were it to be the case that such a deep, meaningful, personal relationship pertained, it should be a minor matter to resolve any ambiguities of a doctrinal nature which clearly exist among the various factions. If they all have the same level of relationship with the same entity, and it is as described, the very reasonable question must be posed as to the different meanings attaching to alleged, words used, sayings said, doings done, and events occurred. Names, dates, places, people and times could all be illuminated and agreed, and in very short order, if the relationships claimed were in fact real as is claimed. The burden of proof argument must again be invoked; those who assert a personal relationship with an imaginary divine entity must do the proving. Otherwise such claims, along with the rest of the nonsense, may be dismissed as mere wishful thinking, and unproven hearsay.

It could be said that Jesus Christ, through his church, effectively, ruled the earth for fifteen hundred years. What is the legacy? Did he do a good job? Would we let him do it again? Or would we put him on trial for crimes against humanity, were he ever to return? These too are problems rational thinkers have with the Christian theist doctrine, the fact of the matter is that there is nothing in our current reality that requires a god figure as a prerequisite mover, or at all for its continued existence, or for its beginning, everything we know and understand can be explained without reference to god whatsoever. Forget where the universe came from for a moment, let's keep it simple, let's confine ourselves to local, and to the manageable, let's content ourselves to the mere immensity of our solar system; gas, dust, electromagnetism and gravity explains everything. We know everything we need to know about the formation of our solar system, from the reasons for a gas cloud collapsing to form a star, to the electrostatic causes of dust particles accreting, to the formation of rubble, gravel, planetoids, planets and the gravitational interactions of the various system bodies. We do not know everything yet, there are discoveries made every day, and no doubt there will be surprises in the future, but we know enough to be able to say, with absolute certainty, there was no need for a god to do it 4.5 billion years ago.

Which of course brings us on to Bible Study, which is unquestionably an intellectual betrayal, and a fraud, and about as profitable as collecting matchboxes or old shoes, it is the intellectual equivalent of apple-polishing. Of course there are two types of bible study, on the one hand there are the objective academic scholars such as, Robert M. Price, see for example his book, *The Case Against The Case For Christ*, who approach the subject from the objective historian perspective with no presuppositions as to the truth or otherwise of biblical content. The second kind are those, and we may number among them such respected scholars as, William Lane Craig who publishes books with titles such as, *Philosophical Foundations for a Christian Worldview*, and who approaches the subject with the subjective axiomatic opinion that the books of the bible contain truth, and cannot be questioned as to their grounding in reality. These comments are entirely directed to biblical scholars of the second kind. In fact, bible study does not constitute any form of serious academic progress, in the same way as reading books regarding accounts of alien abductions is unlikely to be taken seriously as objective academic scholarship, both endeavours are nothing more than naval gazing, the incestuous reworking and reinforcement of apocryphal nonsense. A gathering of believers engaged in bible study is about as productive as a group of teenage girls discussing shoe styles, or a group of bodybuilders discussing food supplements. The people who wrote the books of the bible were of a Bronze Age, agricultural, flat-earth era. They knew nothing. They had no idea of what the world was like or what it was made of. Their go-to explanation for everything was that god, or gods, or daemons, or angels were responsible, they gave, and they took away, they allowed some to live, and caused others to die. People of that time had no idea what caused the sun to shine, or what caused fire to burn or light to emit, or why the sky was blue, or the clouds white or why the leaves of the trees were green. They knew nothing about disease or the human body or where flies or mice came from. They had even less idea about the heavens, why the stars were in the sky, why comets and shooting stars appeared, why certain stars wondered about the sky. They knew nothing about the size of the earth, or its shape, or rotation, they had no appreciation of why the seasons came every year, and lived in constant fear that things would change for the worse if they did not keep on the good side of the gods who were responsible for everything, good gods, for the good things, bad gods, when things went wrong. The notion that there is some deep wisdom or word of god to be found in the nonsense reported in such texts is frankly laughable. There is more wisdom, wonder, glory, grace, and sense of awe in a single photograph taken with the Hubble space telescope of a star field, or a galaxy, or any of the planets, than will ever be found in a thousand biblical volumes, or could ever be in all the Catholic dogma, catechism, or books of revelation, Qur'an or sacred scrolls. There is more beauty in the face of a smiling child, or a flower open to the sun, or a spider building a web than in all the holy bible stories, and nonsense sayings of superstitious, Bronze Age, desert dwellers of antiquity. We have no more an obligation to adhere to the so-called teachings of the bible than we have to follow the precepts of the inhabitants of the Shire or Gondor. Nonsense is nonsense, no matter what wrapper is used to dress it up as the word of some all-powerful superintendent of human activity on earth.

Science and technology do not require to be defended against god, or against those who seek to tear science down, or destroy the technological age, and return the world to a feudal dark age. Science and technology delivered the modern world, and we humans love it, and it is a great success, and you can't argue with success on that level. It is entirely irrelevant if a proportion of people do not believe in evolution, there will always be those among us who decide to take the contrary position simply for the sake of it, to be anti-science, for the sake of it, or anti progress, or simply anti whatever you have. For those who believe, no explanation will suffice. Faith is unshakable, and immune to reasoned argument. There will also be those who lie about the reasons why they believe, or choose to believe. Because it takes a great deal of time, and effort, to achieve in science, the entry fee into any scientific field is significant, to be able to understand science you must do the work, you must spend the time, and for a lot of people it is far too much work, far too much time. There are undoubtedly those for whom a TV documentary, or the word *science*, or the word *technology*, or the word *theory*, become a signal to switch off. As the comedian Billy Connolly put it, referring to politics, *"there are certain things my brain just won't allow me to understand..."* And such people, turned off by all things science, can be blatant in their disregard for the effort that gave us the modern world, and dismissive along with it. It is a form of snobbery, as C. P. Snow observed in his influential essay, *The Two Cultures*, there will always be those who look down on, or sideways at, science, and consider those who practice it for a living to be something less; science and the arts are separated by a gulf even now. There is also the old adage, why be a scientist, when you can be his boss for half the effort. It is a natural dichotomy of the population, with respect to science, those who can and those who can't, those who will and those who won't. Respect for dichotomy is a mark of a mature society, one that can embrace and respect difference, and one whose tolerance extends to the dissenting voice; scientists have lived with it and respected it for years.

It is not at all insulting to suggest that religion is the refuge of a lazy mind. A religious believer, on the face of it, is in possession of a mind too lazy to pay the entry fee necessary to become familiar with the basics of current human knowledge to bring itself up to the required level of reasonable debate. Some people possess a mind content to wallow in feudal nonsense to the exclusion of the modern world. Others are hopeful of the destruction, by fire or flood, of the entire world, and everyone in it, in the sure and certain knowledge that they are already saved, because a fifteen-year-old, ignorant Scottish girl, in the mid nineteenth century had a vision, or because of the lamentations of some saint or other.

There are also the, *don't know, don't want to know*, brigade, the happily ignorant and apathetic mass; the ignorance is bliss faction, who refuse to pay the exorbitant entry fee, as they see it, into any field of science by opening a book. We've moved on. We humans have outgrown the meaning of god, we've transcended the divine, we're better than religion, and we're better than god.

Why believers believe is also a problem that has not quite been solved yet, it is sufficient to say that a belief is a declaration about the state of the world, not some abstract part of the world, but reality itself. Therefore if a belief concerns something contrary to the laws of physics, such as magic, it may be readily dismissed as nonsense. But the very simple fact of existence would appear to be that if you believe something concerning the real physical world in which you live, then you must have a reason or reasons for such a belief. Your reasons may, or may not be well founded, the underlying logic may be faulty or non-existent, or you may possess some latent desire to exist in a totalitarian state, like North Korea or the old East Germany; but there must be some reason. Otherwise why believe at all? It may be, the strong man in the desert, syndrome, perhaps it goes back to our early days when the population pinched to a few thousand individuals, follow me or die, we did follow, and we lived, we are therefore, more than likely, mostly descended from those who followed. Whatever reason is given, from tradition to coercion, all who believe must have a reason for doing so.

The fatalists too, with their nonsense of being on a path stretching from birth to death, and the greater purpose, and the path may be broad or narrow, long or short, one can weave about on the path but that the path takes you inexorably on towards death, and beyond to the greater meaning of life and the universe, and the ultimate meaning of it all. Everything is preordained, no matter how you might weave about on a wide path there is no escaping your fate, no possibility of getting away from the purpose of your existence, or the ultimate meaning of your life which will all become clear when you pass onto the next plain, when you transcend the mere physical of the body, and so on. And in the meantime, you might care to make a donation. Again these believers must be reminded of the burden of proof requirement. This notion of individual human purpose, and some god or other having a grand plan for the species, and an individual plan for every human being, and of there being a correct path, and so on, makes no sense whatsoever unless you accept that we are in fact living within a computer simulation of some kind, and further accept that reality is subjective in some way. That we are, in fact, all part of some program or other, or some vast game or other, or some educational, or custodial scheme or other; a distasteful idea to say the least, if not an impossible one.

All religious cults exist in a permanent state of self-inflicted outrage ready to defend against the slightest perception of onslaught or the least hint of an insult. The underpinnings of their belief system are so weak and tawdry, and so lacking in reasonable argument they have effectively no choice but to adopt the posture of the mortally wounded, and limp away to the darker outskirts of human activity. The lethal combination of a good well-rounded objective education, and an affluent society born of technological mastery really ought to destroy all such indefensible nonsense.

One of the most disturbing aspects of religion is the lack of a sense of humour, the complete inability to see the funny side of anything. We know the obvious, Muslims don't do cartoons, cannot laugh, ever serious, unable and unwilling to see any other side, the legal phrase *audi alterim partim*, compels you to hear the other side, in a legal matter, it seems to mean nothing in Islam, there are no churches in Saudi Arabia, western nations can boast Mosques a plenty, but the other side is not welcome on the holy ground of Islam. There is something profoundly disproportionate in that fact, and symptomatic of the inherent intolerance of religion in general. Next come the Christians, who seem to be in a perpetual state of uproar about humour applied to any aspect of their faith. Jokes of the Jesus, nails, hammer, hotel, 'could you put me up for the night', or the, 'Jesus and St Peter walk into a bar', type, are always met with disgust and derision, and threats of legal action for blasphemy. Jews seem to be the funniest of the big three; they have to be, they might say, New York Jews at any rate. There are many fine examples of Jewish comedians, and there is, of course, a funny side to it. Humour is one of the great human imponderables, like love, one of the great emergent properties of the human brain and of our consciousness, its application to the shoddy portents of religion is worthy of study. We laugh at it because it's laughable, it's funny because it's silly. Religion in all its various forms has been shown to be so silly and nonsensical in the cold white light of human understanding that it is right and fitting that we drag it out into the public arena a apply ridicule and mockery; it deserves nothing more. Billy Connelly has been laughing at religion for many years, see for example, his monologue concerning the preacher Nigel, who is sermonising about football and teamwork, and everybody being on the same team, with Jesus as the manager, and his son Nigel junior, who innocently asks, "*Daddy, Did Jesus play football*"? And, *Monty Python's Life of Brian*, "*it's not meant to be taken literally it refers to any manufacturer of dairy products*". Denis Leary, and his piece about not wanting his children brought up in a church based on, *the size of fuckin hats.* George Carlin's, *but he loves you,* sketch. And many other fine examples of comedy directed at the almighty. This slapping of the public face of religion could not be passed over without reference to the great Irish comedian Dave Allen, who practiced his trade on prime time BBC between 1971 and 1986, and dealt with such jolly japes as farting in church, and the silliness of the creation myths. Some religions even go as far as to get in on the act, they seem to recognise the marketing opportunity, exampled by the Mormons sponsoring of the program for, *The Book Of Mormon,* stage production, and other examples; however, Scientologists and Muslims remain tetchy when it comes to humour at their expense.

Another problem of god and for Christians is justification of aspects of the real world that simply defy their own logic, for example the antics of the Jewel Wasp, so called because it looks jewel like in appearance. Paralyses a cockroach, by stinging it in the head and causing it to become zombie like. The wasp then bites off one of its two antennae, drags the helpless creature down into a prepared burrow in the ground and lays an egg in its body cavity. The wasp larva then eats the cockroach slowly from the inside, in such a way that the cockroach remains alive and healthy, apparently cockroach flesh and innards, would decay very quickly otherwise, and be useless to the larva. Thus the lifecycle of the wasp is completed by the use of the other creature in what to any objective observer appears to be a monstrous and hideous manner. Clearly if there is a god who is responsible for everything, and knows everything, and designed everything, the natural question arises as to what the photon he was thinking when he decided the Jewel Wasp was a good idea, or perhaps someone else, *done it*. The net result is that believers are constantly on the back foot, constantly defending, always trying to add more and better parameters to the committee design that is the image of god, or Jesus, the old adage, *if you're explaining, you're losing*, is apt and wholly applicable here.

There is also further trouble created for Christians by the concept of the contented atheist. The happy non-believer those contented fellows who do not need something more, for whom the search for meaning is no more energetic than the search in the cooler for another beer. The Homer Simpson, blissful ignorance, pig in shit, kind of contentment. It's hard for Christians to come to terms with people who simply reject the concept, or the requirement of faith, and go merrily about their daily business, oblivious and nonchalant, resenting of the time and effort involved in prayer, or church going, or even bothering to think on the supernatural. Many simply don't give a hoot.

Another problem for god and churches comes when non-believers ask questions such as, can god play golf? Are games such as Golf and Chess mere contrivances of Man, the piano too, or are they something more fundamental, projections of the Platonic world into our plane of existence. Abstraction is one of the properties of the human condition. Most Christians don't play golf, it's played on Sunday, and most of Sunday at that. Jews, don't, or can't, not on the Sabbath, Muslims can't, or won't play on Friday. One must assume therefore, god was not a golfer. Einstein reckoned he played dice, making him an indoor kind of fellow, we know he was a hunter too. Provided animal skins for Adam and Eve in the famous garden after they discovered a dislike for nudity.

The very worst thing about religion in general, and the organised Christian religious belief-set and mind-set, in particular, must be is their eternal goddamned pessimism, and their nonsensical end of the world mentality, the walking dead hypothesis, the sinners in waiting proposition, the narrow minded, no future, outlook, and the unforgivable waste of human optimism and hope. God and religion are finally an increasingly unimportant and proven irrelevance. Ignorance arrogance and dogma are the principle components of any organised church religion. It can be simply characterised as a dogmatic belief system held rigidly and arrogantly and protected by a self-imposed, self-perpetuated ignorance. It is maintained against all comers, no challenger is allowed; any utterance to the contrary is silenced, forcibly if required.

Only members of our true church know or can find the true way and even among them only those purified, by the laying on of hands, or the sanctification of the waters, may follow the narrow path, and of those only the chosen few may know the real meaning of the sacred words, and so on. This is plainly arrogant, ignorant and dogmatic to the nauseating extreme. You don't have to be a Christian or even religious to help out. You don't need to be, called to the father, or the faith or, of the water to lend a hand, to get involved, to make a difference, or to give something back. Those who believe and publicly state that only religious persons can be, or are involved in charitable deeds are arrogantly mistaken. There are many fine examples of altruistic behaviour practised by people of all persuasions every day of the year.

Misinformed or ill-informed and selfishly so, lazy and indolent in a blank refusal to expand their learning beyond those texts or branches of learning that reinforce their truths. Bible study and meetings of the faithful to extrapolate meanings from obscurity is not scholarship. It is as redundant as the meetings of the flat earth society convened to reinforce flatness, and dispel notions of the rotund. Religion is the outright refusal to embrace the obvious, and a repudiation of even the simplest facts derived in the most rigorous manner, along with the misrepresentation of such facts and concepts alien to them, or in conflict with their sacred creed.

YouTube is full of idiot preacher types using analogies such as single cells in a test tube being ruptured and expecting the constituents to reform back into a cell once more. Perhaps they might, given three billion years of undisturbed tube time. Another preacher expounds the, jar of peanut butter analogy, it is effectively a jar of chemical bits, and surely if the evolutionists are correct we ought on occasion, because of the laws of large numbers, and why not, find the jar of peanut butter giving rise to life, or at least to an increase in complexity. Clearly the idiot in question has never bothered to ask why the jar has a date stamped on the label indicating a shelf life. Or what particular part of chemical or biological science states that those particular jar contents ought to morph into a living creature on a timescale of days or weeks, and so on. These same people would probably not be surprised if the small change in their pocket morphed into a banknote overnight. The fact that idiots exist and are out there spreading ignorant nonsense across the Internet is not the issue. There will always be idiots at work. The bottom ten percent of the IQ scale will always need something to do. The problem is, some of these people are influential and well-resourced idiots, capable of getting voters out on mass, and they are also effectively corporations, who because of the nature of the products and services they sell are classified as exempt from meaningful taxes or levies. Classifying oneself as a charity, and defining the charitable enterprise, as being the spreading of apocryphal nonsense, along with perversions and distortions of accepted factual science, is hardly what the legislators had in mind when designing the tax code; it could be argued.

Belief is an opinion, and opinion is a form of belief, any separation of them is but a triumph of distinction over difference, or distinction without difference. David Kirkaldy (1820 - 1897) was a Scottish born engineer of the industrial revolution who pioneered the testing of materials such as wrought iron and steel, these materials were new at the time, and their tensile, and other properties, were poorly understood. Over the entrance to Kirkaldy's laboratory in London is his motto, *Facts Not Opinions*, carved into the surround. The motto speaks for itself, an opinion about the tensile strength of a piece of iron to be used in a bridge, for example, cannot be used in place of the facts obtained by testing a sample of the material to destruction.

We are in exactly the same place when it comes to religion; there are no facts on offer, only opinions; though that is not strictly true, there is one single salient fact, which is, there appears to be no evidence whatsoever to support any of the claims. We are asked to believe in opinions only, and in hearsay, and in uncorroborated testimony of revelation, with not the slightest hint of a fact. Those who practice religion have little or no regard for facts. They have clung to the many and various opinions they hold about the universe, the world, its people, and our society since their opinions were formed in the deep recesses of antiquity. And these opinions are rigid, and unyielding, they are given the upgraded status of belief, and called sacred, and become unquestionable, *the faith of our fathers*, as the old hymn puts it. But the facts have changed, they have grown and multiplied, and accumulated to become the sum total of human knowledge, science, technology, art, literature, law, politics, economics, and so on. Yet the unyielding, unsupported, unsupportable opinions persist. This is the irrational tail wagging the logical dog, on the pretext that the holders of such opinions have some moral or other insight; nothing could be further from the truth. It is, however, in the very nature of faith and belief that the critical faculties, must be and are put on hold, for believing without evidence is the very essence of faith, and the more you push the lack of evidence to the side, the stronger is the faith.

What young earth creationists, or followers of John the divine believe in, is irrelevant, their opinions do not matter, but the polluting of a child's education is something that should never be allowed to happen. It is something that ought to be taken seriously, and contaminating the important work of childhood with nonsense is one of the worst abuses imaginable. The net effect is the creation of morons for the future. The creation of unquestioning idiots with no ability to ask why, or without any ability to look for an evidential basis to what they see in the world.

Facts, once established, may be disputed as to the parameters of the fact in question, but in general, facts are facts, and must be accepted as such by all reasonable participants in the process, whatever that process may be. When facts are in dispute it is reasonable to engage in debate, or to mediate, or settle the matter by having it heard by an appointed, recognised and accepted body, expressly established to determine matters of fact. In the case of science, it is the process of peer-review, in the case of law, a court of law, judge or judges, and or a jury. But in the case of religion, the book says so, and the book is true, in fact, the book is absolute truth, and if you do not accept the book, it is because you are not blessed by god, and we may therefore tie you to a tree and set fire to you. Or at least we used to, and might like to do so again sometime.

Clearly philosophical debates concerning relativism, could be explored here, but this is too involved for the intended reader. The physicist David Deutsch in his book, *The Beginning of Infinity*, presents a very good discussion on the nature of scientific investigation, and the nature of proof, and of the fundamentals of inquiry. Along with scientific proof, there is a legal perspective, involving standards of proof, admissibility, and the rules of evidence, most scientists rarely if ever interact with the legal process. Legal logic seems warped, truth can appear variable, facts may be in dispute, waters can become very muddy, and the Law of Evidence contains some very odd rules indeed.

By comparison with both scientific and legal norms, belief in the sanctity and truth of the bible as the revealed word of god, for example, is a mere opinion and nothing more, it is an unfounded opinion, unsupported by any factual nexus, or any evidence whatsoever. The bible as with all texts, scrolls, carvings, writings and records of whatever kind from antiquity are at worst the irrelevant nonsense of a primitive culture, or a certain element of that culture, such as a priest caste, and at best a first approximation at attempting to understand the world, the three dimensions, the observations, the various elements and aspects of the physical world amenable to human perception at the time of writing. They are nothing more! The very best we can ascribe to the bible, taking only the bible, is that it purports to be a first attempt at science, at morality, at explanation, at law, at philosophy. Indeed it is a very poor first attempt at trying to make sense of the world in which its writers found themselves. We do everything so much better, our second, third and subsequent attempts at everything are always so much better. Law for example, what vast improvements there have been in legislating for all eventualities, from minor criminal misdemeanours, to massive corporate insolvencies, to international treaties. And everything else for that matter, in short, we've moved on. Therefore it might easily be said that anyone who clings to the bible, or any other sacred text from the distant past, and holds it up as a source of law, or morality, philosophy, medical practice, dietary preference, or any other important area of human existence, is either living in the past or is wilfully ignorant of the current state of human knowledge.

John, Kenneth Gallbraith, the famous Harvard economist, and adviser to several US Presidents including J. F. Kennedy, famously opined, *when the facts change my opinion changes,* indicating the requirement for opinions to be flexible, and rooted in the reality of fact. In science and in law, complex arguments are settled by an objective adjudication process, trashed out in the literature or at a conference, in the case of science, or in the legal world, mediated, negotiated or litigated in a court. But always, when the facts change the opinions must change. Consider the following parable concerning two elderly gentlemen, let's call them Oscar and Felix, fond of Brandy and cigars. They have been out for a good day's sport, and returned to the comfort of a spacious living room where they fall fast asleep. They wake, one then the other, seated in comfortable chairs, the room is darkened, and they are still a little under the weather; neither gentleman is inclined to leave the comfort of his chair. At length, the clock strikes twelve. A polite discussion begins as to whether the clock indicates midnight or midday. One gentleman maintains it must be midday, the other claims it could only be midnight. They argue back and forth, neither can agree, both gentlemen seem to have a strong case. The argument becomes heated until, as the clock strikes one, a third party enters the room, and draws the curtains back to reveal the light of a fine sunny afternoon. Clearly the fact in this case settles the matter, and any opinions contrary to the observed fact must change. You can't argue with the facts. The third impartial person settles the argument with a sudden infusion of incontrovertible evidence. And the more complex the argument, the more the facts matter to the argument, and the more the requirement for impartial adjudication.

Often the question posed is a lighter shade of grey than can be dealt with by presenting a simple fact or some body of evidence. For example the question as to whether Mozart or Beethoven was the better composer is not a real question, it is simply a matter of subjective opinion. The question is therefore a request to express an opinion. There are many facts that we can bring to the question, they were both Germanic, they were both composers, they were both brilliant composers, and we know that Mozart preceded Beethoven, and many other such facts, they can all be put on the table, but none of them answer or even illuminate the question, because it is not a real question, it is only the request from one person to another person to express an opinion. Is there a god? Or is religion a system of truth? Or does religion contain truth? Or did Jesus exist? Or did Jesus rise from the dead? Are not real questions. They are invitations to offer an opinion posing as questions, and nothing more, and the answers given, carry as much weight as would an opinion about the merits of Mozart over Beethoven.

Religion is now, and has always been in the business of factual denial, regardless of the results, or the consequences for either the world, the species or the individual. This denial of fact has always been prevalent when authoritarian regimes come to power or rely on dysfunctional and superstitious elements to remain in power. Religion is a coercive and corrosive influence, whose intention is to maintain the opinions they hold as absolute truth, or as the revealed word of some vindictive god or other at the expense of the facts, or despite the changing facts. When did religious believers ever allow a set of facts, or a newly discovered fact, to get in the way of a good opinion? They retaliate with defences such as describing the magnitude of their belief, that it is a *core belief*, or a *strong belief*, or a *strongly held belief*, or a *deeply held belief*. Or a *passionate belief*, or that the belief is a gift from a deity, and is therefore a sacred and unquestionable or is an *unshakable belief*. There is also, quite often, the alleged personal relationship with the particular deity involved, the deep and personal involvement up to and including two-way communication. Again, the burden of proof rests with the believer.

Like a lawyer advancing the mitigation that although the client had made a mistake, it was a *genuine* mistake, or it was an *easily made* mistake. On questioning by the learned judge, as to what constituted a genuine mistake, and how this differed in fact from any other kind of mistake, if there were indeed other kinds of mistake, Counsel was required to concur that the client's qualification as to the distinction of the error was wholly irrelevant. By extension this approach may be applied to all religious opinions, it matters not a holy damn, and is entirely irrelevant to the facts of the matter whether or not the opinion is, *sincerely held*, *deeply felt*, or *passionately expressed*, a belief is merely an opinion, and if the facts indicate a requirement to modify or even abandon the opinion then clinging to it becomes manifest nonsense, and intellectual fraud, or in the worst case, the simple profession of wilful ignorance.

Prior to Galileo turning his homemade telescope on Jupiter in August 1609 and revealing its four, hitherto unknown and unsuspected, moons, the Catholic Church was adamant that Ptolemy's view of the earth centred universe was correct in all its erroneous detail, and they clung obstinately to the opinion regardless of what facts were uncovered. The church knew the sun was perfect, the Greeks had said so, when spots were seen on the surface and reported for the first time in 1610, there was consternation, imperfection was contrary to the established facts of the time, but opinions as to solar perfection could not be maintained in the face of the observed facts.

If you profess a belief in some god it is the same as saying that, in your opinion, a god exists, but it is more than mere opinion, it represents a declaration by you as to how the real physical world operates. In other words, an opinion formed in some way or simply arrived at following some experience, perhaps derived in some sense from some perceived external source, or arrived at through discussion with other likeminded people, or from people in whose company you may feel comfortable, becomes reality. Or it might simply be an opinion, because you were brought up to believe. They got to you as a child. You may feel you owe it to your forbearers to believe. To which the rational thinker must simply ask upon what evidence do you base your belief? What fact do you rely upon in support of your opinion? What about other more ephemeral things, belief in love, a case in point, you believe she loves you, and there may be evidence. No one can spell it out, you either know or you don't. It's a binary thing, either present or absent. But she may not love you, there is always room for doubt, or there may be a game afoot. And after a while the intensity may diminish, it may even wither and die. Will it be the same five years from now? Will she even remember your name? Or will it turn to something else? Might there be irreconcilable differences? But the question as to whether someone may or may not love you is not a difficult question, it is not an unreasonable assumption to make, most people have someone, and most people know from the circumstances. Claims of the supernatural, on the other hand, are unreasonable, the explanation reduces to magic, and is therefore nonsense. God so loved the world? God's love is eternal? The Christians, et al claim love in the same way they used to claim light, and thunder and lightning, charity, even humour and music. I believe that joke is funny. Why? It made me laugh. Explain? Because it just is.

The world has never been short of opinions. Everyone has an opinion, or a story, or an excuse, along with other things, and we are clearly now in a place where we have all the opinions we could ever want, but we're also in a place where we have no excuse, because we know better. All opinions are valid, even the most extreme, have the validity given by the right to form and hold the opinion. But not all opinions carry the same weight; my opinion as to the nature of your apparent illness carries no weight when set beside that of a skilled physician. The opinion of someone with a qualification obtained in this or that branch of theology as to the age of the earth has no weight compared with that of a trained geophysicist. In the interest of balance and political correctness, it has become almost mandatory and required to hear the other side of the argument. But this takes the doctrine of, *audi alterim partim,* too far; there are limits, we have run out of excuses for pandering to nonsense. There is now, more than ever before, an objective moral duty on rational thinkers to put nonsense to proof, to demand that the assertions be proved, and to demand that publicly espoused opinions be backed up with evidentiary fact.

The old saying is a truism, you are entitled to your opinion, but you are not entitled to the facts. Facts are independent of opinions, just as money doesn't care who owns it, facts don't care who might hold opinions about them, the facts are immune to the numbers of people holding opinions as to the truth or otherwise of the fact. There is no sanity in numbers, survey after survey shows that only 40% on average of Americans believe in evolution of species by natural selection. It is wholly and completely irrelevant whether or not they believe in it. The fact of evolution is accepted by any rational person capable of reading a rough outline of the theory, the numbers of Christians refusing to believe, or Muslims for that matter, -for they too abhor anything other than the narrow discredited creation myth- is simply irrelevant. Numbers mean nothing. If a fossil is 10m years old then that is the age of the fossil, that's what the physics says, that's what the accepted method of dating tells us, then that is what we must accept as a fact. Stating that it couldn't possibly be 10m years old because the earth is only 6,000 years old, because it says so in the book, is simply crass nonsense of the worst kind, and needs to be exposed as such. And no amount of shouting and hysterical stamping of feet or threatening the fires of Hell will alter the laws of physics to say otherwise. Or using some other asinine retort of the sort used by creationist zealots, who frequently ask, how do you know the rock was laid down 10m years ago? *Were you there?* Or stating, with absolute seriousness, that birds could not possibly be descended from dinosaurs, because birds and fish were created on day five, and creatures of the land were created on day four!! Or was that day three?

If we lived in a time when no one knew or understood what caused thunder and lightning, then each and every one of us who has ever stood staring at stormy clouds, or cowered under a bush or behind a rock, looking at a thunderstorm would be entitled to express an opinion as to what, they believed caused it, and why it should have struck that particular part of town, or why Larry's house was set ablaze. But as soon as scientists make measurements, and determine a proper theory on the subject, such as a theory involving electrostatic forces, and the build-up and discharge of electricity in clouds, those opinions are no longer appropriate, or of interest, or valid. And of course that is precisely and exactly what has happened with thunder and lightning, theories were formulated, and opinions were many and varied, but we now know what it is, and what causes it, and how to predict it, we can even demonstrate it in a laboratory. We are no longer ignorant of thunder or lightning or its cause or its effects, we no longer consider it to be magic or divine. Hence all references to thunder and lightning in the bible or other religious scribbling, for example, contrary to the established facts are simply not true. Or, may be dismissed as nonsense.

Creationists believe, Christians believe, UFO seekers believe. Ancient alien encounter-*ists* believe. Astrology readers believe, psychics believe, dabblers in the occult believe. All of these counter mainstream malcontents along with all religious practitioners believe. They all share one thing in common, they know how to believe, but, as Richard Feynman said, referring to economists, they do not know what it is to know. The only difference between organisations representing the interests of those believing in ancient aliens, or UFO visitations, and organisations representing those who believe in the myths contained in ancient scribbling, is taxation. One lot may be taxed the others may perhaps exist largely tax-free.

A review article by Miron Zuckerman et al, published in 2013, covering some 63 previous scientific studies conducted over several decades concluded that religious believers are less intelligent than those of no faith. The team at the University of Rochester found, "a reliable negative relation between intelligence and religiosity", in 53 of the 63 studies looked at. So no surprise there, the reason is quite simple, religious people clearly have a tendency to simply accept what they are told, they swallow the ancient opinions regarding what the book says, and never raise a question. Belief is the easy part. You can believe in anything. It seems to be a natural part of the human condition to believe. Believing is the lazy way out of thinking for oneself, or thinking while not thinking. One of the many good lines in the 1960 movie, *Inherit The Wind*, delivered by the character of Mathew Harrison Brady, played by Fredric March, is the admission, *I do not think about those things I do not think about*, which sums it up. Believing without question allows a person to do other stuff, and not to have to bother with tricky questions, for example as to existence or purpose. *I believe, leave it at that.* Why? Because, I just do. It is easy to believe, just follow what you are told, and look no further. Never question. Never ask. Or be afraid to ask, or be content not to ask, and delegate thinking to others much as one might outsource the cleaning of a bathroom. Some people are simply content to be, or to remain, ignorant. It seems an incredible, if not ironic fact, that Bible believing Christians in the United States, on the one hand, abhor and resist the influence of state and Federal Government, but on the other hand, welcome, submit, and delegate, their thinking, to a bunch of mammon grabbing, nonsense peddling, backward looking, theocrats. Any objective observer of that particular charade must come away, at the very least, perplexed.

As has been said before, one of the very important maxims of the common law is that of, he who asserts something has the burden of proving his assertion, for example, if I accuse you of breach of contract or claim that you have been negligent toward me, you have the right to, *put me to proof,* that is, you deny everything forcing me to prove my case against you. Consider too the presumption of innocence in the criminal law, the prosecution has the burden of proof, being the party asserting that some criminal offence has been committed. So it is, and must be, with any and all public assertions of religious practitioners, they bear the burden of proof for each and every public assertion they make. Consider the Christian minister who claimed the Berlin wall was caused to fall by the intercession of god via Christian prayer, any such assertion must be met with a request for proof to be provided, of which there is clearly none. It should be noted that no claims are made for divine activity surrounding the Indian Ocean tsunami, or the 911 attacks, which hardly makes for objective reasoning. But you can never expect ignorance, arrogance and dogma to be objective in compiling the catalogue of the stupid club. Nor does shouting about something improve the chance of it being correct. Nor does the writing of great tomes of opinion increase the odds of the assertions being in compliance with the established facts and parameters, of the field to which you wish to contribute.

There have also been, over the centuries, many individual people who, labouring under the mistaken belief that there is a god, and that they have his ear, perform some, objectively very good, works: - some, like Francis of Assisi, it could be agreed went too far. But good works alone do not a belief vindicate, the truth regarding god or the bible or any other sacred holy book, scroll, or rock carving is not proven by good works, it can only be proven by evidence objectively measured against what we all agree to be reality.

Of course the argument back to the rationalist is that belief in divine things is a truth of a higher nature than, *'mere truth'*, or the ordinary truth that one might discern in a physics lab, for example, or that might be presented as forensic evidence in a murder trial. Higher level means, of course, a level not requiring the usual necessity of evidence to be presented. Higher means that only certain special people, righteous people, are amenable to the truth, it is esoteric, only to be revealed to the true believer. And, of course, the true believer gets his faith directly from god. In fact, faith has always been seen as a gift from god, and if you are lucky enough to have received this gift then you are fortunate indeed, for you have been singled out by god himself to be blessed and favoured by him with the gift of understanding and knowledge. As any physicist will tell you, we have no idea what this higher level means, nor do we subscribe to the notion that evidence may be dispensed with, and one's trust placed in some metaphysical argument that belongs in the dark ages, when ignorance and fear of the dark were commonplace. Lawyers too will put a brace of question marks against this notion of a higher level of truth, which may not be questioned or examined in any manner whatsoever, but must simply be accepted for what it is; try getting that past a Superior Court Judge.

If magic is to be accepted as a valid explanation for any physical phenomena, then all bets are off, anything goes, and it is then clearly possible to, for example, to load a jury in a criminal matter to the extent that their beliefs are coincident with those of the defendant. Consider, for example a murder case where the victim is brutally murdered and the suspect on trial is allowed to enter the defence that, at the time of the killing, he was possessed by some form of evil spirit, and was not in control of his actions. In other words the defence of *'god did it'* is attempted, in this case we are talking about the bad god. If the defence was allowed, a jury of like-minded people carefully selected might find in the defendants favour. The logic being that once you rely on the supernatural for explanations of something as grand as the creation of the entire universe, and humanity, and everything else, it is a small matter to allow a supernatural explanation to be put forward for the mere death of a single human being. If you are comfortable with, *'god did it'*, as an explanation for the air we breathe and the water we drink then explaining away a killing is not such a huge leap. In fact if you accept that, *'god did it'*, is a valid explanation for any observable physical phenomena, then you must accept that it would be entirely possible for the opposite of god to intervene, and in such a case a jury of believers would have no choice but to acquit the defendant. Modern judicial practice would preclude any such trial ever going ahead, the defendant might well be declared mentally unfit to stand trial, or on advice might seek to rely on some other avenue such as insanity, in which case he would be required to offer proof of his assertion to the court. It is also unlikely, for among other reasons, public policy considerations, that any judge would allow such a defence to be offered, or any such evidence to be admitted. There was one attempt to have such a defence placed before a court, it happened in 1981 before the Connecticut Superior Court in a case where a defendant was accused of killing his landlord, the Judge rejected the attempt of the defendant's lawyer to admit evidence of demonic possession describing it as, among other things, *unscientific*. And if magic is wholly unacceptable in a court of law, civil or criminal, anywhere in the world, we might well ask the question why should magic, or the publicly stated belief in magic, be accepted or acceptable in a political campaign, or in the public utterances of candidates, or elected representatives?

Every human person has a right to be an idiot, or to hold a belief; and a right to hold it deeply, passionately, intransigently, or in spite of all the available evidence. People have the right to close their minds and look the other way; they have the right not to acquaint themselves with the facts of any given situation, in short they have the right to be ignorant, even deliberately and wilfully ignorant. You may believe what you will. You also have a right to express your deeply held belief. But, you have no right to the facts. You also have no right whatsoever to have others take your belief seriously. Nor do you have a right to be respected for holding any particular belief. There are no absolute rights. All rights have proportionate limitations, and are generally qualified. Your right to property, for example, does not mean that your small strategically located home, however many memories it contains, or regardless of the emotional attachment you may have for it, will be allowed to stand in the way of a much needed, and long in the planning, freeway or military firing range. Your right to bear arms does not mean you can own an arsenal of nuclear weapons, which you claim are for home defence purposes. And your right to life does not mean that if a police sniper needs to shoot straight through your sternum, or brain stem, to take out the bad guy he will not take that shot.

The sanity of numbers is another hinge joining the two arms of ignorance and dogma, which deserves mention. It works as follows, consider the opinions and beliefs of those engaging in psychic prediction, or those believing in the premonitions and ruminations of Nostradamus, and consider too, the vast numbers of books and TV programs, and the hordes of credulous individuals who arrive at the conclusion, or form the opinion that, because there is an industry surrounding the myth, there must be, *something in it*. The Nostradamus industry is a particularly nauseating retrospective application of the inane drivel of a long dead scribbler, which has as much relevance in 2016 as any of the biblical prophets, and is more useful to psychologists studying group dynamics and mass hysteria, than it is in prediction. Of course it is possible to find anything in the quatrains if one looks hard enough and plays the, word, letter, character games that are played. But nothing is ever seen in advance, it is on the face of it, simply an example of the old truism that if you look hard enough at the Moon you will see a face, and if you stare long enough into a fire you might see the devil, it is as empty and pointless as bible study. The fallacy of prediction gives us nothing to hang on to; precision in predicting the future is non-existent, or dependent on many spurious parameters. Information is everything, and the facts must always be primary. Not only are we going to be saved, but we have the bumper stickers to prove it, just because it's on a bumper sticker don't make it so. A Neilsen survey found that 70% of Americans believe in Angels. So What? Similar surveys, by many and various polling organisations in many different countries have many and various results concerning such phenomena as: Miracles, Astrology, UFOs, ET, God, Satan, the Supernatural, Spoon Bending, the Tooth Fairy, Leprechauns, and so on. So what? We are none the wiser.

There seems to be a quaint numerical sanity involved, the greater the number of ardent supporters, espousing deeply held beliefs, the more rational the belief seems to become. The fact that the vast majority of people on earth once believed the world was flat did not alter the fact; that the world *'is'* flat? The fact that the Catholic Church burned Giordano Bruno at the stake for suggesting the sun might be at the centre and not the earth, and that stars might be other suns far away, did not alter the finally uncovered fact. Simply believing in something, and seeking the safety of numbers to justify that belief with the slogan, *'we can't all be wrong'*, is insufficient, because sometimes we *are* all wrong. Many Germans believed in the principles of National Socialism, they passionately believed they were the master race, the *ubermench*; they were tragically wrong. Scientists too get it wrong, sometimes it is simply wrong because the instrument is wrong or cannot measure what is being looked for, as with the Michelson Morley experiment, and the search for the ether. Sometimes we are victims of a fraud, like the famous Piltdown man hoax. Sometimes it is simply wrong science, like cold fusion. In any sporting contest a great many of the supporters believe their team, or their man, dog, horse, fly, frog, etc, will triumph in the contest, and are often prepared to bet substantial, often ruinous, amounts on the outcome, the amount staked often being directly proportionate to the strength of belief; they are frequently wrong. The magnitude of numbers holding a particular opinion is simply irrelevant to, and independent of, the facts.

The creation evolution debate is the name given to an ongoing attempt by extreme American Christian groups to convince rational people that there is such a thing. It is an attempt to have the teachings of the Book of Genesis accepted as factual occurrence, or accepted as a balanced alternative explanation for the formation of the earth, the origin of life, and the evolution of species. It is also an attempt, particularly in America to have creationism and Intelligent Design inserted into the public school curriculum. It is rare indeed, to ever come across a more wanton waste of time or a more futile intellectual pursuit than the so-called creation, evolution debate. Nor would it be fair to call it a debate of any variety whatsoever. On the face of it, it would appear to be little more than a pointless rant on the part of a few irrational individuals hell bent of the crazy notion that every scientist on the planet has misread the evidence, the laws of physics, the laws chemistry, the laws of biology, and that in fact, the entire planet and everything upon its surface was created by magic. To use the scientific vernacular, the ideas are, bonkers! Why every scientist on the planet should thus conspire or why the laws of physics should be discontinuous in such a glaringly illogical manner is, of course, irrelevant to the proponents of these bizarre beliefs. The only argument they can muster to put against the entire weight of scientific scholarship accumulated over the last three thousand years is that they have a book, and the book must be correct, therefore all evidence to the contrary must, and can be, dismissed out of hand. The book says so, and we need not question further. More to the point, to question further is to anger the very creator of the magic, which gave rise to the universe in the first place. And so on. What annoys many scientists is that there exist serious individuals with a scientific training who, for reasons entirely of their own, sign up to these irrational beliefs, and openly profess them in public. The recent debate between Ken Ham of Creation Museum fame, and Bill Nye, which interested readers will easily find on YouTube, is a case in point.

Another presumption of the Creation and Intelligent Design movement, and they are the same movement, is that science is some kind of a finished product, that the project has been completed, and reached a conclusion, and hence they will attempt to argue that there is no understanding of mass, or the origins of life, or the beginning of this or the formation of that, statements and arguments such as, the origin of life remains a mystery, and therefore god did it. Science is no more finished than is medicine, no one is going to suggest that medicine is a completed and closed subject, that everything we need to know on the subject is known, and the book can be forever closed? We still have no cure for spinal injuries, organ replacement, kidneys for example, nor is there a cure for cancer. Intelligent Design is simply a god of the gaps theory where the gaps become ever smaller. It is the last desperate gasp of an intellectually bankrupt enterprise. Neil deGrasse Tyson, the astronomer, and great populariser of science, is of the opinion that intelligent design ought to be thought in the science classroom in schools in America. It is, after all, very much part of the history of science, from the Greeks to Galileo, to Newton and Huygens. There has always been a point where the scientist gives up, where the science stops, and the speculation begins. When the giants of early science were faced with the unknown, they frequently resorted to god as the answer, god was very much part of their world, they attributed the unknown or the complex or the unfathomable to god, to intelligent design. Intelligent design is therefore a part of the rich history and culture of science, it is part of the process of scientific thought that has developed over the last five hundred years. It is therefore not unreasonable that creationism and Intelligent Design be objectively thought as part of the history of science.

Some Christians have a saying that if you scratch the surface of an atheist you will uncover an agnostic. This cannot be the case, for the simple reason that an atheist is an atheist for a very good reason, usually they have thought about religion seriously for more than just a passing few moments, and come to the only conclusion possible, that, as the late Christopher Hitchens opined, *it's an empty sack*. Agnostic, believer, or atheist, golfer or non-golfer, we cannot escape the facts of our existence. It is a simple statement of fact that our lives are but a series of fragments, we are all therefore *'fragmentists'*, we experience fragments of reality that comprise each life, the elemental, essential essence, that grand list of experiences and episodes, good, bad and indifferent, memorable and forgettable, pleasurable and loathsome, that make us human, and we store some of them in our imperfect memories. In the end it was said, or implied by Samuel Beckett, the essence of the human remains human, whatever else is taken from them that will always remain. Reduced to the absolute minimum, to a voice of the faintest whisper, with only single words stabbing the dark, the human essence remains, and is human, *Unless it goes on beyond the grave*, as The Unnameable inquires. And no god is required for explanation. The hydrogen gas and dust cloud hypostasis will do the job just fine. There seems, on the basis of non-acceptance of evidence regardless of its strength, little or no difference between Christian fundamentalist adherents to creation myths, and the rump of National Socialist sympathisers refusing to accept the facts of the holocaust as reported, and as discovered by US troops in Germany.

Christian fundamentalists are content to wallow in dogma with the certain knowledge that they preach and screech the revealed word of god. The very words ushered forth from the very lips of the Son of God. So too were the SS certain, and Hitler was certain, they knew in their hearts that what they did was an act of purification, they were working on behalf of humanity. Jacob Bronowski's famous appeal to reason and understanding from the pond outside Auschwitz throws such certainties into sharp relief, the Jews did not perish because of gas rather they perished because men believed that what they did was based on the certainty of their belief system. The ignorance arrogance and dogma, as Bronowski labelled it, was the cause peddled by men who were certain. The many and different flavours of Christianity that have run the world for sixteen hundred, of the last two thousand years, did not serve us well, they have now been pushed away from the table, and should never again have a seat, they have nothing to offer.

In the United States, the Supreme Court banned the teaching of creationism in schools because it was in breach of the first amendment separating church and state. But the Court in its wisdom allowed that alternatives to evolution may be taught in schools if they were scientifically based, in other words, other scientific theories regarding the origin and development of life on earth which did not rely on, or adhere to evolution, as understood by the Court, could be taught in schools. To date no such theory has been advanced, nor is it likely that one will emerge; evolution is a simple proven set of facts, which best fits the observed structure of all life on earth, and explains the origin of species. The Christians remain scratching their heads and looking to the sky for inspiration, but thus far, came there none. No new theory, no one to save the day. No contradiction of the fact that every Christian on the planet today is an Ape, an upright walking, talking Ape, whether they like it or not.

What they fail to realise or accept is the simple fact that there is no argument. Creationists and members of the Intelligent Design movement are so far beyond the pale there is no point in engaging in a debate. They have lost the argument in the face of overwhelming evidence to the contrary, and the matter ought to simply be closed. In exactly the same way as there is simply no point in discussing whether or not the earth is flat. Would any respectable scientist take seriously an offer to appear in a debate with a representative of the flat earth society as to the particular topological properties of the earth? Well, probably, for fun, or at the right price?

Evolution is a clear example of a closed argument. But that still leaves the question of why otherwise intelligent men, and women believe or claim they have faith, in esoteric principles and tenets of religious doctrine quite clearly at odds with the reality we observe, and the laws of nature we have discovered? In times past, everyone was compelled to have faith or at least the outward trappings of faith, it was a matter of life and death, literally, believe or perish. They knew no better. Newton a case in point, he devoted a large part of his existence to the occult and alchemy, and to the

Book of Revelation; even arriving at the year 2060 for the end of the world. But now, in our time, when there is no compulsion, it seems a strange position for men and women of science, or any kind of learning for that matter, to adopt. Although quite often for the sake of social inclusion non-believers will conform to what they see as the normal patterns of suburban life. It's an open question as to when, if ever, an atheist could be elected as president of the USA, for example. One thing is certain; the future will be non-theistic, for the simple reason that objective education is now truly free and available to all who care to partake.

In reading Richard Dawkins book, *The God Delusion,* you come across his bafflement at otherwise intelligent people such as scientists, for example, the eminent theoretical physicist John Polkinghorn, a man not content to hold a private belief in some superintendent of the universe he required to take holy orders, and become an Anglican priest. And the eminent Oxford mathematician John Lennox, a bible believing Christian, who has debated both Richard Dawkins and Christopher Hitchens on the subject of the existence of god. There must be a sense of bafflement, not just with why a physicist or a mathematician should come to such conclusions given the available evidence, but also some mild bafflement as to why someone such as Richard Dawkins would enter into debate with creationists of any ilk, or with any irrational believer, a Cardinal of the Roman Catholic Church, for instance. Certain people seem to be simply beyond argument, and cannot be reached; they appear immune to reasoned debate or logical, or factual argument. In the same way alcoholics cannot be open to seeking help or recognising the existence of a problem until they hit rock bottom. Otherwise they are simply beyond argument, beyond logic, beyond any attempt at rational discourse. There was a time when all scientists were Christians; there was a time when it was not safe to be otherwise. But then as well, there was also a time when there was no difference between astronomy and astrology. The study of both was the norm for a learned individual. There was no distinction, and the practitioners of the period saw none. Now of course, we know better, the facts have changed, our opinions have changed with the factual alteration. We've moved on.

An example of this flight from the rational, and the complete refusal to accept that any belief about the physical world, not grounded in reality, or not capable of being tested in reality, is simply a baseless opinion is illustrated in the following scene. Picture if you will, the laughable site of a creationist Pastor standing on the stage of a revival meeting, with a silicon-chip based laptop computer, connected to a projector containing a photon emitting lens, and a cavity containing, photon amplifying, laser pointer. The laser pointer is used create a green dot of light; to point out on a projected graphic, generated from a file stored on a solid state drive, various facts concerning the planet earth, and the grand conspiracy perpetrated upon Christians by the scientific community, and against the holy revealed word. The slide show reveals, the fact that evolution is false, that the earth is only six thousand years old, and that carbon dating is discredited, and therefore woolly Mammoths could not have lived twenty thousand years ago; they all evidently perished in Noah's flood. The Pastor further asserts carbon dating does not work, because god said it couldn't work, and continues to flash the laser pointer across the illuminated screen; there are photons everywhere. What the Pastor fails to see, and clearly seems not to know, or is too ignorant to appreciate, or is in denial about, or is simply fraudulently concealing from the congregation, is that the laptop, the laser pointer and carbon dating are all interconnected by the most successful scientific theory in history: that of Quantum Mechanics. The theory formulated by learned men and women, and modified for over one hundred years, describes the behaviour of matter at the atomic and sub atomic level. The computer and laser pointer could not have been fabricated without a detailed knowledge of quantum theory and its application. The Pastor would not have the use of a laptop with its PN junctions, and uncountable numbers of electrons moving according to the principles of the greatest theory ever formulated. Nor would the laser pump out its amplified photon beam without its makers having had a very detailed understanding of the energy levels of the electrons in the atoms comprising the laser cavity, not to mention the behaviour of photons within the cavity. It is therefore a very reasonable conclusion to reach by any objective observer that in using the laser pointer, the light emitting projector, and the knowledge-storing computer, here is a Preacher, who not only comfortably accepts the validity of quantum theory but also places more than a little trust in it. Rejection of the carbon dating evidence is a rejection of the very same theory that has already been implicitly accepted. And this is what passes for debate, and makes sense, among Creationist Pastors. You cannot *pick n choose* which of the laws of physics you will obey and which you will discard; not that the laws of physics give a damn either way. You can't have it both ways. Even the most fervent Creationist would have to agree, *you can't be a little bit pregnant.* It's a deployment of logic more at home in the kindergarten.

Creation and evolution in America has a somewhat funny history bordering on the nonsensical. In the United States of America, the stark fact is that over the years in poll after poll vast numbers of the American public still cling hopelessly to the silly notion that the world and America were both created by god, and that evolution of life on earth never happened. Such polls are easy to find on the net and need no particular reference here, but the figures are alarming at around the 40% mark, year in and year out. It seems astounding that the greatest nation in the world has an education system that, on the face of it, would appear to be such a dismal failure. Well-funded right wing groups and organisations are determined to have evolution inserted into the classroom in some form, with many states being urged to allow the teaching of alternatives and objections to evolution, along with plugs for what are seen as various gaps in accepted evolutionary theory. With numerous attempts at having some form of intelligent design considered, with the designer, of course, being some external creator, as found in Genesis. All of course leads on to Jesus. Those involved in education and the scientific community are, of course, concerned by creeping theological intrusions into the education system. It must appear an odd fact that nearly 90 years after the Scopes Monkey Trial, organisations such as, Answers in Genesis, the Discovery Institute, Creation World View and the Creation Studies Institute, should not only be in business but find themselves extremely well-funded, and very well supported.

Ever since President Bush, caused consternation among scientists, teachers and anyone else who had given the matter the slightest thought, by declaring that, *the jury is still out on evolution*, such organisations along with the entire American bible belt has been on the march with the stated aim of getting creation back into the public school system. The tactics and stunts are those we have come to expect from the Christian right in America, and involve lawsuits, the loading of school boards, and votes in state legislatures, and all with the clear aim of casting doubt on evolution by declaring it to be an unproven theory, and nothing more. And of course the language is emotive, with for example, people in Kansas quick to affirm that they did not descend from monkeys. A point the evolutionists might well concede, as there is a clear species distinction between monkeys, and Apes.

Fundamentalist Christians in America are also fond of martial language such as fighting a war for the soul of the nation, as if it had one, opening a new front in the war for America's culture. They see it as a battle to preserve freedom, or try to make it one; the freedom to speak has instead become the freedom to pollute the minds of children with arrant nonsense. Christian claims that there are growing numbers of scientists who doubt evolution are simply erroneous; no scientist properly trained in a scientific discipline could dismiss the available evidence, which increases every year. As Karl Popper noted, it is intellectual fraud to do so.

The debate, such as it is, is clearly a political debate rather than a scientific one, every poll conducted shows that a majority of the American population believes in god, therefore that belief, being the majority must be respected and given prominence. Politicians, always fearful of offending the majority are typically cagey on the subject, but on the right, the prevailing view is that there are defects in evolution and therefore there are legitimate concerns, and so a debate must be had, and an alternative view must be allowed. Hence the approach seems to have evolved, *yes evolved*, into one of trying to get creation in via the side door by insisting that evolution is open to an alternative viewpoint, and the kids have a right to hear it. But that has always been insufficient both in science and in law. There is no sanity in numbers, just because the majority believe something to be true, does not mean it is true.

It can therefore be reasonably inferred that movements attempting to have the teaching of evolution in schools companioned by some set of hints and assertions indicating it might be an unproven theory, and open to reasonable question are theological in nature. The intelligent designer hinted at, is none other than the god of the Old Testament. And there are powerful conservative lobbyists at work, for example, the wealthy conservative Howard Ahmanson Jr., who holds as a stated goal "*the total integration of biblical law into our lives.*" There is also the Maclellan Foundation, an organisation that believes in, "*the infallibility of the Scripture.*" But regardless of who is behind the attacks on science and the anti-Darwin rants, it remains a good old-fashioned, '*god did it*', solution firmly grounded in magic. And, as with all such attempts to explain something by blaming god, it fails, being nothing more than the same empty sack complained of by Christopher Hitchens.

There can be no freedom of choice if one is ignorant of the alternatives. One cannot make a choice unless it is an informed choice. Children forced to blindly follow a particular faith are not informed. The choice is being made for them. Religion only survives by forcibly brainwashing the children into it. Forcing children to memorise vast tracts of obscure texts is tantamount to child abuse; not even grown up jobbing actors are required to learn lines to that extent. Fairy stories, and threats of hell and damnation are the oldest tricks in the book. To be ignorant is to be a captive and to be constrained. Scientists know that all too well. To be ignorant of the cure for lung cancer makes us prisoners of lung cancer, forces us to be constrained to suffer it, and watch others suffer and die from it. We are prisoners of our ignorance. The true path to freedom is towards the light of knowledge; this is what gives rise to the right to an education for all humans, and ancillary to that right must be the right to an unbiased education, unburdened by superstition or magic.

All Christians believe in evolution, is a simple statement of fact backed up by masses of good supporting evidence. Evolution as an existing process that occurs all around us in our everyday experience, and most Christians, even the most extreme creationists have no choice but to agree. And not only does the process exist but most Christians believe it when they see it. Cell phones, are a good example, the 1987 movie, *Wall Street*, contains a scene where Michael Douglas wanders along a beach at sunrise with a cell phone the size of a cereal box. Now we have the *iPhone*, there is no comparison, the former clearly and unmistakably evolved into the latter via a host of intermediate forms. The same may be said about the personal computer, jet engines, and any other piece of technology you can name, hence a Christian seeing such wonders, simply cannot deny that evolution, as a process, exists. But any attempt to apply the process to life is not allowed by Christian rules, evolution cannot possibly operate on life because the book says so. There simply was not enough time, because the earth was created 6,000 years ago.

Even the most fatuous and intractable creationist, and there are many, would have to agree that things in general have a tendency to evolve over time. Whether it is an economic system, a political ideology, a rock band, or individual human people and their opinions. Time and a variety of different external influences have a noticeable effect. Things grow and alter with time, and external influences, new blood joins the party, or the church for that matter, and over time subtle effects and shifts are seen to occur. Cityscapes change over time as new buildings and infrastructure are added. People alter as they live their lives, become educated and have children. Organisations, such as companies change with time as successive generations of management shape the structure, or new laws or regulations are applied. In fact taking the early unreformed Catholic Church and comparing it with the situation today where numerous distinct species of Christianity now abound. If that is not a clear example of evolution giving rise to speciation, the Pope is no catholic. Systems of economy or political thought change, and are refined and adjusted as new opinions and results are added. In fact, wherever we look, and whatever we choose, whatever aspect of our reality you would care to examine. Everything changes with time, so why not life itself? Why should life be immune to the type of natural alteration we see everywhere? And the simple facts made known by even the most cursory glance reveals that life is not immune to the pressures of time and external natural influences. And if the changes over time, which Darwin called *'natural selection'*, are so obvious to us now, and there are so many examples of the process, even in our own individual daily lives; why do people exist in the world who deny the most simple facts amenable to even the most disinterested schoolchild? The very simple, very alarming, and potentially very dangerous fact of antibiotic resistance is the starkest proof that evolution of life is a proven fact, and one likely to affect a great many bible believing Christians whether or not they chose to see it as something else. Pathogenic microbes evolve resistance to the drugs we humans have evolved to kill them; that is the simple fact. We are running out of drugs to treat disease. A world without the ability to fight disease-causing organisms is not a pleasant prospect. Although Christians may simply see it as another sign that we are living in the *'end times'*. Why do otherwise intelligent human beings engage in pointless campaigns to discredit science in the hope that creation will be accepted? This kind of intransigence and ignorance is what causes most rational people to look at religion with the utmost distrust, and curious bafflement, and conclude, that money must be, and is a factor.

The argument of Intelligent Design, is also countered by the contention that something as perfect and sublime as a grand piano could not possibly have come from the mind of a man that it must have been created by god, yet we know it evolved over a great many generations of precursors to the instrument. And indeed the piano has no inventor, no one actually sat down with a piece of paper and said, *'I think I shall invent something along the following lines'*, and then proceeded to draw the inner workings of a Steinway grand. No! The instrument we know and love today came about over centuries of invention, improvement, refinement and adjustment until a consensus was reached in materials, construction, design and function to give us what concert pianists the world over have come to expect. There was no creator, no initial idea, no grand design, simply evolution brought about by many generations of musical instrument makers working upon a concept over time. Time, complexity and human problem solving gave us the piano via an evolutionary process, just as Natural Selection gave us all the species, from the humblest antibiotic resistant strain of gonorrhoea up to and including humans.

Education is the key to a great many aspects of human existence, and is the only known cure for ignorance. Education is the great philosophical and political solvent of humanity. Like water acting on rock, education has the capacity to dissolve mythology and other nonsense such as religion over time, perhaps only over generational periods of time. Like a persistent stream, education can erode stubborn ideas and concepts acting as it does to replace the eroded surface with a fresh one that can then reveal the truth of the underlying structure. Education is the dissemination of objective, unbiased, factual knowledge, usually through a process of learning and testing, resulting in some form of accreditation in recognition of the work done. The knowledge must be honest and true, that is, the accepted truth of a particular position must be given, as opposed to opinions or rigid doctrine, like consent it must be fully informed and freely given, without prejudice. There must also be no propaganda along the lines of George Orwell's, Ministry of Truth, from the book *1984*. Proper education must allow the recipient to get to the very heart of the matter, with all aspects explored and discussed, and all questions answered with facts where appropriate. Notwithstanding the rights believers have to know in detail the parameters of their belief, the study of Theology or divinity, or study of the Bible or Koran, or any other sacred scrolls or texts, from the presupposition their contents are true, is hardly worthy of being described as an education.

There is a basic human right not to be ignorant. The right to education is enshrined in many constitutional and charter documents, but usually implemented in a rudimentary manner, or not at all in some cases, the education of women and girls in certain parts of the world, for example is despised and discouraged to the great detriment of those societies. Ignorance is no defence at law, is an old common law maxim, and is true in all jurisdictions, the alternative to education is ignorance, pure and simple, and ignorance is as vile and miserable an existence as it was in the time of Dickens. The alternative view to biological evolution, as an explanation for the existence of animal species including humans is: nothing. The anti-creationist website, No Answers in Genesis, puts it quite simply, *Creationism is not the alternative to Evolution – ignorance is.*

There is no possibility of a theocracy being a state of society capable of producing a technological world, and ultimately a Purple world. The probability would be very small indeed, because conjecture and the need to place arguments is a requirement, and in a theocracy the religious doctrines, and not the people, are supreme. Hence all conjecture and all argument must by necessity defer to the doctrines of the theocracy. One would require permission to refute, permission to contradict. Not easily obtained, especially if those from whom permission is sought have not paid the entry fee into the field of debate. Theocracies vary in strength, Iran and Tibet are different in that their religion is not the same, but similar in what constitutes freedom of thought. Saudi Arabia is to all intents and purposes a theocracy; thinking outside the box is something the proponent must consider very carefully. The great strides of the Arab empire in the Middle Ages in mathematics and astronomical observation, were tempered by the prohibition on drawing the human form, progress but only so far and in specified and approved directions. It could be argued the establishment of a theocracy overthrew the Roman Empire. It emerged from the chaos, and ruled for fifteen hundred years. The rump of it remains to this day in Vatican City and in the myriad of offshoots all following the same ragged dogma. But the European enlightenment grew out of that theocracy, by the brave questioning of the few scientists and philosophers who discovered that all was not quite as the book said it was. They questioned, what they observed, they explored, and they threw off the chains. Everything we know tells us that a mature exploring society can only develop unfettered by chains of dogma and ignorance. Everything must be questioned; everything must be explored and reported on, in the blinding light of discovery. Evangelical Christians and creationists would have it another way, they would like to impose a theocracy in America, and anywhere else they could get away with it, a theocracy every bit as rancid as that in Iran, replace the Bible by the Koran, and that is exactly what they would wish. Government by prayer, is their stated aim, rule by the bible imposed by god himself, non-believers punished, science rolled back, a new dark age of superstition and mythology. If the catalogue of achievement of a society, or a group of people for that matter, is to be written down, there can be no better way than to measure the total number of Nobel Prizes won for science and literature. The entire Muslim world, almost one quarter of the World's population has managed to collect a total of four, one in Physics, one in Chemistry, and two in Literature. Contrast that to the very significant fraction of the total number of Nobel Prizes won by members of the Jewish community, numbering only about fifteen million people worldwide. Even a place as small as Cambridge University, or a country as small as Ireland, has more Nobel Prizes than the entire Muslim world. This is not meant to be a racist slur or an attack on people in the Arab world or elsewhere, it is simply a stark fact, and ought to be a matter of deep concern to educators and governments in the Muslim world. On the face of it an explanation is required for the fact that to a first approximation there has been no contribution whatsoever to human culture in the twentieth century, from that quarter of the world's population. Theocracy simply cannot compete. Only the inhabitants of the Muslim world can explain this glaring anomaly, because it requires explanation.

Any fool can be a preacher, or claim to be, you need only say you have been anointed by some supernatural being, and you're there. But only a theoretical physicist can get to the bottom of a problem in theoretical physics. No amount of reading alone, or working at home will get you there. You have to go through the process. And you have to do it properly. The origin of the universe is a problem in theoretical cosmology. It is not metaphysical. It is not theological or ontological; it is mathematics at the front end of the envelope, and theoretical conjecture in multidimensional arenas, and not for the faint hearted. The detailed studies of rock strata, and chemical and physical analysis, and classification of the rocks of the earth is also a discipline for those who pay an entry fee in time and effort, and expertise gained through long years of experience. Following the immutable laws of physics and chemistry arrives at determined ages of whatever they find in the ground, they did not need to be there to see the rocks form. We know what we know because we have done the work, we have earned the right to know what we know. It's a matter of established and accepted fact, not opinion and unfounded belief or revealed wisdom. People are not stupid, might be an over simplification, but they can be very poorly informed, and the effect of the one often contrives to give the appearance of the other. There are very few stupid people in the world. But certainly there are ignorant people, and there are apathetic people. Even the best golfer would have to admit to having a great deal of ignorance, about golf; they know it for a fact because they are exposed to their obvious limitations every time they play. Likewise, many a golfer would also have to confess ignorance, perhaps even apathy, towards soccer and basketball. Sometimes both ignorance and apathy operate in parallel. Some people find learning a chore. There's nothing new in that, we have always had the, *don't know; don't want to know*, people, sometimes with alarming levels of intelligence and ability clearly present. Science does not know everything: that is abundantly clear to anyone who has ever put a penny's worth of effort into studying it. But for some learned possessor of a holy book to step forward, and claim, that his book and his book alone, contains the solution to some pressing problem borders on, if not trespasses into, the laughable. For anointed ones to claim they know what science does not yet know, or that they know why science is ignorant on some point, is arrant nonsense. It is ignorant, arrogant, dogmatic verbiage of the basest degree, for the god-inspired to undertake to fill in the gaps, as it were, regardless of width. Every measurement is made *on the edge of error*, Jacob Bronowski said, again in his famous deliberation on science and humanity delivered, with great human feeling, while standing in the pond of the Auschwitz death camp, and it is worth repeating here, *Science is a very human form of knowledge. We are always at the brink of the known, we always feel forward for what is to be hoped. Every judgment in science stands on the edge of error, and is personal. Science is a tribute to what we can know although we are fallible. In the end the words were said by Oliver Cromwell: 'I beseech you, in the bowels of Christ, think it possible you may be mistaken'.*

Stephen Hawking claims science has made Philosophy redundant, and therefore, the laws of Science and Nature may explain everything we are capable of knowing. It's hard to think of an end of philosophy, because people love to argue, and Science is still a minority pursuit. It is certainly very reasonable to propose that everything we do should be grounded in reality, and founded on fact as opposed to magic and superstition. Nothing else will do. We humans are better than that. Anything else is insufficient; anyone accepting any aspect of our modern technological world implicitly places his or her trust in Science. We pay for Science, and we do so for good reason, because we have made a collective and irrevocable decision that we wish to understand everything. It really is that simple. And just as we no longer make our own clothes, or kill our own food, we leave it to others to discover those things we need to know, or would like to know.

But that still leaves us with a fundamental problem. If religion is clearly such a discredited man made mediocrity, and so on, and the study of theology such a gross perversion of the concept of learning. Why are so many otherwise intelligent people involved, engaged, and indeed wholly committed to its study, practice, profession and continuation? Jesuits are hardly fools, professors of divinity are not idiots, and physicist, and biologist, and chemist converts to Christianity don't suddenly lose all power of rational thought. So what's going on? What do they know, that we don't? Belief? Delusion? Mental illness? Old Age? Fear? What, you may ask, is the cause? Is there some deep spiritual thing in human beings that calls to them? The George Harrison effect, George, of all the Beatles, was known as the *'spiritual one'*, whatever that may mean. Richard Dawkins confesses, if that's the correct term, to being baffled by the phenomenon of, otherwise intelligent, rational people, even the odd scientist becoming believers, it seems to be, like other quirks of the human condition, somewhat beyond current understanding, although as dutiful scientists we may freely speculate, inquire and seek to establish the facts, as a matter of public duty.

They only want your money; might be the cry from the true cynic. Money is the great human motivator the lubricator, James Joyce summed up the entire motivational relationship between money and religion when he wrote in Ulysses. *From Cahill's corner the reverend Hugh C. Love, M.A., made obeisance unperceived, mindful of lords deputies whose hands benignant had held of yore rich advowsons.* The sale of indulgences, the buying of future goodies in the afterlife, provisioning the path into the dark beyond, prayers for the departed a penny a pop, the relic of the true cross, of the true crown of thorns, the one true shroud, and so on. It's the oldest trick in the book. *"Salvation by the shilling"*, as Robert Bolt put it. Separating the sheep from their wool. Religion has always been a racket, a store, selling redemption and revelation, and the certainty of knowledge in the spirit realm, and the unlocking of ancient secrets, which purport to lead the believer on the only true path to eternal bliss or enlightenment, everlasting contentment, riches and success. All packaged and presented in a modern format, tailored to suit your busy lifestyle. You become our partner, we put you in control, sign up now, make the commitment, *'it's all about you'*. Like all good salesmen they have a wide variety of tried and trusted techniques, proven and refined over the years to yield the desired result. There's nothing new, nothing mysterious, nothing sacred. It's just like selling chocolate to kids, or cold beer, or sunny vacations. They pander to the supposed hunger that exists within some ignorant people for answers to naturally arising, if naive, questions. Anybody wishing to delve deep into the subject of the economics of religion might take a look at Rachel Mc Cleary's book, *The Oxford Handbook of the Economics of Religion*. Religion is like any other product or service, companies exist to supply the product, and serve the space; the market is fluid, and behaves much like any other markets. Adam Smith commented upon the religious space in his, *Wealth of Nations*, it is therefore nothing new that there is money in faith. The laws of economics, and rules of business apply, companies form, from one man, dog and pony, outfits, to large multinationals with a global reach. Wealth and excess are commonplace in modern religious practice, the profit motive, rules all, the accumulation of money, and growing the customer base; the price of faith and hope. If ever evidence was required that the sole purpose of religion is to operate a cash business, and extract as much money as possible from adherents for the sole purpose of amassing a vast pile of cash, they need look no further than the Church of Scientology. The HBO documentary, *Going Clear*, produced by Alex Gibney, is a disturbing testimony of the cult and its obsession with money. Like the movie *Marjoe*, before it, which showed us the tricks of the trade of the evangelical preacher class.

Religion never had right on its side, only might, the scurrilous drivelling of perverts and disillusioned would-be intellectuals reduced to the navel gazing antics of theology. They have been seen off by scientific heresy by the profanity of truth seekers, by the emergence of a body of knowledge, both of the world we live in and the planet we share with the other animals. They are rendered moot by the careful observation using optical and electric instrumentation never dreamt of by the cave dwelling, blood-burning peasantry who invented the few myths and fables upon which the entire rancid mess is based. God bless the telescope at the one end, and the microscope at the other, their illuminations have given us a world of wonder. Seek yee then the truth in Hubble, and in the eyeglass, and the spyglass, and the prism, and the grating, and the electron beam, because you will not find it in the grubby pages of a bible or any other holy scribbling. The oldest cartoon stereotype is surely that of the delusional poor moron replete with sandwich board declaring *'the end is neigh'*, the end has always been neigh, and will probably always be neigh or close to neigh. As long as there are followers ready to part with their hard earned Dollar, Euro or Yuan to support these cabbage-like fossils, the end will always be neigh. America will always be under attack from the enemies of Satan, and the Lord will always be on his way, his coming imminent. And it will always be a cliché. As it undoubtedly was, in Europe, in the Middle Ages, at the turn of the first millennium.

Just like tobacco executives defending a product shown to be overwhelmingly harmful to humans, religious people continue to peddle their nonsense, especially to children. Otherwise intelligent human people clinging to nonsense, leads to only one possible explanation, that of money. It's all about the money, *follow the money*, Woodward and Bernstein were told by Deep Throat. Religion is a business, make no mistake, religion is a going concern, hiding behind the thin veil of charitable giving each and every small religious enterprise is a profit seeking commercial entity, competing with other commercial entities in the religion space. These people have invested in their business, and cannot walk away from it. What else would they do? They are, for the most part, otherwise unemployable. Samuel Beckett, the famous Irish writer and Nobel laureate, was very rarely wrong in any matter concerning the human condition. But his assertion, that being a poet, was, *the last ditch for an unemployable man*, must be called into some question as he very clearly and evidently failed to consider those of his fellow creatures whose only employment outlet is to be found in the religious space. Poetry, by contrast, is a noble, if underpaid, profession, though there are significant exceptions. Seamus Heaney, the late Nobel laureate, also from Ireland, went from being merely, famous Seamus, to being rich and famous Seamus. Anybody interested in evaluating the collective achievements of the profession of poet over the years need look no further than *The Oxford Book of English Verse*, arguably one of, if not *the* greatest books in print. Compare that work to the mountains of holy gibberish published under the headings of theology, divinity, biblical scholarship, and even eschatology; Beckett was surely wrong. The last ditch for an unemployable man has always been, and must surely remain today, the occupation of, religious professional. What skills do they possess other than the ability to preach nonsense, and lie to children? At some deep internal level, they obviously know, or at least it would be reasonable to impute that they know, that the opinions they hold, and pass on to the children they teach, and the ignorant members of their congregations, are manifestly and undeniably false. And yet knowing this they continue, with the pretence, with the sham, with the mean practice of leading vulnerable people along a dead end. It baffles most rational people that seemingly otherwise intelligent people allow themselves to continue with such a charade. Apart from the obvious one, of financial gain, it is hard to understand their motive. Of course they can never bring themselves to spill the beans, as it were, to admit to the fact that their finest wares are no more than threadbare rags. Few salesmen would stand in the public arena and declare that what they are, in fact, selling is *crap*, well, there was this one salesman, CEO in fact, of a leading British Jewellery Store chain, the company was Ratner's, and his name was Gerald Ratner, at a speech to the British Institute of Directors in 1991, he described certain of his company's products as being *'total crap'*, the publicity associated with the remarks caused the group to collapse in value by £500m, and cost Ratner his job, and led him into gaff of the century charts, and business textbooks as an example of how not to treat customers.

Creationism is a business just like any other branch of religion. The way to look at the creationists, and other biblical, *'preach and screech'* individuals, and organisations is to look at them as a business subgroup, the creationist sector of the religion space. They have a product, and they sell it into a market place. Noah and the flood, why is there a persistence in this myth among these people when it is manifestly mistaken to hold such opinions according to all the evidence. Where do holders of such beliefs acquire the extensive funding necessary to conduct archaeological digs in remote regions of the world where their belief system tells them they will find proof of the resting place of the Noah's wooden boat? The answer is that they record their exploits and make TV programs, DVDs, and write books, create detailed slideshows that can be taken on the road, they tour the church circuit, $10 a head, it all adds up. It is every bit, a form of entertainment, going to a church to hear such a lecture perhaps accompanied by a prayer session, and some mediocre Christian music, might in some people's minds constitute a cheap night out. Fun for all the family. A trip to a burger joint on the way home, and the entire evening might well come in at less than $50 for a family of four, not at all bad if you are on a tight budget. And if you are such a family that believes, or needs to be seen to believe, or have neither the time nor the inclination to ask questions then you probably require frequent reinforcement, and therein lies the business model. That principle is above all what keeps the show on the road.

Why do we tolerate religion, is another obvious question we need to ask, after all, atheistic, humanist, socially minded, scientists are a tolerant lot, but there are limits. We very rarely if ever tolerate conjectures, theories or explanations we consider to be simply bad, as in magical, mystical, and wholly unsupported and unsupportable, along the lines of, the earth is flat, because it's flat, because the book says its flat, and that's good enough for us, and so forth. So why should we tolerate religion? Our modern technological society is founded upon science and learning, and all that science and technology has delivered. What particular reason therefore, should we have for allowing the public practice of such obvious nonsense in our societies across the Western World, why should we not simply recommend a ban on all religion, and suggest the closure of all Mosques, Churches, Temples and Synagogues?

Because in a Purple world, which is by definition the world we aspire to evolve into, we tolerate almost everything, with the long list of obvious exceptions, such as intolerance, dishonesty, violence, hatred, and so on. And that must imply that we expressly include those beliefs, or opinion sets, grounded in obscure mythological nonsense. We tolerate religion because we can afford to tolerate, because it does not damage us to tolerate it. More damage would be done to our society by not tolerating it. Because we are strong enough to tolerate, in the same way that strength in a totalitarian regime is seen as a means to oppress, in our western enlightenment based civilisation, the opposite is true; strength is used to absorb and include. We tolerate because of our indifferent and nonchalant approach to religion, and superstition in general, in the same way that UFO spotters, and conspiracy theorists are tolerated, along with astrology and palm reading. There is no problem to which that kind of intolerance is ever likely to provide a solution.

Toleration always comes with wealth, the more wealth a society has, the more it can tolerate the dissenting voice, it was true in the Holland of Spinoza, and it's true today. But there are exceptions to every generality, and we have the exception of the Middle Eastern theocracies, whose vast wealth is rooted in oil, and accident of geography. Perhaps the generality that tolerance seems to be associated with wealthy societies should be better phrased to distinguish societies based on earned, rather than found, wealth.

Toleration is that aspect of our existence that allows us to accept the validity of other opinions or activities even though we find them either disagreeable, of little value, or simply incorrect. Our tolerance seems to be a mandatory requirement regardless of the opinion expressed or the act complained of, to the extent that every voice acquires the right to be heard. Yet there are limits, and we can include the obvious candidates here, supporters of Stalin, Hitler, Pol Pot, and the like, and those who deny the Holocaust ever happened. Our principle seems to be, and is, that we dislike the act but insist that the culture of freedom and pluralism that allows the individual to develop the ideas or carry out the act is to be cherished and protected at all costs. Our atmosphere of freedom and tolerance will always be open to those who wish to take advantage of it, and use it as a way to do us harm, and evil extremes committed by states, companies and individuals will probably always be part of our free world, no matter what kind of technology we develop. There will always be people who are *anti*, those who oppose, with or without reason, they protest, they oppose, they draw the line. Simply because they do, reason usually has little to do with it. Like Marlon Brando's character in the movie, *The Wild One*, when asked what he was protesting about, he replied, *what have you got?* Our toleration of religion is one of those aspects of the cult of modern freedom that is a great source of academic debate. It is clear that religion or rather people impressed or obsessed by religion are responsible for a great many atrocities, and again the reader is referred to Dawkins and Hitchens for a catalogue of such incidents. In his book, *Why Tolerate Religion?* For example, Brian Leiter draws on sources such as Thomas Hobbes, John Stuart Mill, John Locke, and modern legal thinkers such as John Rawls to make his case that tolerance of religion should not be a special case, he writes, "Toleration may be a virtue, both in individuals and in states, but its selective application to the conscience only of religious believers is not morally defensible." He also argues from a Nietzschean perspective that because false beliefs are sometimes considered necessary to life, simply because a belief system is based on falsehood, untruths or is a proven nonsense does not mean that it is not deserving of our toleration.

Essentially religion is nothing special, and therefore should not, and does not, deserve any special kind of tolerance above and beyond the ordinary tolerance we reserve for other believers of the many, and various irrationalities at large today, to mention only, Ancient Aliens, UFOs, Astrology, those who believe they are destined to win the lottery, politicians who believe in the correctness of a clearly failed position, those who follow them, the list could go on. Irrational belief seems to be as human as eating and sleeping, and we simply have to tolerate it. We tolerate it because we tolerate everything, with the exception of intolerance, and those self-evidently adverse aspects and mores of human behaviour, which, largely by consensus, governments the world over have seen fit to outlaw.

Is it ever possible, as the Christians put it, to love the sinner but hate the sin? Can we retain a respect for the person, as a person, as is that person's natural right, while at the same time having neither regard, nor respect for the beliefs held by that person. Voltaire's alleged famous line about defending, to the death, the right to hold an opinion, is all very well, but does that include ideas that are obviously and manifestly incorrect and untrue, or contrary to observed fact, or just downright bonkers, for example the absurd biblical belief that the earth is a mere six thousand years old, and that the entire observable universe was created within the time frame of a literal week, on the whim of an infinite being whose parameters, motives or intentions are forever to remain a mystery? What about downright evil, or divisive, or dishonest beliefs? Must the idea itself carry the same degree of respect as the person holding the idea, regardless of the idea, or can the two be separated. And of course the two are frequently separated, you may very well love, as a dear friend, the fellow who passionately believes the earth to be flat, he is as entitled to his opinion as the next deluded soul, but idiotic beliefs remain idiotic, regardless of the underlying passion. And you may well despise the ranting Nazi sympathiser standing on a corner, sailing close to the wind with his rhetoric, but as long as he remains within the law he may continue, and confrontation by reasoned argument may prove to be the best counter to such a person. There is no intellectual grace, if that is the correct term, in describing people as morons or idiots, the terms are insulting at best, but polite disagreement with someone professing manifestly repugnant beliefs, such as Creationism, or Nazism, perhaps ought to be along the lines of: I respect you too much to believe you would be capable of holding such a nonsensical idea as truth. If, that is, you actually care about the person's feelings, although regardless of that, it is clearly important not to depart from the reasonable norms of rational debate, and stray into insult. For example, a statement indicating that it might be a good thing for the world to see the back of Judaism, should not be confused with some, right wing inspired, desire to rid the world of Jewish people. On the one hand, the religion is a silly and dangerous Bronze Age nonsense written by peasants while, on the other, we have a proud and prosperous collection of individuals who have given the world among others, Albert Einstein and Stephen Spielberg, to mention only two. Sam Harris has pointed this out in his many pronouncements on the subject, and observes that no other field of human endeavour do we allow such a wide latitude as in the profession of religious faith. He also holds this taboo of religious moderation responsible for the extremism of the world. The extremist hides behind the moderate religious practitioner because religious moderates have evolved a cult of non-questioning when it comes to faith.

It is reasonable therefore to suggest that there are clearly limits to our tolerance. There are obvious self-evident limits to our tolerance when it comes to protecting our children, from sexual exploitation, for example, or society in general, from being poisoned, by imposing smoking bans, or banning the use of asbestos, or lead or a host of other toxins. The practice of religion ought to be no different. We ought to consider carefully our tolerance, of religious freedom, if the denial of incontrovertible facts or adherence to a position of wilful ignorance, and intellectual fraud, for the sake of personal gain is the result of that tolerance. Society should also reasonably seek to prevent the pollution of education by outlawing certain manifestly untrue teachings. It would therefore seem, not unreasonable; to impose a criminal sanction for evolution denial, in the same way that holocaust denial is a criminal matter in certain jurisdictions. That is, for the wilful spreading or dissemination of proven lies in a matter of the utmost public importance.

However, this area of legislation is difficult and controversial; the notion that what people think, or hold as opinion may be proscribed goes against our very nature, and our enlightenment value system. Surely silly, or corrupt, or dangerous, or malevolent ideas otherwise repugnant to reasonable people will fall under the weight of their own stupidity or absurdity, surely an educated population will soundly reject foolish nonsense? We may well think like that, but the facts do seem to bear another interpretation; adherence to nonsense sometimes has nothing to do with levels of education. There are to date seventeen countries, where holocaust denial has been made an offence attracting a criminal sanction, among them being those countries most affected by the holocaust, Israel, Germany, and Austria, as one might expect. However, in Spain a similar proposal was found to be unconstitutional, and proposals to create such an offence in British law as well as in Denmark and Sweden have been soundly rejected. Interested readers may read more in Deborah Lipstadt's book, *Denying the Holocaust: The Growing Assault on Truth and Memory*. The author and her publisher were famously sued by historian David Irving who lost the case, and in the process, was described by the judge, in a long and notable judgment as being a "holocaust denier" as well as "right-wing pro-Nazi polemicist". There is a long list of usual suspects who deny the holocaust for many and various reasons, the former Iranian President, Mahmoud Ahmadinejad, is a frequent holocaust denier, for probably obvious reasons. Others deny on the basis of historical revision, or lack of evidence or other concocted reasons, but try as they might they cannot deny the facts of the matter. The overwhelming and accepted factual evidence in the public domain is that the holocaust occurred, and pretty much as described. Hence there is little or no point in denying it, and those that do engage in such denial do so for purely selfish, malevolent or financial reasons. It can be argued there is no point in prosecuting holocaust denial as a crime, or that no problem is solved by such actions, and that prosecutions such as that of French literature professor Robert Faurisson, who was convicted and punished under the Gayssot Act in 1990, do not address the perceived problem. It is, on the other hand, understandable for a variety of reasons, such as prevention of some future rise of fascism or racial tolerance or the prevention of incitement to racial or, ethnic hatred; already crimes in most countries, that holocaust denial should be prosecuted. Prosecution also sends the message that we, as a society, do not tolerate such activity that we hold such views, to be at such variance with the facts, and so repugnant to our reasonable society that criminal prosecution is the necessary remedy. The magnitude of the holocaust itself also demands that we ought reasonably to continue with its denunciation forever more as a special case. Most people who apply any objective attention to the issue will incline to the view that holocaust denial is a silly and mean activity entirely devoid of any human feeling or consideration for the documented suffering of the millions involved, and that perhaps its status as a crime should be retained as a special case as a symbol of our continued revulsion.

So what should we do with those who deny the facts of evolution? There is arguably more evidence for species evolution, and in particular for human evolution that there is for the some of the events of the Second World War. Should evolution denial be made a crime? It is arguable that we should certainly seek to ban the teaching of creationism, or the doctrine of intelligent design, as scientific fact in any, so-called, faith based school, and confine the teaching of any religion, or religious text, to the history, or language class in secular schools. All right thinking members of any reasonable enlightened society should also be in agreement in seeking an outright ban on the surgical interference with the genitals any human person without that person's freely given and fully informed consent, and explicitly in the case of a child where not strictly necessary for sound medical reasons.

To paraphrase an old truism, religious tolerance is to tolerance as religious music is to music. Another view is that perhaps we are far too easy on believers and their mythologies. Not every opinion is worthy of opposition, not every view requires to be balanced by considered response, calling it what it is perhaps serves the better, higher, purpose in the long run. There is no right to be an idiot, no right to be wrong headed, and no right to pollute the education of children, a criminal sanction for evolution denial, even if only as a token, is perhaps taking things a bit too far. But it is arguable that evolution denial, and evolution deniers ought to be given the same status in our catalogue of the lunatic fringe as Holocaust denial. Everything we need to know about religious intolerance can be summed up in the French phrase, *Je Suis Charlie*.

When we seek to limit long held freedoms, we embark on a dangerous road, especially where human thought is concerned, the idea of thought police or religious police rightly turns the stomach of all right thinking democrats. But the law imposes limits on everything we do; there is no reason why religious practice should be singled out as different and distinct. Society has a duty to protect itself from dangerous idiots as much as from pollution, or from dangerous drivers, or from illicit substances, or from terrorists. Idiots, though, have a tendency to be laughed off the stage, nonsense tends to evaporate and diffuse away in the face of reason and intelligence, honed by objective education. Hence there should be no need for legislative provision to counter bad ideas; they will surely fall on their own, without any judicial assistance. Creationist Christians have a favourite taunt reserved for those they collectively lump together as '*Evolutionists*', which essentially accuses rational scientists of invoking a doctrine of, '*millions of years*', to explain everything, a kind of scientific version of magic; given millions of years anything can happen. In the case of religion, however, time is the great erasure; time is all that is required to see the back of religion, in the face of rationality and relentless questioning, it is hard to see religion, of any hue, surviving many more generations. G.K. Chesterton, famously stated, when people stop believing in a god they don't believe in nothing they believe in anything.

Perhaps science is a religion? Sometimes this question is touted as reality, frequently the retort that scientists have evolved a form of religion called scientism, with its very own contrived dogma, which is every bit as pernicious and dishonest as those available in mainstream religions. The essential tenets of this dogma are alleged to be its maintenance that science, and only science has the answers. As any scientist knows all too well, we do not have all the answers, and we frequently alter opinions as new facts come to light, it is the very basis of what has become known as the scientific method, the facts change the theory needs to be altered to suit the new facts, sometimes radically so. Science is falsifiable, a better theory wins the day. Suppose we lived in a time when the prevailing theory regarding the moon was that it was made of cheese, and everyone on earth believed in this cheese theory, and debates were had as to the type of cheese, and so on, and learned papers are published, and factions formed. Now suppose that someone actually goes to the Moon and returns with a bag of rocks, and says that in fact the Moon is not made out of cheese, but is in fact, made from rocks. Then the cheese theory is out. It matters not a damn who believes in it, or how long the theory has been around, or how many supporters it has, or what the king says, or how emotionally attached to it, or psychologically dependent on it, people have become, or how much comfort it brings them in their old age. If the cheese theory is not in line with the observation then it is worthless as a theory, and it's out; '*cheeses has to go*'. Richard Feynman spelled out the method in one of his famous lectures, we start with a guess, he said, then form that into a theory, then test it against experiment, and if it's wrong we throw it out, no matter whose theory it is, or how respected the scientist, if it fails to agree with the experiment, out it goes. Radical theories too, are always acceptable, but only if they are a better fit to the data; they are rarely accepted just because they are radical. There are gaps too, gaping holes in some places, and we have no idea what the full extent of science is, or is likely to be. Science is a work in progress, and is far from complete. But just because we don't know conveys no right whatsoever upon, so called, men of faith to pollute the enterprise with magic. Listening to the pronouncements of some of these anointed ones often conveys the impression that they wish nothing but failure on every scientific or technological enterprise. In response, we should not waste any opportunity to express the view, or rather to state the fact, that all religious belief is hopeless nonsense, devoid of any evidential basis or grounding in reality. As was said before, if the claim contradicts the known laws of nature then it is nonsense; there is no need to be polite, or mince words.

Religion and science are simply incompatible, the one cannot explain the other, there is no peaceful coexistence, religious texts are not scientific explanations of anything, and science publications do not concern themselves with religious concepts. People who claim to be scientists as well as being believers in a god or practitioners of a religion are either not scientists or are deluding themselves, or are just plain awkward, and professing a belief just for the hell of it. Or have been handsomely paid to lend their support or perhaps even blackmailed into supporting a belief in magic. The best description of the dichotomy, is that of the comparison between science and religion as being akin to a Tiger and a Shark, as espoused, by the evolutionary biologist Professor Steve Jones, both animals are formidable predators in their own environments, but neither could match the other on its own particular terms. There of course, people, including scientists, who like the atmosphere of a church, and see attending a service as a community activity, a social gathering, they like the music, the gold and marble, and the general ambiance, and of course they are perfectly at liberty to do so. A great many scientists have been married in a church, and funerals are not quite the same without the ceremony. But that conveys no truth whatsoever to the doctrines, magic remains incompatible with reason, and nonsense remains nonsense.

It is not really a serious question whether or not science pretends to be a religion, or even if science could ever be seen as any kind of belief system. Science is not based on any system of belief; it is simply a collection of observations and resulting facts, collected over time. Science is nothing more than a list of stuff. Occasionally in the history of science there have been periods where belief in a particular theory was a matter of fierce debate, and factions formed on different sides, belief in the atom, was such a case, there was a time when the atom was a questionable entity, there were those who supported the atom, and those who denounced the concept as nonsense. These opposing opinions were only finally resolved with the actual discovery of the atom in the experiments of Rutherford and others. The ether too, was a matter of debate as to whether or not is existed, and again there were those in favour and those against. The famous Michelson and Morley experiment finally resolved the matter. Sometimes arguments can only be settled finally by revealing the factual truth, which cannot be disputed. There have been many examples, from technology, over the years, the debate, for example, as to whether the overshot or the undershot waterwheel was the more efficient waterwheel, was only settled when the engineer John Smeaton (1724-1792) Harnessed the two types to the same axel in and showed that the overshot wheel caused the undershot wheel to rotate backwards. He had calculated that the overshot wheel was 63% efficient, while the undershot wheel was only 22% efficient, no mean difference to an eighteenth century mill owner. Another very public demonstration occurred in 1845 when the screw propeller was introduced and suggested as a new more efficient method of driving a ship, a famous encounter took place between the screw-driven HMS *Rattler,* and the paddle steamer HMS *Alecto,* in which the *Rattler* managed to pull the *Alecto* backwards, at a speed of about 3 knots, along the river Themes, thus demonstrating the efficiency of the propeller over the paddle wheel. And we have the same thing in many other branches of science to this day, lively debate among scientists is nothing new, and as before, when the facts change the opinions change. Even today Christian apologists speaking against Darwinism as if it was a belief system, often speak of those who *believe* in evolution, as if it were something optional. Evolution is a mechanism, which explains the observations; it is the current best fit to the available data. We call it a theory because that is the way scientists speak, it effectively means that we are leaving the door open, if a better explanation of the observed facts is presented, and this better fit explains more, and predicts more, then evolution will be discarded, and replaced with this better fit model. But to date, there is none.

There is however one other aspect of science that deserves mention here because it is frequently cited as having religious overtones, and that is those aspects of theoretical physics and cosmology that are so remote to almost all human minds as to be, or appear, magical and religious. When a theory such as super string theory or M-theory is presented, as being the best explanation of the origin of the universe, yet can never be proved, because individual strings can never be observed, and the theory does not predict anything that can ever be revealed by experiment. And when there are many competing theories, then it almost becomes a matter of belief, and almost a faith issue as to which scientific theory one accepts as the best explanation of the observed universe. We have a similar situation with the scientific basis for human induced climate change, competing opinions on the same data. And sometimes it comes down to a matter of whether one is a believer, or one is waiting to be convinced; till such a time that the weight of evidence leaves no choice. But on the other hand science has always passed through such phases, and will continue to grow and evolve our long list of things we know. There will always be opinions on everything science discovers, and hopefully healthy debate too, because we live in a world of no excuses, no one can claim ignorance of science, it is simply too easy to find the facts in a format that even the most ardent non-scientists can master with little or no effort. If you can understand a newspaper, you can understand science; there is no excuse.

But there is no theocracy of science, nor a deity of technology, or personification of progress, nor do we see the general anthropomorphising of knowledge. There is such a thing as scientific faith, a belief in progress, and a future that will be, and must be better than the past, but it's not blind faith, and it is not misplaced trust. We should never allow ourselves to arrive at a point where we are compelled to trust, or where we must have faith, such notions belong to religion. Trust in science, faith in progress, belief in technology arise because of what has been achieved to date, because of the world that has already been delivered by science and technology, and because of what we observe with our own senses all around us in the real world.

It is easy to make the comparison, and see similarities between faith in a god, and faith in science, technology and progress, we place our faith in a group of strange remote and distant people, that they will deliver us a world of plenty, and a bright and peaceful future, bountiful with the fruits of progress, benevolent as the kindness of a loving parent, flowing in the milk of human kindness, and so forth. Our superstitious predecessors yearned for eternal life after death? We yearn for eternal life prior to death, if a tree can live for 600 years, why not a human?

'Oh tall and graceful splendid tree
Would I could live as long as thee'

And we could be accused of praying to science, and to technology, and to progress, personifying them as gods as the ancients did with the wind, and incense vapours, and the like. We give them names, corporate symbols and logos, colour schemes and designs. We venerate the high priests, the tech CEOs, and the Silicon Valley venture capitalist heroes, and the rock star scientists, constantly present as TV talking heads, endlessly touring, debating, and preaching the eternal message from the book of progress that science is our salvation, our refuge and our strength, the worthy epistle of hope for a gleaming future in an often depressing modern post-industrial society. We might even scribble the odd prayer?

> *Oh Science most high, most knowing, most grand,*
> *give us this day our daily progress*
> *Lead on to the future promised land, allow us no regress*

That is probably taking things too far, science is a very ordinary pursuit, and a very human pursuit, for the most part it involves the mundane collection and presentation of data, and the monitoring of instruments, most PhD theses are unreadable gibberish to anyone outside of the field, and most science is not understood by the people who pay for it. Very few scientists have any kind of a media profile, and most of the time what they do has no effect on anyone whatsoever. As the astrophysicist Lawrence Krauss puts it, most scientists don't think enough on the subject to even know if they are atheists.

It is sometimes hard to fathom the interactions between scientists and god, or certain attitudes of scientists towards religion, and how some technology writers view religion and god. But on closer inspection, it's not really that hard to fathom, god is a creation of humans, he therefore exists, religion is the hijacking of this anachronism by groups of believers for their own ends, there's nothing difficult in it. There seems to be still some need, on the part of some scientists and technologists, to involve the religious believer, to pander to those who insist on a magical, mystical or superstitious approach. Einstein was fond of speaking about god, famously stating that, *god does not play dice*, to the extent that Niels Bohr told him to, *stop telling god what to do*, Stephen Hawking too, mentions god, and the mind of god, but these are only humorous, passing references, almost an expression of non-belief. More worrying, for some, was the acceptance by Martin Rees, arguably the most eminent scientist in Britain of the Tempelton Prize, for which Richard Dawkins called him, *a compliant quisling*. In his book, *The Singularity Is Near*, Ray Kurzweil, ponders that the evolution of technology, and the growth of thinking machines effectively leads humanity closer to the ideal that is god. While Kevin Kelly in, *What Technology Wants*, spends the last chapter, entitled, "Playing the Infinite Game", meandering about technology and god. Perhaps it would be more to our purpose of advancing human technology into the Purple age, to treat god and religion as we do a piece of antiquated technology, like the telegraph, or the first PC, or the Betamax video player, they still exist, and we remember them, sometimes with fondness, but we neither need or use them today. Religion, like some obsolete technological product, sits on a metaphorical shelf, it retains a loyal user base transitioning to the next generation feature rich version. Some users will never upgrade, some swear the bugs in the new version are too dangerous, and they will stick to what they know, and so on.

A note on future magic is also warranted here, in his book, *The God Delusion*, Richard Dawkins speculates about possible existence of god-like aliens, and how their technology might be as far beyond ours as to appear to be magic, in accordance with Clark's, so called, third law, which is dealt with elsewhere in this book, in the same way that a Dark Age peasant might view our technological enterprise. It needs to be said here that many of the people wandering about the malls and streets of our cities, have all the qualities of limitation one might expect to encounter in examples of Dark Age peasantry, yet clearly have the capacity to interact, even control, very advanced technology. It is submitted, the Dark Age peasant, were he or she to make an appearance, would exhibit shock and surprise at first, followed by a period of acclimatisation, followed by interaction, and possibly even mastery of any technology placed before them. The matter is one of capacity, there is in all humans the innate capacity to understand anything, once it is properly explained, and of course, mastery of anything will require payment of the requisite entry fee of commitment, assuming also the person has some ability. Notwithstanding these requirements there is nothing aliens could know about our universe that we humans could never know or understand or master eventually, if shown to us and properly explained, for the very simple reason that we are both made of the same stuff, and inhabit the same universe, and with that in mind, we would instantly reject the proposition that anything an alien might have would involve magic. However, the antics of this god fellow, we are told, are entirely magical, and to a degree that mere humans, not only don't understand, but also could never understand, and to be even outside of both space and time. This we must find objectionable with the contempt granted to us by 400 years of scientific endeavour.

Is atheism a religion? Is that even a proper a question? Religious people are fond of referring to atheists as having belief in this or that, belief in evolution for example, or a belief in nothing, or a belief in the opposite of god, which they see as anti-god, or antichrist, or evil, as it suits, and according to the holy scribbling. Atheism has evolved, and that is the correct term, into a movement of human beings wishing to embrace a future devoid of the baggage that religion of necessity brings to any situation. Think how easy the many and various Middle East conflicts might be solved if the holy books were to be closed, and left permanently upon the shelf. In a non-religious atmosphere, there ought to be as much difference between Israel, Palestine, Lebanon and Jordan as there is between Holland, Belgium and Luxemburg. Atheism has become a movement, and a pressure group because atheism solves problems, and as a bulwark against the repugnance to logic and knowledge that religion has shown itself to be. Some people go further, and appear to behave as though atheism might be a religion, by aping the behaviours of religious practitioners. Only one example is needed. An atheist *congregation* in England, if congregation is the correct, and not an offensive term, calling themselves the Sunday Assembly, hold regular Sunday, *services*, again, services may not be the correct term, in old unused churches in London. The model is the traditional Christian Sunday service but without the reference to god or religion, instead they indulge in science, and other talks, sing or play music, watch stand-up comedians, and generally hang out. Why, you may ask, organise such an event, and who, you might ask, would attend. The aim seems to be to provide non-believers with a sense of belonging to a community they feel is missing if they simply abandon organised religion. Sanderson Jones, a stand-up comic who began the event claims the reason is all summed up in the movement's motto, *live better, help often and wonder more*. Aping religious practice hardly seems a profitable use of the time reclaimed from Sunday, following a conversion to atheism, these people are strongly advised to join a golf club; a Sunday game of golf lasts four hours, provides a great deal of community spirit, and works up one hell of an appetite. And, if it is felt, there is some spiritual deficit, then there can be no more fitting spiritual pursuit than that of perfecting your short game. It can only be espoused here that as long as '*Theism*' persists in any form, there will be a role for *Atheism* as an '*ism*', and as long as religion protrudes and pollutes as it does, there may be a role for assertive, aggressive, even militant atheism. Because for all the love and peace pronounced by religions, they are still quick to threaten violence, and take up arms, and denounce those of their fellow humans who refuse to believe as they do, or who may wish to exercise their right to change their mind. Obsolete it may be, but it is not quite as obsolete as astrology.

Is Purple a religion? Could it be so described? Exactly the same may be said for Purple as for science. Purple is an assertion; nothing more, an assertion that if we continue as we are, we will eventually arrive at a point where all significant human problems will be solved, and every human will have a share of the infinite wealth we know we will create. While it is an opinion and may be a belief, it is also based on a sound set of self-evident facts, that we humans exist, and are capable of solving problems, and have created the world we now have, and that nothing we propose in future is contrary to the laws of physics, or beyond human ingenuity or understanding; so hardly a religion.

Religion and the state is a contentious old chestnut, hot potato, or hand grenade, depending on individual taste. International recognition and widespread state acceptance, and practice, or support for religion, organised or otherwise, has no place in the modern empire of technology. We are secular through and through, secular and humanist, and nothing more; that is how we got here. Freedom of religion is, of course, guaranteed, but this right only applies to one's private beliefs, and the freedom to hold and practice, express and discuss a chosen belief in a private capacity. Religion ought not be state sponsored, directly or indirectly, especially not through tax exemption, there needs to be a recognition that religion, like the practice of astrology, is a private and personal matter, and a business transaction, and has no place in the workings or machinations of the state.

Religion is a product and a service, to be sold to consumers of religious products and services, by religious corporations, and organisations, positioned in the religious space, in the same way as consumers are sold TV programming, or packaged food. There ought to be a disassociation of the state from the practice of religion, or any form of state sponsored superstition, for example, no swearing on sacred religious texts, in court or anywhere else, for example, the swearing in or politicians, or judges. There should be no invocation of superstition or, of non-existent entities living in the ether. Abandonment of religion as a feature, or as an adornment at state occasions, no more kneeling, or supplication of any kind, by politicians, for example. Or kissing of rings! *"For god sake, we've moved on!"*, as the Irish comedian Dara O'Briain puts it. And no more excuses for religious extremists. Religion is a proven nonsense, the claims made are contrary to the laws of physics and the laws of nature, and are therefore nonsense, -that cannot be said often enough- and politicians should call it as it is. So when extremist Christians kill doctors working in family planning clinics, or extremist Jewish settlers kill Palestinian farmers, or extremist Muslims kill cartoonists and policemen in Paris, or tourists on a beach in Tunisia, or college kids attending a rock concert on a Friday night, the political outcry should be against religion; the real cause of the atrocity. Religion provides the underlying reasoning, it is the religious nonsense, the contents of the holy scribbles, that provides the licence, and the authority that mandates the believer to kill the unbeliever. Religion solves no problem, and serves no purpose in our modern society, all the claims made by religion are contrary to the laws of nature, and are therefore fair game to be denounced as nonsense, and should not be condoned, or excused, or pandered to, or publically respected, and politicians in the Western World really need to acquire the requisite backbone to plainly say so. And if they refuse, they should be shamed into doing so by irrefutable argument. It was interesting to see in the aftermath of the 2015 Paris attacks, the officially sanctioned gathering in Notre Dame Cathedral comprising the French political establishment, the diplomatic corps and grieving family members and friends, along with a few dozen assorted bishops and priests, all adorned in ecclesiastical purple. It had all the hallmarks of a statement that the Christian god was better than the Islamic god. Notwithstanding the obvious grief, and unnecessary destruction of lives and families. It is hard to understand the purpose of such a gathering, is it to give thanks for the lives lived, or offer comfort to the bereaved, or to be thankful that only those few were taken, and not more? Of course as usual the official line was that the cause lay in *'warped ideology'*, and was the work of deranged, politically misguided terrorists. The *'R'* word was not used, there is no linkage, no one saw the elephant in the phone booth.

What we want as we move toward a Purple world is that religion effectively becomes abolished, laughed out of society for the nonsense it has clearly been shown to be. No more prayers to open parliaments, or blessing of buildings to declare them open, no more involvement of religion in state sponsored funerals or swearing in of presidents or politicians. It ought not be the case that swearing on some sacred holy tome should convey any more meaning than the making of a solemn declaration, in fact, how could it possibly mean more, now that we know what lies behind the text is laughable nonsense? In which case, it should not be offered at all. There need not be any more oaths of office involving fealty to god. There should be no further recognition by any state, that there is something higher or better than humans, we now know there is not, so why continue with the practice and the pretence. In the final analysis the theocrats have had their day in the sun, they ran this world for over 1600 years, and they blew it. James Madison knew that back in 1785 when he addressed the Virginia Legislature. Religious believers now need to take their place on the side lines beside the druids, and the Oracle of Delphi; they have no place in public life alongside rational thinkers in the new republic, they simply have nothing to offer, we know better. It is probably high time to stop pandering to what we know to be nonsense of the worst kind, as though arrogant and dogmatic believers are ever likely to suddenly present the world with some fragment of proof. Religion has no place in public life, it solves no problem, serves no purpose, it is fractious and divisive, and we are better than that.

And it is for the above reasons that politicians and public officials must be challenged on their beliefs, and on where they stand in the public space. Just as there is, belief in god, Theism, there must also be non-belief, Atheism, however, it is not as straightforward a dichotomy as that of golfers and non-golfers, as noted by Neil deGrasse Tyson; non-golfers have no need to congregate, with the possible exception of golfing widows or widowers, who may, for all we know, wish to gather for Bridge, or light refreshments, or even sexual recreation, while their respective partners are engaged in the black arts of iron play and putting. Atheists, otherwise simple non-believers, akin to non-golfers, have no choice but to become members of a movement called Atheism. Unlike non golfers, they even evolve militant tendencies, and become militant Atheists, and congeal into groups, in particular into lobby groups, and groups seeking to have god and religion forever removed from the public space and relegated to the side lines or to the funny pages. The feel the need to do it for the simple fact that religion in all its forms is so irrational, and so distorting, and so corrupting, and so polluting, that there is no alternative other than embarking on such radical action. Because the religious believer is not a happy or contented individual, and as has been noted by many authors over the years, will not be, until every knee genuflects, and every head bows, or every convert prostrates before the lord almighty. Despite such concepts as love, and joy, and grace, and effulgent coalescence with the divine, they cannot be happy or content until you believe as they do, and are joined up, and signed up; they are at war with believers and non-believers alike.

Human society has reached the point after four hundred years, and in particular the last one hundred years of study, and research, and scientific endeavour, and observation. We have formulated the most spectacularly accurate picture of Nature and the Cosmos, and the immutable natural laws that govern everything that has ever been assembled. Yet the religious adherents simply choose to ignore the enterprise, and give pride of place to their own irrational, discredited, and ultimately disconnected opinions. Further, we now know so very much about the workings of the universe, and the natural world that we can say with certainty, that all of the many and various beliefs held by the four thousand flavours of religion are simply false. They are simply wrong. Science can make that statement with certainty, because, while science is famous for its principle of uncertainty, and the fact that every measurement has an associated error, and mathematics has its incompleteness theorem, and scientists use words like theory, and deal in probabilities, and state our findings in terms of what may be more likely or what may be less likely. Science can however, state with absolute certainty when something is just plain wrong. Not simply unlikely or doubtful, or of dubious tenure, but wrong, in other words, so manifestly contrary to the immutable laws established at the formation of the universe as to be clearly impossible. Or so completely contrary to all of the observed evidence obtained from direct observation and measurement as to be simply wrong. And the list is a long one, almost infinite, and would obviously include, the old perennials such as, perpetual motion machines, time machines, antigravity devices, along with the denial of evolution, or denial of the existence of tectonic plates, and continental drift, or any claim that the earth was recently created. Or that miracles can occur, that water, for example, composed of molecules containing a single oxygen atom, and a pair of hydrogen atoms, could ever physically transmute into a solution containing a variety of complex organic molecules, including ethanol. Or that bread can become flesh of some kind that residual plant matter can become meat. As stated, the list is a long one.

It is therefore not unreasonable to pose a question to a person running for public office or to an incumbent office holder as to the nature of the beliefs they profess publicly to hold. In fact, it might almost be said to be compulsory to determine the extent of wilful ignorance on display. It must also be incumbent upon any public official or person running for public office to explain, if asked, where they stand on the matter of their publicly stated belief. If they publicly state adherence to a belief system clearly contrary to the immutable laws of the universe, then rational and reasonable voters, have every right, we might even go as far as a duty to ask, why that should be so. It is simply insufficient to state that they have a right to believe whatever they wish, public statements of belief in the irrational, and in the unreasonable, are the same as dismissing the established and agreed certainties of science as discovered over the preceding four centuries. It is the same as saying that personal opinions, however odd, weird, strange or nonsensical they might be trump irrefutable facts. The rational citizen must be outraged and complain that such an attitude is simply not good enough. We live in the twenty-first century; this is not 1535, when men like Sir Thomas More stood against tyrannical Kings on a matter of conscience. Today's political class do not have the luxury of More's ignorance. In the cold reality of our technological society, the maxim of the law that ignorance is no defence is doubly true, ignorance on any important matter such as the age of the earth, or whether it is flat or round, or the truth of evolution, is easily cured with the movement of a finger over a screen. There is no possibility of that excuse being even remotely valid or acceptable, like the excuse of the child involving a domestic pet for the non-production of homework. Nor is it a practical proposition for a politician today to assert that perhaps there is a possibility that his irrational belief might be, in some respect, valid, for there can be no doubt, science is every bit that strong. Therefore the matter of belief in nonsense now becomes one of deliberate wilful ignorance, and reduces to either of two possibilities, neither of which is pleasant for the politician involved. On the one hand, the politician is incompetent, and obstinate for no good reason, holding onto an irrational and unsupportable opinion against an insurmountable weight of factual evidence to the contrary, is simply the action of an incompetent, or worse. On the other hand, the politician is engaged in a deliberate public demonstration of the irrational for the sole purpose of alignment with those of the voting class possessed of the same, arrogant and irrational opinion, in which case, the politician is blatantly and openly dishonest. In either case it would appear we have reached a point in time when the holding, or public profession of, irrational religious belief is no longer reasonably acceptable in the public space. If a candidate ran for public office, and when asked, claimed and asserted a belief that the earth was flat, and behaved in a manner that could only be categorised as being wilfully ignorant, for example, refusing to accept photographic evidence for the spherical nature of the planet, and insisting that the earth was flat because, some holy text or ancient authority, to which he subscribes, declares it to be flat. Such a candidate would, quite rightly, be asked to offer proof of any such beliefs and assertions, beyond mere opinion, and evidence, beyond mere sacred text, and in the absence of proof the candidate would be ridiculed in public, and rightly so, both candidate and opinions would be questioned, dissected and finally laughed out of the public arena. Needless to say the campaign would be of marginal success, at best, and certainly it would be hard to envisage a term of office; with the possible exception of an otherwise wholly disillusioned electorate. Similarly, were you to ask someone what day it was, and they replied Saturday, regardless of which day it actually was, and regardless of what evidence you adduced, and continued in the assertion regardless of any and all information to the contrary, you would be free to form your opinion as to their suitability to hold public office or represent you in any matter whatsoever. But such beliefs are equally as absurd as any on offer by the mainstream religions, and every bit as arbitrary, easy to disprove, or simply, diametrically opposed to common knowledge. Wilful ignorance has never been tolerated in law or in professional conduct, and many books have been written on the subject, and many cases won and lost on the point. If your lawyer, accountant doctor, dentist, architect or other professional acting on your behalf for a fee, deliberately closed his eyes, or she deliberately and wilfully looked the other way or failed, refused, or neglected to see some fact, or deliberately chose to ignore some body of evidence widely held to be common knowledge, you would probably sue. If their action caused you material damage of some kind, you would have a cause of action, and be entitled to recover damages. Similarly, were you to sit on a jury and hear a defendant claim that while he knew it was wrong to commit the offence in question, he deliberately ignored that knowledge, and went ahead anyway, you might be inclined, in fact you might feel duty bound, to return a guilty verdict? Yet the absurdities acceptable to people of faith, concerning their politicians of faith seem to beggar belief. To give but one example, that Hurricane Katerina was the wrath of almighty god for the sins of homosexuality and gay marriage in America. One wonders what might happen to the career of a politician were he to blame the same event on the wrath of some god from the ancient world, or place the blame before evil imps or witches. Sam Harris notes that religion and faith gets a free pass, if such nonsense were translated into the sphere of UFOs or aliens, such people would not be elected, '*they do not find themselves on the boards of universities and public companies*'.

But religious belief among the political leadership goes largely unchallenged, incompetent and dishonest as they might be shown to be, such beliefs enjoy the protection of a complexity of rights and laws stretching back into antiquity. Freedom of religion and of conscience, and of thought, and freedom of worship are, of course, fundamental tenets of all human rights proclamations, and it is not suggested for a single moment that any human person would ever be forced to relinquish any such right.

However, all rights, and that must include freedom of religion, and freedom of conscience, as mentioned before, are proportional, and are necessarily qualified. The right to believe in the absurd, and the irrational, and the manifestly untrue, is a given, but public pronouncements of the absurd, or public stands taken as a direct result of absurd and irrational beliefs, are by the same rights and freedoms, open to the sternest possible dissection and questioning. If you are approached in private by someone expressing that old Christian belief that the, *End is Neigh,* you might congratulate him on his insight and move on, or recommend that he see someone more professionally qualified than yourself. If, however, you were to encounter the same person in the public square, replete with sandwich board making the same declaration, you must surely be entitled to challenge that individual with questions of the type. Why? How do you know? Where does this information come from? How can you be so sure? What evidence do you have? The right to ask why is every bit as fundamental as the right to believe as one chooses, and in particular, to know why a political leader, or public official, will choose to believe in the irrational, and accept the absurd as true when the overwhelming evidence, beyond all reasonable doubt, is to the contrary and sufficiently so, as to leave the candidate open to the charge of being wilfully ignorant and intellectually fraudulent. And this must be especially so when such a public figure relies upon the irrational to justify or defend a position in rational society. If ignorance is no defence, and it is not, then wilful ignorance must be positively detrimental. We can no longer claim to have the luxury of ignorance, we have come too far, we know too much, there is nowhere to hide, and there can be no excuse.

Irrational and absurd beliefs will always be with us, in some form, as long as there exists a single human believer, a true believer. It can probably be compared to smoking, a filthy rancid habit that we all know we ought never to take up, and vast amounts of wealth have been expended in trying to make people stop, and to prevent new smokers emerging. Yet smokers persist, and new converts join, and swell their ranks every year, even some, otherwise quite intelligent, people smoke, doctors, lawyers, scientists, and politicians among them, and we may well wonder why they would engage in something so entirely irrational, absurd and downright stupid, as well as injurious to one's health, when all of the evidence points to the contrary; addiction alone, while providing an answer, is insufficient, the reasons people smoke are deeper and more complex than addiction alone. And it is the same pointless question one may ask of adherents to religion. Why do they believe? And the answer is the same, and every bit as complex as why smokers smoke, faith alone, or comfort alone, is insufficient. Humans believe, because they are human, and they can believe in whatever they wish. And in the same way as no one gives a damn if you choose to smoke in private, no one cares, or ought to care, if you pray or otherwise follow or practice some absurd and irrational belief system in your private life. Bringing it out into the public domain is altogether a different matter. Just try smoking in public, try lighting up in a restaurant or a cinema or other public forum. Just as smokers have restrictions so too must religion. There was a time when you would visit your doctor, and he might offer you a cigarette, now we know better. The questions to be asked for those seeking public office, and who declare a public belief in the irrational or the absurd must include questions as to their suitability for office.

But if any little bit of advice were to be imparted to young Christian, or Muslim, or Jewish, or Hindu, or Buddhist people looking for, or searching for something spiritual in their lives. It would be this. Stop looking to the clouds, stop looking in books, and texts, and scrolls, written by peasants with quills, they couldn't possibly know more than we do. Start looking at starting a rock band, or a dance group, or take up golf, or football, or needlecraft, stretch and bend your intellect, and your abilities as far as they will go. Or find someone to really love, and plan a family; those things are as spiritual as you will ever need. Develop your brain to the fullest extent possible for you. There is no evidence for anything in the clouds, there is no virgin infested heaven, no loving Jesus to guide your every action, no answers to payers stuck in jars, or spinning on wheels, or placed in a hole in a wall. It is what it is, we are where we are, you're on your own, and it's fun.

The redundant god, is all we are left with, we may surmise that once upon a time, there was a god and probably did, once upon a time, serve a purpose, as Professor Daniel Dennett suggests. *God has not left the world, but the world has evolved past god.* There is no longer a requirement for god, the problem to which god is a solution is undefined, and indefinable, and so, like god, cannot possibly exist. God was entirely the creation of the minds of backward, unsophisticated humans, living a largely pastoral, Bronze Age existence, and ultimately at the mercy of a hostile biosphere, and of blind chance, and the whims of indifferent weather patterns. The purpose of this god was that of a place to go, a refuge, a reason for things being the way they were perceived, simple people require simple explanations; or rather simple people evolve simple explanations. The first people to think about the problem came up with the first approximation; god is nothing more than a first tentative guess at how the universe might work. God is therefore very much like a childhood toy; it has, for some, the very real comforting effect of a favoured Teddy bear or cuddly bunny, regardless of how worrying it may appear that grown up people still converse with imaginary friends. How much more important for a mature civilisation to be able to put the toys away. God equals the building of monuments to something else, to something beyond, to something intangible, and out of reach, to something ephemeral and unreal. God is a magical representative of ignorant men staring into the dark abyss. However, the abyss is not as dark as it once was, we have discovered light, and seem to have moved on. We are on our way to a Purple world. There is no more need for magic. There is no more a place for gods. We're better than magic. We're better than religion. We're better than god.

Chapter 8   We Love Problems

Unforeseen consequences stand in the way of all those who think they can see clearly the direction in which a new technology will take us.

Neil Postman, Technopoly

Humans are the supreme problem solving species; the only species nature evolved to define a problem then set about solving it, record the method used, then store and transmit that knowledge to every other member of the species. In short, humans generate and manipulate information, and we do it effortlessly. In fact we have no choice, we are compelled to do it, and do it as easily as other species hunt, graze, swim or fly. We call this ability to solve problems and the history of our problem solving, our technology.

A very important, though not obvious principle to be noted here is that once a problem is solved, it stays solved, and its solution is recorded and disseminated. It should also be noted that if there are ultimately, even in an infinite universe, only a finite number of problems, and we continue to solve them in sufficient numbers then eventually we reach a point where, to a first approximation, all problems will have been solved. We even go as far as to create problems just so we might solve them, we invent games such as chess and golf so we can practice.

Given a problem which is not contrary to the laws of physics, and a willingness that the problem in question should be solved, then it can be solved. We even have rules for solving problems, for example, there must be a solution, that is, the problem must be defined, there must be a willingness to solve the problem, including overcoming any opposition or resistance to solving it. And of course, the old mathematical adage, if you don't know what you're looking for you'll never find it, comes into play, along with the proviso that you should know the solution when you see it.

It's in the nature of individual humans to become obsessed with solving problems. The entire human project can be summed up in terms of problems and solutions. There have always been problems to which the human animal has found solutions, and these solutions include workarounds, Band-Aids and ignoring the problem altogether.

Now in 2016 the problems we face are bigger, they are not of themselves more difficult, or more intractable, just bigger on a different scale, they are, however, nonetheless soluble; we have always been able to meet the challenge. Our success is a testament to that fact. As the character Mr Macy says in the 1947 movie classic, *Miracle on 34th Street,* referring to the store's new Santa, *"you can't argue with success"*. At least, it's a difficult thing to do, success proves itself, and one of those self-evident and undeniable attributes of human existence is the undoubted success we've enjoyed as a species so far.

While we are the ultimate problem solving Species, we are not the only problem solving species we take our place alongside chimps, parrots and spiders, and many other species that dwell in three dimensions. Humans do it better than any other species, we do it professionally, it's what we do, it's why we're born, any problem we can imagine, from food shortages, to the climate crisis, and right the way down to why marmalade sticks to everything; we solve them all.

Our science and technology, and our open-minded approach gives us a visibility of the future never before available, we see the future of humans on earth and in space. We see further because we have built the necessary tools to allow us to see further. We have reached the limits in some areas, for example, we know there are only 92 naturally occurring elements, so there is no point looking for others, although it remains a vibrant area of research, and very heavy stable elements are a theoretical possibility. We see further than any of our predecessors, and with greater clarity about what we see and why, we have nothing to fear, we have a duty to be optimistic, and we have earned the right to dismiss the negative.

The problem with problems, if it might be put that way, is that in the best tradition of the human species, those problems perceived as being both huge and pressing are often neither huge nor pressing, while those problems we refuse to acknowledge as being problems are potentially highly dangerous to our project. Further, there are also some alleged problems we continue to attempt to solve even though everyone knows that the problem does not require to be solved.

Peak oil is one of those problems that has been predicted for over a century, and we are told it will have a detrimental effect on the entire planet. We are told oil is a finite resource and must be conserved; our thirst for oil drives wars and greed, pollutes the world and kills people. If it is finite and running out, then one day, regardless of what we do it will be gone. What therefore is the point in conserving it? What problem does cutting back and holding on to diminishing reserves solve? When it's gone it's gone. Its imminent going should kick the markets into a higher gear, and drive technology to find a solution. Its imminent demise will hasten the development of alternatives. There is no point in being concerned. We either sink with it, or find the alternative. Necessity being the mother of invention will assure a solution, otherwise all bets are off, and it's not going to matter. It is clearly too big of a problem not to solve. The mind of the world is upon it. A solution, or a very wide range of solutions, must and will come from our endeavours. And peak oil is not the only problem seen as being huge and pressing while being in fact neither.

Consider the problem almost every politician on the planet thinks is the biggest problem humanity has ever faced or is likely to face, global warming, or climate change if you prefer. Carbon dioxide is the ash of our fires, and has been since we learned to rub two sticks together or strike a flint. Our entire world depends for its very existence on combustion, we burn stuff to make the world go, and to keep it going. Everything we have that burns is composed of carbon and hydrogen, and when we burn it we create carbon dioxide and water. The water is fine, but the carbon dioxide has the potential to cause real problems. And there is clear and compelling evidence that the excess of both carbon dioxide, methane and other so-called greenhouse gases has caused the mean global temperature to rise. But just think of how this entire problem would vanish almost overnight were we to evolve some method of capturing carbon dioxide and methane faster than we could put it into the atmosphere. And further, consider the situation if this capturing were in fact a profitable and easy undertaking. For example, consider a machine that could turn these products into a useable fuel for home heating or into gasoline. We can't do it today, but very soon you will be able to buy a machine, for a very few dollars, that will do just that. It's an example of a problem that is well defined, and one we wish to solve, regardless of the opposition or any resistance. Companies such as Liquid Light, a Princeton University campus enterprise have produced catalysts that use carbon dioxide as the main input with some near-commercial and interesting results. And a thousand more start-ups and spin-outs are working on the problem round the clock. Every oil company is pumping money and chemistry graduates into micro-enterprise, spin-off companies that have the single goal of converting the output of your tail pipe into the contents of your gas tank. There is also a $20m X-Prize. All of which means the problem doesn't stand a chance.

Problems don't become real problems, and don't get solved until enough people want to solve them. The so-called drugs problem must be the classic example of a non-problem posing as a problem, and an intractable human problem if there ever was such a thing, perhaps even a problem everyone has agreed not to solve. Drugs are both universal and immemorial, they have been with us since our prehistory, and every culture on the planet indulges. In our modern western democratic model we have competing sets of rights and responsibilities. On the one hand it is very clear and self-evident that individuals have the right to alter their state of consciousness, from the default state of sobriety to some other state, euphoria, or anger for example, or enhanced contentment, even happiness, and of course drunkenness. Although in many jurisdictions inebriation is a criminal matter. And this alteration is accomplished by the ingestion of a very wide range of substances, such as caffeine, chocolate, alcohol or any one of a million prescribed medications. There are other substances contained on a list which very many people wish to use as their drug of choice. But these are banned by law, and may not be used, those using or supplying are breaking the law, and very serious penalties await those apprehended by the enforcement agencies upon conviction, enormous amounts of money also continue to be spent each and every year maintaining this state of affairs. However, despite the penalties, even including death in some instances, the range of banned substances continues to be very widely and freely available, to the extent that it is generally agreed by all who examine the issue, that the money spent, and the legal sanctions imposed make no difference whatsoever to solving the perceived problem. People still use what they prefer to use, and have no problem getting hold of any amount they require, for example, the drug of choice at the Woodstock rock festival in 1969 was cannabis, had it been alcohol, there might have been a different outcome given the underfoot conditions. So governments the world over spend inordinate amounts of taxpayer's money restricting the valid choices of citizens expressing what is a universally agreed human right. Of course, the reason they do it is all wrapped up in what lawyers call public policy considerations, and the old chestnut of child protection. There is so much nonsense surrounding this entire, so called problem, that it is the subject of vast numbers of books and journalistic output, each and every year. The bottom line is that the so called drugs problem is an example of a problem that cannot be solved because it is not a real problem, rather it is a deliberate public policy of control posing as a problem. It is also a public manifestation of the human ego, the last thing any politician wants to do is admit that they have got it wrong, or that the policy has been wrong for sixty years, and all the money spent on it has in effect been wasted. The obvious solution to the problem is to legalise all natural substances, such as Coca leaves, opium, cannabis, magic mushrooms, and all other herbs, spices and assorted plants that people have used, or might wish to use to exercise the basic human right of altering their state of consciousness. Let them be sold and taxed. What you might do with it when you get it home, is of course, entirely your own business, in the same way that no one cares what you might do with a bag of coffee beans or a crate of Irish whiskey. The taxes collected and the savings in criminal justice overhead ought to be more than enough to educate, heal and rehabilitate any citizen straying too far into the darker places. But all reasonable people know that such a solution will never be willingly introduced, there are too many vested interests and political reputations at stake.

Some problems are thoughtful enough to solve themselves. Or at least the perception that there ever was a problem simply evaporates. The great horse excrement debate of the late nineteenth century is an example of one such problem that was a problem, then faded and disappeared from the world elbowed out of the way by human technology. The debate concerned the perception that in a horse drawn world everything moved at the speed of the horse, and that was not fast enough. The solution was more horses, and more horses again. Concerns began to be raised that the increase in horse numbers was having an effect on the environment. Horses needed to be housed, they needed to be fed, they required land for stabling, and land for the growing of oats and hay, and they also generated copious amounts of manure. As usual calculations and projections were carried out, and the numbers showed very clearly that a problem of enormous proportions was on the horizon. The cities of the developed world would be under many feet of horse manure by the year 1910 unless something drastic was done. But of course nothing was done, and then again something was done. The internal combustion engine was developed, and went on to replace the horse. Nothing was done in the sense that the inventors of the internal combustion engine did not set out to solve the problem of cities disappearing under mountains of steaming manure. They set out to make a more efficient engine than the steam engine, and the gas-powered internal combustion engine. But the net result was that the horse problem went away with the horse to the extent that no one is concerned about it today.

Once a problem is solved it stays solved, no one need solve it again, the solution becomes part of the list of things we do, and can be copied by anyone who wishes to use it. The law of technological convergence, if there is one, may be stated as follows, that when a problem is identified and solved, further work in the field simply results in variations on the theme. Most of the time. A simple example of this might be the invention of the table and chairs. The origin of this technological marvel probably came about from the perception of the problem of where to have a meal in peace. Sitting on the ground is clearly an option. Standing is also clearly and option. However, the act of sitting at a bench, or table to eat clearly arose in certain parts of the world, and in the absence of records on the matter we may suppose the origin lies in a scenario where some individual rolled a log against a rock and commenced the practise. Once the solution was discovered the application of the technology to other aspects of life followed, two people could sit, and hold a meeting, or play a game, and so on. In due course many different variations on the same theme began to appear until we eventually arrived at the Chippendale dining suite, or a deckchair and folding table. In effect the solution of the problem was found and the field of research effectively closed. Any further work in the field is design, the innovation having been complete.

We don't know if all branches of knowledge converge to a solution or not. Certain fields will remain open, mathematics, or music perhaps, but others may close, medicine for example being concerned with a finite limited system might become closed, in that there might come a point when everything we need to know about the workings of the human body are known, and nothing further needs to be done. It may be added here that sometimes apparently new technologies are no more than old technologies repackaged and reworked. For example, the MP3 player was around long before they became portable, it was called a PC, with music files stored on a hard drive and earphones plugged into the sound board. Making it portable created a new product, but it didn't solve a new problem. The rest is marketing. Of course some design is clearly sufficiently transcendent to be considered art in and of itself, the coke bottle for example, and design can solve problems of function, but often the same thing is just repackaged and resold to a new generation, flared trousers for example, or Cherry Coke. Original shampoo becomes new classic shampoo. Now in new convenient packaging, or now with a new great taste, new meaning repackaged with a new convenient weight. We've taken some out, increased the price, but hey! It's all in a great new package, all for your convenience, and it's all about you. It might look good, but in effect this is all sales fluff, and has very little to do with solving problems or making progress.

Energy is a problem we must solve, and will solve, oil will be displaced and eventually the entire energy supply of the planet will probably come from solar panels on the lunar surface, until we get fusion reactors that work and are cheap, and safe. Many reasonable physicists believe nuclear fusion is the key to all of our energy needs. They believe fusion is doable, fusion is also clean, and will provide potentially limitless energy obtained from sea water, and clearly once it is made to work it can replace all other power sources. They also decry the fact that more money is spent on cat grooming in the USA than is spent worldwide on the nuclear fusion project, or some such statistic. Like the internal combustion engine of the late 1800s fusion may come to the rescue, and save us from the energy problem, the peak oil problem, and the climate change problem, all at the same time. In the same way the gasoline powered engine saved us all from a plague of horse manure. And once the problem is solved it will stay, well and truly solved.

Consider also the problem of dark matter and dark energy. This is something that has cosmologists and particle physicists scratching their heads, and seems no nearer to being solved than it did last year. These two types of stuff make up about 97% of the universe and we have no idea what they are. Opinions abound. There are a clear set of observations and experimental results. What causes the observed effects seen in the experimental data remains a mystery. A problem to be solved. In this case more technology is not going to help much, we appear to have all of the technology we need. What we are missing is a theory that explains what we see, new physics and new theoretical methods, and that will only come after many physicists have banged their heads on the problem, and someone comes up with something that works; a lucky guess perhaps. But it's not a pressing problem, and makes no difference to the human project whether it is solved today or in one hundred years. Like quantum gravity, the vast majority of humans will not lose sleep over it, but once solved, it will stay solved, and we will be in a new era.

The old saying that necessity the mother of invention is probably universal, and as old as the hills, it is also true. When backed into a corner and faced with calamity human ingenuity tends to save the day. We have never come across a problem that could not be solved eventually. Given the willingness to solve it, and so on, as discussed. Given what we know now, we can look back at the past and see that none of the predictions of calamity have ever come to pass. In the 1930s there were concerns that the population of the planet would drop to very low levels. Or predictions by climatologists that an Ice Age was on the way. Human necessity has always overcome the Malthus conjecture, no matter what the resource, no matter how doomed our human project seems to be.

Sure the carbon problem is a big one, but it has a solution, and is in good hands, it will be solved, and the solution will be refined and optimised. The solution will then be rolled out, and we will look back with fondness to a time when huge conferences and gatherings of international leaders were held to discuss the problem of climate change that went the same way as the horse manure crisis of 1890. Yet climate change and global warming seem to be the biggest problems on the political agenda, at least with political leaders. Most ordinary people feel more mundane things to be huge and pressing problems. Credit card fraud, identity theft, fear of losing a job, fear of crime, fear of economic collapse, terrorist attacks, and the like.

It could be argued that our most pressing problem, the one we ought to solve as soon as we can is probably the problem of ignorance. Because ignorance causes so many more problems for people and for nations. Apart from the obvious manifestation of wilful ignorance on the part of politicians in refusing to recognise that there is no problem faced by humans to which religion is the solution, all other problems faced by humans may be in some way laid at the door of ignorance. Ignorance of science gives us a lack of technology to solve some of the big problems. Ignorance of each other leads to wars and misunderstandings. Ignorance of what our politicians are up to leads to corruption wherever it is allowed to fester. Ignorance is at the back of disease, and famine, and any other human misery you might care to name. If we solve the problem of ignorance we will be on our way to a purple world.

The low hanging fruit theory suggests all the easy problems have been solved the more difficult problems are beyond us. The Purple approach disagrees with the negative outlook and maintains that all problems we wish to solve will be solved eventually, and we have an infinity of time. A core proposition of Purple is that no problem or crisis facing humanity can be solved negativity, or by pessimism, or by giving up, but all potential problems benefit from the application of technology, and a rational scientific approach by intelligent, enlightened humans.

Chapter 9   The Fallacy Of Prediction

> *We have no idea where we are going, and sweeping, confident articles on the future seem to me, intellectually, the most disreputable of all forms of public utterance.*
>
> *Kenneth Clark, Civilisation.*

When Marty McFly, the hero of the *Back to the Future* movies went forward thirty years to 21 Oct 2015, his future contained, hover-boards, flying cars, 3D adverts, and a whole array of feature packed gizmos, most, if not all, of which have yet to make an appearance. The stuff they did get right, or appeared to get right was stuff already visible in 1985, or obvious, or inevitable, based on the technology of the time, two-way communication involving video, for example. More important though than the technological correctness was the generally correct philosophical extrapolation that the future was going to be bright, and filled with largely contented, agreeable people, and their convenient, easy to use technology. In a sense we are now officially the far future, as far as people in 1985 are concerned, and in every respect it is bigger, brighter and stranger than they could ever have imagined, and the world is generally populated with contented agreeable people. We have yet to reach the Soylent Green date of 2022, but chances are that too will be bigger, brighter and better again. These days, as in 1973 Hollywood prefers its futures to be dark dystopian affairs, understandable when you consider the drama to be extracted when the characters spend the entire movie fighting for their lives. The Soylent Green date of 2022 is half way between now; 2016, and the Singularity date according to Kurzweil et al of 2030, and no doubt those dates too will be closely watched for the accuracy of past pronouncements.

There can be no shorter route to idiocy than attempting to predict the future. The game of Golf has the capacity to make fools of grown men, especially with putter in hand, however, predicting the future is the more potent fool rendering mechanism because its effects can be as latent as they are cumulative. In other words, predicting the future is a spectacularly hard thing to do, while at the same time appearing to be a very simple thing to do, after all, you don't require any special training. The history of prediction is littered with failure, and is densely populated with red faced individuals who get it wrong but still get busy with *'Oh that's not what I meant at all'*, articles. It is also the reason why politicians never actually say anything, instead speaking in meaningless stock phrases, and canned platitudes, such as when asked what they intend to do, they will always reply that they intend, to leave no stone unturned, or do whatever is necessary, you get the point. You can never be accused of having said anything if you never actually say anything. One suit looks very much like another suit, and if the phrases are sufficiently bland they cover all eventualities.

There is a very long list of those who got it wrong, a long list of famous predictors from the Greeks, through the Romans, right the way down to our time. In fact, it is probably true to say that almost anyone who has ever tried to predict the future is on that list. There are some spectacular, *'got it wrong'* entries worthy of mention. The great horseshit problem, of the last few years of the nineteenth century is a case in point. Serious minds were set to considering the obvious problem of manure, but as we know 1910 came and went, and there was no great manure issue. In fact the problem peaked and began to decline, the internal combustion engine saw to it that a very serious and intractable problem simply disappeared to nothing, along with all the effort put into trying to solve it. A massive leap forward in technology solved the problem and made it go away, the horse population was replaced by trucks and cars.

And while the subject of horse manure occupied the agendas of 1899 the Commissioner of the US Office of Patents claimed with authority that *"everything that can be invented has been invented"*, clearly a prediction very quickly shown to be wide of the mark. Equally worthy of mention is the comment by Jack Warner, the founder of Warner Bros Studios, posed as a now famous rhetorical question, *"who the hell wants to hear an actor speak?"*. Who indeed. We also must note that many flawed predictions are made by people of eminence, who perhaps ought to have known better, the Chairman of IBM in 1943, for example, declaring that there was perhaps only sufficient market for *"maybe five computers"*. Computers of the day being large multimillion dollar installations requiring entire buildings, and a staff of a few dozen. Finally, cconsider the book *Dow 36,000: The New Strategy for Profiting from the Coming Rise in the Stock Market*, by James K. Glassman and Kevin Hassett, published in 1999, just before the famous dot com bubble went pop. The thing speaks for itself, for even today the Dow is only at 16,000 or there about, having attained the dizzy heights of 18,000 in June of

2015. We can also add to the list the long line of doom seers predicting the end of the world to be neigh for some or other religious reason; it's still here, far as we know. The end of civilisation has been predicted more than once, the old Mark Twain aphorism may be applied, reports of our demise have been greatly exaggerated.

There is also a long list of those who got it right, mostly through serendipity, or by stating the obvious that was, at the moment of prediction, technically impossible, such as Jules Verne predicting that people would eventually get to the Moon. Or Arthur C. Clarke and his prediction of communications satellites, long before rockets existed to place them in orbit. The notion that inputs in a complex system may be traced to outputs has long been known to be a non-starter, therefore no specific dates or time frames, can be predicted given the role of chaos and serendipity in technological advance. The role of serendipity cannot be underestimated, the right man in the right place at the right time, and the right combination of random chemicals left for the correct interval on a sunny window ledge while on vacation. We are dealing with the unknown and unknowable, the unmeasured and un-quantifiable. The system changes but remains the system, the system continues to exist in a changed form in the new state. Perturbing the system causes a change from one state to another. Noise keeps the system bubbling like a pot on the stove. And every state of the system even the simplest is a complex chaotic system, and prediction is therefore impossible.

Nor can we forego the opportunity to examine the long list of things that no one saw coming. The fall of the Berlin Wall, was not predicted, and no one quite foresaw the rise of China, the one party totalitarian state crammed to the brim with capitalists, most commentary expected it to eventually fail, and go the way of the Soviet Union. No one really predicted the rise of the internet, it had been around and in use by academics, but only became something usable once the World Wide Web was invented at CERN by Tim Berners Lee in 1991. The millennium bug, was predicted to be a disaster but nothing happened. The dotcom boom, the subsequent crash, and the old economy versus new economy were largely unpredicted, and most predictive commentary was wildly wrong. Those who did know, kept quiet and made money. No one really foresaw the importance of search or the rise of Google, or the demise of newspapers, and who knows, perhaps even the rise of newspapers again, and the decline of Google. No one predicted the emergence of social media or its importance in people's lives, and the many and various disruptions it brought about. Lastly, no physicist predicted high temperature superconductivity, no one thought it was even remotely possible until it was demonstrated.

People also have a habit of backing the wrong technological horse, because the ultimate choice of which technology is adopted, sometimes depends on luck, or circumstance. There is a long history of failure on this point too, all the way back to the introduction of draft animals; whether manpower or horsepower was to be preferred. In the nineteenth century there were people and companies who thought steam was the future, and backed it with hard cash while ignoring the internal combustion engine, and as a result paid a heavy price. One hundred years later, VHS became the adopted domestic video standard, despite the fact that Betamax was considered the better technology, DOS became the preferred operating system for the early PC, while arguably, better systems were available, and there are many other examples.

And given our spectacular inability to forecast the future the question needs to be asked why should environmentalists and Green Party activists be singled out as the one section of human society, and the scientific community capable of predicting the future with any degree of accuracy? When everyone else throughout the entire course of human history has been so wrong. If we know anything, or have learned anything from our failed history of prediction it is probably reasonable to predict that the carbon dioxide problem taxing the minds of the scientific and political elite of today will probably be solved by a leap of technology every bit as radical as that which spared the cities of 1910 from the calamity of two feet of horse manure.

If anything can be drawn from our previous prediction record, it is that the future continues to be better than the past on every metric, like the game of golf, the future is a game of surprises. The one thing we can predict with any degree of accuracy you like, is that we cannot predict but we will be surprised. Purple does not attempt to put dates on specific events or predict specific outcomes, the approach is to discuss the initial conditions of the system, and to use these to show that certain inevitabilities must, or ought to follow as they did at any period in the past. Our world and the many interleaved systems comprising it is far too complex for that kind of exactitude. Things have a tendency to improve, people generally want their lives to be comfortable and agreeable, they like to possess new things, and the world of technology generally obliges.

Prediction of technology is sometimes easy and straightforward, because the constraints are such that the facts lead to only one inevitable conclusion. It is the low hanging fruit theory, and is the difference between pure speculation, and prediction of the glaringly obvious, and the almost inevitable. For example, in the days of sailing ships it was not uncommon for a ship to be stuck in a port sometimes for weeks because the prevailing wind was so strongly against that they simply could not sail. The inevitability of a better way drove innovation to install steam engines in iron ships driven by propellers. Similarly the facts of climate change, and the inevitable exhaustion of sources of fossil fuels coupled with the fact that the sun is an infinite source of power compel us to conclude that if our society is to continue to grow and expand, as it has in the past, we must and will close the information deficit preventing us from utilising this infinity of energy. The complete exploitation of the sun must therefore be considered an obvious inevitability. But when, is the question, and the only answer that can be given is that it the transition is beginning to happen now, and will be complete long before fossil fuels are exhausted. Necessity, after all is the mother of invention. When faced with problems we have to solve, they generally get solved.

There is however, an inherent fallacy of attempting to predict the future specifically, otherwise we would have a much more accurate history of prediction than we have. The history of prediction is littered with failures, and notwithstanding the earlier discussion regarding certain inevitabilities, it is acknowledged that prediction of specific events if fraught with the usual pitfalls. Prediction based on the laws of physics, for example, solar eclipses, space craft velocities, and so on, is entirely different form speculative prediction of some future state of a technology based on conjecture, and an understanding of the current state of the art. However, such prediction is possible as long as it is understood that the margin of error increases the further one gets from the starting point. We know all too well that one thing leads to another. Once a steam engine was invented, and shown to be capable of doing useful work, making it work better was an inevitable next step, and predictably so, because many more capable and intelligent people, saw the potential and became involved. Also inevitable was the subsequent displacement of waterwheels and wind power to the status of secondary power sources, as was the use of steam to achieve locomotion. And if these wonderful things have come to pass, can the development of large cities be far behind. The further discoveries that electricity and magnetism were connected, and that electrical energy could be usefully generated by rotation of a magnet within a coil meant it was a small matter to predict the connection of a generator to a steam engine, to effect the production of instantaneous power. And the requirement for more power, and the inherent limits of the steam engine led to the development of the steam turbine, which is where we are today. The fact that birds and other animals can fly leads to the inevitability of powered flight, and by extension to air travel; because we see that it can be done, we set our minds to determining how it can be done, and then we do it. Similarly, because the human brain exists, and is the size it is, and can clearly and obviously do the things it does, artificial intelligence is a predictable future occurrence, though the particular forms it may take are clearly the subject of much speculation, as are the various legal, moral, and ethical issues.

There is a claim among the Singularity proponents that the material world inherently evolves, and each generation increments upon the last, and so on, in a blind process apparently towards something, and that this evolutionary process is what drives us inexorably towards the Singularity. As one paradigm is exhausted another paradigm emerges to allow the exponential explosion to proceed undeterred as before. This is not how evolution is seen by biologists, and those who study the fossil record. Evolution can either reach a dead end, or it may converge to an acceptable solution that remains effectively stagnant until acted upon by some external selection mechanism; a plateau of development that represents a kind of completeness. Sharks would be a good example of something evolved in nature very long ago, and unchanged throughout geologic time, they were swimming in the sea hundreds of millions of years before the dinosaurs, and are the same today as they were then. Sharks and internal combustion engines would therefore appear to have something in common. The internal combustion engine has not effectively increased in efficiency in the last fifty years, there have been minor modifications and improvements in materials, along with a few design improvements, and so forth, but in essence we may say that like the shark, the internal combustion engine has converged to an evolutionary solution. There has been no doubling of performance, no exponential increase in speed. The internal combustion engine, for all its pollution issues is one of the finest mechanical engineering solutions ever arrived at by the human species. And if a specific prediction were to be made it might be that the internal combustion engine will endure, and still be part of the fabric of human society in one hundred, and possibly one thousand years from now. So evolution sometimes stops, because there is no further requirement to make progress, the problem has been solved, and there are no external selection forces acting.

Moore's law is said to be the driving force behind all advances in computation, but Moore's law is a bit like investing your hard earned cash, will the graph continue to rise, will it flat-line for years, or will it tank? Suppose that in May 2008 Larry liquidates his portfolio, and is advised to use the proceeds to purchase Lehman shares. We know the rest. Larry complains. He is told two things by the advisors, both of which are to be found in the advisor's standard terms of business; the small print. Our rating of the stock is nothing more than an opinion, and not to be taken literally. The second caveat is that past performance is no guarantee of future gains. Clearly both of these caveats must also be applicable to any reliance on Moore's law, being as it is, a speculative future bet, and every bit as speculative as the stock market. Moore's law is extrapolation along the line of the curve taking the few points we have to date and extending it as if everything that happened in the past is likely to happen again. At its worst it is blatant wishful thinking, the vane hope that the path continues as it did before, and like the bankrupt investor the disappointed techno-futurist is a factual outcome.

Another factor influencing the inability to predict the future direction of technology is the fact that just because something is possible does not mean it will occur. There is a long list of things that are possible, and were long predicted, but have yet to come to pass, the paperless office, long predicted, and yet for all the technology we surround ourselves with, we still have a place for pencils and bits of paper. The cashless society is still something seen as being an optional part of the future, the usefulness and anonymity of cash remains, and personal preference is a big retaining factor. The promise of virtual reality has been around since the 1980s, the full emersion, glove and goggles experience, we have yet to see a proper working version, there are hints and first attempts, of course, but noting to take your breath away, nothing that has parents complaining, or people writing strong letters.

There is an old saying in corporate culture, and politics too, which declares that a rumour is merely a premature fact; we might rephrase this to state that today's problems are just premature technology. Inevitabilities occur when we already know something can be done, because we have seen it done, we just need to know how it was done, what the details are, we require the information. We are engaged in a kind of industrial espionage against nature, for example, we know a very sophisticated computer the size of a human brain can be built because it already exists. We don't know how it works in detail, but day by day we are working on it, and forcing the problem to a solution. We are not, on the other hand, working on ways to travel faster than light, because we know it cannot be done, nor do we pursue the illusion of the time machine of the, *Back To The Future* type. Despite the flawed nature of prediction and its discredited history, all of the facts, and a little speculation would seem to indicate technology is improving, growing, evolving and developing in all areas, and there is a certain amount of acceleration involved in some strands. It's a statistical effect, there are more of us involved in doing it than ever before, and we don't give up easily, of course we're going to make progress.

We must, however, proceed with a great deal of scepticism when referring to those who make very specific future predictions. Ray Kurzweil's, prediction of the Singularity happening in or about 2030, is a case in point. And it must be stressed here that Kurzweil's prediction for the emergence of human level general intelligence is an optimistic one, there are many opinions on the matter, and a good summary of the general expert consensus can be found in Nick Bostrom's book, *Superintelligence: paths, Dangers Strategies*, it would appear that the general consensus of those active in the field is closer to 2130 than 2030. In his Book, *The Singularity Is Near*, Kurzweil deals with the criticisms of his proposition, in a chapter entitled *Response to Critics*, he provides a long list of challenges to the Singularity and in every case shows how these challenges will be met and diffused by technology. However, many of the objections lack conviction, and are reduced to comparisons such as to the political and theocratic objections to stem cell research as being akin to pebbles in a stream, with technological progress washing over and around them like water; such analogues are trite and arrogant, and ultimately meaningless. AI is very much different, because as has been pointed out before it is largely the preserve of corporations, and the first impact it is likely to have is to take away people's employment, and that will almost certainly get their attention. The old maxim that there is nothing like a death sentence to concentrate the mind might be invoked here, there is nothing quite as effective as taking away people's money to cause them to sit up and ask why, or to get them to take to the streets in vast numbers demanding action.

If we continue as we are AI of varying strengths must be considered an inevitable consequence of the work currently underway, and the fierce competition among technology companies to get to the prize, and claim the first mover advantage. But putting a specific date upon it is problematic. The closer we get to the proposed date, and the more intelligent the intermediate developments released into the public domain are shown to be, the stronger will be the opposition, and the more visible the potential downside. As AI is deployed with a vengeance across the commercial and industrial landscape, the bigger will be the mob of economically displaced citizens standing outside the door, or taking to the streets, demanding that something be done. AI and the Singularity then become the subject of legal and political discussion with potential strong regulation, even prohibition becoming a reality.

There are of course limits to technological advance, which may well serve to block the ability to predict. Limits can be imposed for a variety of reasons, either self-imposed by ourselves, or by agreement, or by prescription. For example, nuclear bomb testing, and the further development of nuclear weapons technology, still continuing, but grossly scaled back, and limited by treaty, there is also the agreed ban on nuclear weapons in space, and so on. Other WMD technology is also constrained by international treaty, nerve gas and other chemical weapons, along with bio weapons controlled, or at least partially controlled by international agreement. Sometimes technology fails to advance further because the natural limit of the technology itself is reached, further development yields little if any further improvement, the steam engine is a perfect example of such a technology, the end of the paradigm was reached in the early twentieth century. Anyone predicting that steam power would continue to develop, and become ever more compact and efficient was on a looser.

Human cloning is an area of technology where the ban is by agreement, it could probably be done, but the ethical issues of doing so are outweighed by the scientific or technological advantages. Sometimes technological research is discontinued for religious reasons, for example, the White House ban on human embryonic stem cell research under the Bush Presidency. There was no effect other than to divert the research funding to Europe and other more enlightened countries; it's never a good idea to allow church people to make important decisions, those who believe in magic rarely act rationally. In the case of the White House ban, the reason given was that the three day old human embryos used in the research, and consisting of no more than 150 cells, had souls, no evidence was adduced to support the claim; faith was seen as sufficient reason.

Sometimes further technological development runs into an insoluble problem, something so intractable that it acts to stall development, such as a fundamental law of physics which limits the size or the number of components on a chip, or limits the speed of operation because of heat generation. The insoluble problem may also be economic, perhaps the cost of transitioning to the next generation of technology is prohibitive, or perhaps there will always be a negative return, regardless of how many you build or how many you can sell.

Another form of self-imposed prohibition comes about when we decide that we do not want or need further advance in technology, for example, in the game of golf. If Moore's law was applied to golf, the average golfer would now be able to drive the ball onto the green of every par four. Technological advance is limited to design and perhaps changes in materials used. The rules of golf set strict limits on ball design, club design, and on how far a ball can travel in the air after being hit by a given club. All manufacturers must adhere to these rules in order for their equipment to be approved for play. Chess too is an area of human endeavour where technology is restricted, because computers play better chess than any human, when human tournaments are played, machines are barred by agreement.

There are a host of other technologies restricted, regulated or banned, GM Food in Europe, for example. Sometimes public outrage and protest swings the day, sometimes the prevailing moral and ethical climate, makes it impossible to pass or to repeal certain laws. There are often very good reasons why we may seek to prescribe, or limit technological advance in one or many areas, and sometimes the reasons make no sense at all. Drugs used by people for recreational purposes remain illegal, despite being, as John Gray notes in his book, *Straw Dogs*, both universal and immemorial. The policy continues despite the fact that it has been shown to be unworkable, and a waste of taxpayer's money; not to mention the blatant hypocrisy of alcohol being the acceptable method of altering your state of consciousness. There are vast numbers of people in prisons throughout the Western World for flouting the prohibition, there are many more sentenced to death in a few unsavoury countries around the world. Yet the drugs war rages on. Politically acceptable and technically possible are often conflicting terms, anyone who thinks that artificial intelligence research and development, for example, could not be prohibited does not understand political pressure. As has been pointed out before, AI is a corporate activity, it would be easier to prohibit AI globally than drugs. Just because we can do the research doesn't mean we will be allowed to do it or be permitted to continue. Vested interests, corporate entities, intellectual property rights, restrictive practices, established apple carts, and so on, can all be swept aside if political expediency dictates.

It is not, on the whole possible to be exact and precise about future events or even trends, the Singularity may or may not happen by 2030, many commentators think not. Kevin Kelly calls it a myth, every bit as mythical as Superman or Utopia. The economist Jeffrey Sachs may or may not be right when he declares in his 2005 book *The End of Poverty*, that extreme poverty will be driven from the world by 2025. The only thing we can predict with any degree of certainty, is that we cannot predict with any degree of certainty.

Chapter 10   Scepticism And The Myth Of Progress

I believe that this world is newly made: its origin is a recent event, not one of remote antiquity. That is why even now some arts are still being perfected: the process of development is still going on. Many improvements have just been introduced in ships. It is no time since organists gave birth to their tuneful harmonies. Yes, and it is not long since the truth about nature was first discovered, and I myself am even now the first who has been found to render this revelation into my native speech.

Lucretius 99-55 BCE. De Rerum Natura.

Purple is not about a specific future but the general inevitable future, eminently predictable from the current starting point. More specifically Purple is about our attitude to the future. When we think of the future we think of making progress, we are engaged in a process of gradual improvement and progression, from where we are now to any point in the future. As far as this book is concerned that future point is the point when the human world achieves a Purple state. Purple means that we have come together as a species, and have done all the necessary big stuff, we have acquired all the information necessary to unleash the infinities of matter and energy, and everyone alive has pretty much everything they could possibly want.

If that sounds like a kind of utopian wet dream then perhaps it is, when stated succinctly, but the fact remains that the only thing missing is the information necessary to achieve that dream, we have the energy, we have the matter, and more than enough educated reasonable people willing to embrace a future of abundance. In his New Statesman essay, *The Myth of Progress*, John Gray, Professor of European Thought, at the London School of Economics, asserts that faith in progress is a superstition, akin to a religion. It is hard to argue that blind faith in progress is clearly on the same level as superstition, however, while faith is defined as belief without evidence, there is no faith in progress because there is a great deal of evidence, simply put, we have the evidence of where we have been and the evidence of those things we cannot yet do, and our self-evident ability. More than sufficient to render our belief valid. But not unconditional.

However, almost nothing happens instantly in our world, most things of any real worth seem to take time, which casts a great deal of doubt on certain futurist claims both past and present. The greatest attribute you can possess in this modern empire of technology is a vigorous scepticism regarding claims made by anyone who states with any degree of certainty that they know what is going to happen, or when. Purple requires the reader to be sceptical, and challenge the current and predicted rates of technological growth, and the exponential theories of many technological writers and prophets. Illustrations from medicine, computer science, power generation and other areas amply demonstrate that rates of growth and technological advance are not necessarily what are predicted. And even if technological progress does continue at the optimistic pace of many of the more exuberant commentators, that is no guarantee of a Purple world delivered, because so much depends on what people do, and what they do when they form groups, and what those groups do when they form larger groups.

So called laws, such as Moore's law, Clarke's Law, and Asimov's Laws are more akin to Murphy's Law than to Newton's laws, and of themselves have no validation or merit other than as aspirations, guesswork, or rules of thumb. The low hanging fruit theory must also be considered, along with the fact that in the real world of resistance, greasy poles and slippery slopes, many problems are insoluble for a variety of reasons, inability to understand, unwillingness to solve the problem, the prohibitive cost of solving the problem, and so on. And of course, we always come up against the law of unintended consequences. And we must never forget the law of diminishing returns. Scraping the last from a tub of ice cream, or a pot of honey, at some time you just give up. Or writing a book. At some point in time the end is reached the rest is simply polishing. The concept of apple polishing too, is well known. There is a point at which further effort is just tinkering, and achieves nothing, the problem is solved the plateau is reached. It has been suggested, if not demonstrated that progress in many of the fields of science and technology is neither linear nor exponential, even if it appears to be, it is perhaps better described as being sigmoidal, which is essentially to say that progress stumbles haphazardly along the curve from one plateau to the next as each burst of progress happens, often by accident. The intervals between advances may be very long indeed.

If we continue as we are, and technology continues to evolve at the current pace there is a chance we can overwhelm all other problems, and the world may become Purple; eventually. Quickly or slowly, it doesn't matter.

It ought to happen, because there are more people working to make progress than there are resisting, and those resisting have nothing to offer by comparison. There are no alternatives to making progress. But the wild promises of certain techno-utopians require to be tempered with a great deal of caution. For example, useful and applicable AI may be ten years off, or it may be fifty or more, no one knows for sure, those who say they do know are guessing or engaged in wishful thinking. But what is certain is that because we know it can be done means that sooner or later we will understand it, and we will eventually make it work, hopefully in a form we can control and apply to our problems, that fact alone will liberate a great many people to do other things. Or such achievement may serve to create an angry mob to add further to the resistance.

Some things in our modern world don't change, or change so slowly that we humans hardly notice. You could be sitting in the garden of a house built in 1865 in an antique garden chair, staring at a plant that might have been growing there since the garden was first laid out. You could be, to all intents and purposes, at any time between 1865 and today, nothing has changed. You might look to the sky and see a vapour trail, that will quickly put a date on it, or your cell phone suddenly rings with a quaint musical tone, and gives the game away. The only thing that separates us today from the builders of the house is our technology, and the liberation from the mundane drudgery of existence that technology has brought us. Our homes are crammed with it, and while our 1865 home might superficially look as it did when it was constructed, for aesthetic reasons, it is now crammed with modern technology, our persons are adorned with the same stuff, from synthetic fibre clothing to electronic biosensors and communications. And we love our technology, and we all want more.

Progress is very much a process and our progress is entirely linked with our ability to evolve and deploy technology that solves problems better than the previous generation of technology or solves new and different problems. It is therefore appropriate to ask if the rate of technological development is increasing or decreasing or reaching some plateau. Is it perhaps stagnant? Is there what we might describe the illusion of progress, whereby new versions of the same technology are released with a new design, or the same features are simply rearranged? In the same way that detergent is repackaged with a 'NEW' label emboldened on the fresh new packaging containing the same old detergent. Clearly, in certain areas technology has entered a phase of what we might call stagnation. A period when nothing happens or nothing appears to happen, where all is boring, and has ground to a halt or at least appears to have ground to a halt from the point of view of a casual observer on the outside.

Aeroplanes, for example, to the casual observer there has been no appreciable change in air travel since the jet engine was introduced to make air travel possible. To a first approximation the planes look the same, superficially at least, they all have similar looking wings, with two or more engines, the cabins all look the same as they ever did, with seats numbers and tight spaces, the food, or lack of it. Boarding by a single entrance, pressure changes, and so on. The flying experience has not changed in sixty years. Nor has the time taken to get from Europe to North America. However, the impact of technology has been incredible, and has resulted in massive numbers of people taking to the air, and increasing every year. The casual observer will not see the technology, the computer systems keeping everything working, the changes in materials technology that make modern jet engines the most efficient ever, the flight systems and navigation controls that allow planes to fly in all weather, and arrive safely. And there are enormous numbers of planes in the sky. The massive increase in safety records globally leading to the fact that pilot error is almost a thing of the past. If a plane crashes it is probably, more often than not, a deliberate act. Notwithstanding the 9-11 disaster, three recent examples will suffice. The Malaysia Airlines Flight 17 brought down by a Russian ground to air missile over Ukraine, the Germanwings Flight 9525 crashed by a suicidal co-pilot over the French Alps, and the Malaysia Airlines Flight MH 370 which disappeared, probably deliberately, over the Indian Ocean. Despite these and other dreadful incidents, flying is still the safest way to travel. But it is not what we were promised. Way back in the 1960s and 1970s futurists and science fiction writers looked forward to the future of air travel, and saw supersonic flight, even hypersonic extra atmosphere low earth orbit flights, allowing us to get to from New York to London in two hours, or even less. They saw enormous sky cruisers, and trips to the Moon, and to orbiting space stations for vacation. From their perspective, and the techno-optimism of the time it was both obvious and inevitable. None of it ever happened. The rate of progress stagnated. It was a false start, with a premature attempt at supersonic flight, the technology was simply not up to the job. Concord, the only ever supersonic passenger jet, was desperately kept flying by Britain and France as a fervent nationalistic symbol of the new white heat of a technological future, a hope that finally crashed and burned on primetime TV in a field North of Paris.

Just because progress has been exponential in some field to date is no guarantee that the future trend will hold. As your investment advisor is legally bound to point out, past performance is no guarantee of future growth, and they are frequently correct. The markets might outperform or underperform, conditions may change. Moore's law might hold or fold. Exponential progress in science or computation is as unpredictable as exponential growth in carbon emissions. The two sets of data have the same thing in common, limitations in predictability. If Moore's law holds, computability will double every eighteen months. That being the case an average personal computing device of two years hence will be twice as powerful as the one you have today. And in fifty years, or thirty-three iterations of doubling, the same device will be two to the power of thirty three times, more powerful. And in one hundred years, or in five hundred years, or in two thousand years, and so on. Or will Moore's law break down, and will we reach a plateau of development from where we can no longer make progress? Or perhaps we will call a halt on reaching some point where we say that no further progress is desirable, or required?

Once the problem is solved it is solved, and stays solved. As was said before considering the humble table and chair, any further development in the space of tables and chairs is purely a matter of aesthetics and design, different shapes, other materials, and so on. The difference between innovation, inventing and design, has been debated by academics for centuries, and sometimes good design can be an invention of itself, but solving the problem is what makes progress. Moore's law applied to problems already solved leads inexorably to endemic feature creep. The many and varied advances in computer technology have led to word processor programs, and other software becoming ever more complex. With the end result that the feature set is so complex, and the learning curve so steep that most users simply bypass the lot. The additional features and speed of machines simply does not translate into a faster more efficient method of creating a document, for example. Once the problem is solved we seem only capable of adding features, certainly over the last twenty years office software has seen no exponential advance. In the book *Abundance: The Future is Better Than You Think,* Peter Diamandis and Steven Kotler provide us with the example of the *Osborne Executive Portable* mobile computer, from 1982 and compare it with the modern *iPhone* as an example of exponential growth in computing power over the intervening years. And it is an impressive difference by all metrics. On most modern uses there is simply no comparison, but in terms of utility, the 1982 machine was one of the best for what it did at the time. Even today, if all you wished to do was type a 1000 word document, the Osborne's robust keyboard may well be the better machine. The modern smart phone does a great many things, and we all love them and use them, but if all you want is a phone, then most of it is redundant, and if all you wish to do is type the occasional few pages then it is probably useless. In a way it is not really progress, there are a great many features, and a great many devices all rolled into one, with more colours, and nicer graphics, they are a great pleasure to use and a great pleasure to own. However, there has been no exponential increase in the ability to answer emails, or make phone calls, or create documents, or get your work done, it is part of the myth of progress, part of the illusion of futurist claims that seem to elevate faith in future technological progress to the status of a religion, with all its false promises of magic and solutionism.

Similarly very little, if any, human endeavour advances exponentially, art, for example, the ability to draw and render the world in pencil or any other medium, has not changed. You might also ask if art peaked with Picasso. Or with Rembrandt? Or Leonardo? Perhaps with Jackson Pollock? Again it all depends on your perspective. The functional sketch of an engineer, or a portrait of Rembrandt's mother, have not changed in hundreds of years, either in conception or in execution. Literature has also remained static and manageable, the modern novel may have different themes, but could well have be read by Proust, or Henry James. And let's not forget music, perhaps you might think it peaked with Beethoven, or Mozart, or Chopin, or even Elvis, Sinatra and the Beatles. All depends on taste. And a good tune is timeless. But there are no huge advances taking place, the only difference between a Beatles melody and a Beethoven melody is on a subjective level. So there has been no exponential advance in these areas. Beethoven might not know what to do with an electric guitar or an iPhone, but he would certainly be capable of operating a Steinway Grand.

Of course progress in computation has been exponential to date, and might well continue through the Singularity, and well beyond, or so we are led to believe. Is it likely that everything necessary will continue to advance exponentially until we reach a point where strong AI takes over, and continues the exponential advance towards infinity? Or perhaps we are in the middle of an occurrence of the low hanging fruit phenomenon, chomping our way through the easily discovered good ideas. As with the periodic table, it was easy to discover new elements when only a few were known, later it became much more difficult. We have no way of knowing exactly where we are on the curve until we pass the point.

We saw similar spurts of growth in the era of the steam engine, following a slow start, there were many decades of continuous growth and progress, as the science was better understood, and new methods and materials were deployed. Steam engine technology progressed throughout the nineteenth century, the field converged to a solution and reached a plateau, known as the golden age, when the most efficient engines ever built were in service throughout the world, enormous, stationary engines used for pumping and winding, and locomotives puffing their way across continents. But how many corporations today have a research division working on steam technology, perhaps one or two, some minor effort on turbine efficiency which is hasn't improved in decades. Diesel and other electrical technology took over from steam to become the second paradigm for heavy power usage, and has converged to its own plateau where it remains to this day. And there are no new paradigms of technology about to subvert the diesel engine, or the electric motor, or the steam turbine.

The first ever steam engine was built by the blacksmith Thomas Newcomen in 1712, and if he could be flicked through time to 2016, and be brought to a car dealership he would not feel out of place. If the workings of the internal combustion engine, and the standard transmission were explained to him he would no doubt understand. His familiarity with the machine would not lead him to suspect magic, as with Clark's law. He would understand that instead of coal, oil was used, that the oil was refined to enable us to use only the finest portion of the oil, and that the oil thus refined was burned, not outside of the engine as he had made in his workings, but in fact within the engine. It could be explained to him that it was found by the art of those men who had come after him that this was the better way to burn the oil, and that it was the burning of the oil within the engine that produced a kind of steam which forced the piston to move within the cylinder as it did in his engine of so long ago. He would understand that he did after all use the fire of coal to convert liquid into gas, and thereby generate motion, and by doing so pump the water from the mine at Dudley. While there has been some exponential growth, and one or two paradigm shifts in 300 years, it has not gone so far as to be beyond the capacity of the first workers in the field. While Newcomen would no doubt be impressed it would not be as magic to him. He would understand it all, and very quickly.

Where then does this inevitable Moore's law bring us, what lies at the end of the path if it simply continues along on its exponential way? Clearly if it holds then we get progress, and lots of it. Eventually Kurtzweil, and Diamandis et al, declare that the Singularity has been reached, and you can go to the store and purchase a computing device for under $1,000 with apps operating with human level intelligence, capable of passing the Turing test very comfortably. It does everything you require, and lives in your pocket. Five years later we have for the same $1,000 a machine that is ten times smarter than a person. This mass produced entity is super intelligent, immortal, capable of out-thinking the average human by a factor of ten, yet still resides in a pocket, and obeys its owner. Oh what a charmed life we would all lead thereafter. But is there a human on earth who would actually need such a device? What possible problem could the average human face that might require such a powerful device to solve? That is probably an irrelevant point, we will of course as dutiful consumers purchase them simply to have one in a pocket. And all of this to occur by 2035 according to the Singularity believers, if Moore's law holds, and no one steps in to call a halt.

Perhaps someone will step in and cry halt, there are precedents in other areas of technology, where companies could produce technology to exponentially overwhelm the problem set, but are prevented from doing so by governing bodies, or regulators with an agreed set of rules. When Jack Nicklaus, arguably the greatest golfer ever, won his first Masters at Augusta, The legendary amateur golfer and designer of the Augusta National Golf Course, Bobby Jones, told him, *"You play a game with which I am not familiar,"* so convincing was his victory. When Tiger Woods won his first Master's tournament people began to talk about how golf courses would, in future, have par fives of 700 yards and more, and how all the golf courses in America would have to be redesigned to accommodate these new golfers with their new overpowering technology. Apart from being the ultimate game of surprises, golf has always been a happy hunting ground for technologists, and every year since the beginning of the twentieth century there has been new and different technology on offer to amateur and professional alike. The game is saturated with technology. The golf ball is the biggest source of technological input, as you might expect being the game's primary consumable, and every year new dimple patterns, new coverings, and new innards are promised to help the golfer keep them straighter, and hit them further. And none of it works, because if it did the average golfer could hit the ball five hundred yards, and dead straight. The mowers, and the variety of grass used, on the fairways and greens, are also a big factor in the improvements made to the way golf is played compared to how it was played in the early twentieth century. The golf clubs, grooves, shafts, grips, and so on all help, expert fitting of clubs to individual golfers also makes a difference. The fitness of the players, attention to diet and exercise, the psychology of competition, and greater understanding of the interaction of these inputs is also important. But there has been no exponential increase in performance, either by amateur golfers or in the professional ranks. The 1947 Open Championship was played on the links at Hoylake in Liverpool England, the course was 7,000 yards in length, with a par of 68, and the Irishman Fred Daly won with a score of 293, with only a handful of shots separating the top ten players. 20 years later in 1967 Roberto De Vicenzo won the Open on the same course with a par of 72 and a total of 278. In 2006 the Open returned to Hoylake, and Tiger Woods was the victor with a score of 270, eight years later Rory McIlroy won with a total of 271 shots. In the interval of 67 years between the first Irishman winning the Open at Hoylake to the next Irishman winning the same tournament at the same venue there has been no exponential increase in performance or in technology. There have been improvements, of course, but there is no Moore's law operating in golf equipment design, or in performance.

Golf magazines from 1947 promised 10 yards extra from the new design of fairway wood, magazines of the mid 2007 were promising exactly the same, in fact, every year between the two dates, the advertising promised an incremental improvement of a few percent over the previous technology. And similar sales-fluff claims were made for golf balls, putters, drivers, irons, wedges, shoes, gloves, umbrellas and waterproofs. But the stark fact is that Fred Daly could hit the ball every bit as far as today's player; if he teed it up today with modern equipment, no one would give him any shots. Nor is there any appreciable increase in distance by using the modern ball, and the modern putter certainly does not get the ball in the hole any better. Of course there have been vast improvements, in materials and technology, clubs are easier to hit, requiring less skill than in the past, the poor shot is perhaps not as poor, the modern ball has a computer evolved dimple pattern to promote a better flight, you can now find a ball to suit your particular swing, and that might help. The machinery used to prepare and maintain the golf course has improved, the ball will run further on the better prepared surface, and hold its line better on the greens, irrigation and agronomy are better, and golf course management is better understood. The reason you can't hit the ball twice as far as you could in

1947 is that the governing bodies of the sport set performance limits on balls, clubs and on technology. These limits are set out in the technical specifications of the Royal and Ancient Golf Club of St. Andrews, and the United States Golf Association, with only very minor differences between the two codes. So while there is progress all around the golfer, in materials technology and new devices, such as GPS course guides, there is no progress in the one thing that matters, hitting the ball further and more accurately, and in one hundred years from now the same game will be played on the same courses with pretty much the same equipment, and scores will be pretty much as they are today. The regulating bodes seek, in their own words, "... *to prevent an over-reliance on technological advance rather than skill, and to ensure that skill is the dominant element.......*" So there is no Moore's law for golf, and practitioners in the field of golf like things as they are, no one feels the need to exponentially grow the technological capabilities of the game into the future far beyond the spirit and traditions of the game. As a result there are many golfers with identical skills, which means that scoring in any major tournament finds the top 25 players often within a handful of strokes. The game is healthy, popular and has a future.

So perhaps regulation of technological developments in computing and AI will stop Moore's law in its tracks, perhaps we will say, thus far and no further, as was discussed in previous chapters. There are areas where we could use a little Moore's law action, areas of knowledge where we are desperate for some exponential growth and an explosion of knowledge. Medicine as a whole, but in particular that stubborn collection of genetic diseases we collectively call cancer, and the general problem of human aging.

The awful and depressing facts of cancer are all around us, nearly everyone on the planet is affected. A diagnosis of cancer strikes terror into the human heart. And still there's no cure for cancer, Denis Leary was right on that score. All the promises of all the futurists, and the optimistic predictions of the past fifty years, and all the extrapolations seem to come to nothing when faced with the complexity of the human genetic code conspiring to destroy the human body. Of course there are advances, and of course there is progress, and of course there is a better understanding, and of course there is far better treatment. Above all there is that vast reservoir of money collected for cancer research all across the world, and the numerous billion dollar biotechnology corporations spun off from the research with their constant outpouring of semi-favourable clinical trials. So it is possible that a major breakthrough is just around the corner. But the cynic might retort that we've been hearing it in one form or another for the last fifty years, funding the research seems to be an end in itself, harsh treatments by chemotherapy and radiation remain the only weapons, and the only real cure is prevention.

However, our skill in chemistry is such that we can isolate individual components of complex amalgams of neurotoxins found in snakes, spiders, frogs, scorpions, and, the many and various, poisonous sea creatures found in our environment. It is not a difficult task to read the popular review literature on the subject and conclude that medical breakthroughs are on the way as a result of the work being done, it is one of the inevitabilities of the work that some the vast list of problems in medicine will yield to the new drugs that will come from this world wide effort. It is unlikely that in twenty years or so the review literature on the subject will lament the wasted years, and the false promises of the work, and the billions of research dollars all of which resulted in a single partially successful pain relief treatment, or a moderately better anticoagulant. But it is not exponential growth in ability, and bold statements such as, the work will result in cancer being a thing of the past, are clearly out of place. Vast amounts of time people and money thrown at any problem will usually result in a partial solution, or at the very least a far better definition of the problem and its parameters, for it is in the nature of complex problems, such as those associated with the human body that there are very often hidden components to the complexity. Reading, for example Siddhartha Mukherjee's book, *The Emperor of All Maladies*, or *The Immortal Life of Henrietta Lacks*, by Rebecca Skloot, shows quite clearly the nature and depth of complexity involved in tackling the moving target that is cancer. Mukherjee's conclusion, is a very depressing admission that cancer is here to stay and that we may never really be able to claim a cure. The ever-patient public awaiting a breakthrough in the field exist in a constant state of disappointment, nothing seems to change, every family in the West is affected, the body count seems to remain the same, and little or nothing seems to change. It seems to be a field absent any hint of Moore's law.

On a more trivial note, there are other less dramatic failures to make progress, areas where the myth of progress comes into sharp relief. The cure for baldness is a case in point. There is no apparent progress, no exponential increments discernible to the reasonable external observer. Nothing to cause one to comment that perhaps there will be a cure in the near future, nothing to suggest that a prominent bald person in the public eye might suddenly appear sporting a mane of healthy new hair, or that the hideous comb-over, as sported by Donald Trump, for example, might be consigned to the past. And don't think for one moment that the lack of progress in curing baldness is some natural consequence of medical priority. Because it is very much seen as a priority. Baldness affects forty percent of men in the Western World, and so the research has impressive if not obscene amounts of money attached to it, which makes it a priority. No price is too great to put on male vanity, the vast number of wealthy bald men will make it a going concern at any price. Yet despite the unlimited amounts of money thrown at the problem it remains stubbornly unsolved. Nor is it a lack of understanding, apparently everything about the condition is completely understood, and has been studied in the minutest detail by some of the best, yet solution comes there none. Again the cynic might exclaim, progress what progress? If baldness can't be cured, what chance a cure for anything else? If these medical people cannot resolve issues with the humble hair follicle, what chance with the complexity of a tumour. There is probably as much money spent on the one as on the other in any given year.

Technology is defined as the long list of problems solved, we must therefore ask questions such as how many problems exist, and how many have been solved? How many remain to be solved? The number of remaining problems can be quantified and prioritised into big, small, important, and so on. Cancer is obviously a big problem and an important problem. Feeding the world, on the other hand, is a solved problem, there is more than enough food in the world to provide every human person with 3000 KCals per day, and that's despite nearly fifty percent of the food being lost between field and table. The problem is getting the food to those in need. Water is another problem that has effectively been solved, again the problem is not producing the clean water and lots of it, the problem is distribution, and getting the water to where it's needed. These kinds of problems are political, and can be solved given the political will, but they lie beyond technology, and you cannot stop religious hatred or civil wars by solving technological problems. Getting to the stars is also a problem to be solved, long term of course, it's not of such immediate concern or of any vital importance as some other more pressing problems. And so you might conclude that in our attempts at getting to a Purple world we have only to solve the outstanding list of problems, and we are there, of course not, that would be a techno-utopian approach. Purple is not a branch of techno-utopianism, Purple recognises that not only do we have to solve the technical problems, and advance the technology, we also need to solve the equally long list of political and self-interest problems. Hence the requirement for a great deal of scepticism for any suggestion that technology alone equates to progress. In fact it is this insistence by certain activists in the field of prediction that technology has any capacity to be of itself progress that is the myth of progress.

Techno-utopianism is, according to those familiar with the subject the proposition that technology can solve all of the problems faced by humanity leading to a perfectly moderated world populated by well-educated, and well-adjusted individuals all living a very agreeable extended life in great comfort with all their wants and needs met by technology. So far so good.

There's nothing new in it, of course, it has probably been around in one form or another for as long as humans have been upright, certainly in the modern era the dream of a final political settlement of man's affairs dates back at least to the French Revolution, we then had Marxism on the extreme left, and Hitler and his like on the extreme right, all promising similar outcomes as soon as the few wars were won, or the dictatorship of the proletariat was concluded. Today our mistrust of politicians has elevated technology to the status of hero, leader, and problem solver for humanity. We ask a great deal of it and thus far it has delivered. It is, however, also seen as the most likely source of apocalyptic failure, whether nuclear, nano, viral or AI.

Trying to predict the future is nothing new, even Cicero had something to say about those making predictions when he opined that he found it hard to understand why a couple of soothsayers passing each other in a public place would not be compelled to burst into laughter. The Victorians too were filled with awe and wonder at all the marvellous accomplishments their technology delivered, over the sixty years of Victoria's reign. Their achievements of steam travel and industrial development, horticulture and farming, filled them with pride, created vast wealth, and gave them the belief that they were just getting started, and anything was possible. And they were right. Any educated Victorian lady or gentleman arriving in a modern city today would be amazed and perplexed, they would be immediately struck by what we have achieved, the use of electricity, air travel, automobiles and trucks, and the road networks, computers and TV, and all the rest. And while there were two major global mechanised wars, and several economic calamities, and political upheavals in the interim, the plod forward of technological achievement continued, the production of wealth continued, the taxation of that wealth by governments continued, and paid for all of the visible civic improvements. Our Victorian visitor would be justly proud of the foundation they and their fellows had built. They would also be required to admit that we have a great deal more fun than was ever possible in Victorian times.

The techno-utopian movement, if it can be so described, though this term is disliked by many of the active participants, preferring the term futurist, or transhumanist, or the abundance movement, is led by people like Peter Diamandis who along with Stephen Kotler has produced a seminal text of the techno-utopian movement, *Abundance: The Future Is Better Than You Think*, Diamandis is, among other things, the chairman and CEO of the X-PRIZE foundation, and a co-founder of the Singularity University. More recently he has announced an X-Prize for, *Making CO2 Something New*, or finding uses for all of the carbon dioxide complained of by the climate change people. Ray Kurzweil, author of *The Singularity is Near*, and *The Age of Spiritual Machines*, is also a very prominent member, as is the MIT based Ethan Zuckermann, whose book, *Rewire: Digital Cosmopolitans in the Age of Connections*, urges us, *"to take control of our technologies"* and, *"start the process of rewiring the world"*. Among the more exuberant titles on offer by this modern band of soothsayers is the book by Byron Reese, *Infinite Progress: How the Internet and Technology Will End Ignorance, Disease, Poverty, Hunger and War*. Strong stuff. We might also include in this category the book by the Oxford based physicist and thinker, David Deutsch, *The Beginning of Infinity*, which effectively makes the claim that we are just getting going. For most of the claims made in the numerous books the starting point is the presumption that Moore's law will hold true for each and every eventuality and area of technology, specifically in the development of computers and software. Moore's law originally applied only to transistors on a chip, but as Ray Kurzweil points out in his various books, it effectively applies to our ability to compute. This exponential growth eventually leads to the point in time known as the Singularity, when human level intellect is achieved by an artificial general intelligence, thereafter all fields of technology and human endeavour progress exponentially with the aid of ever more intelligent, ever helpful, compliant and controllable smarter than human machines. Purple also belongs firmly in this canon of work, though with a much more tempered exuberance. As has been described in other chapters of this work, it seems not only reasonable but inevitable that the great wealth to be unleashed upon the world when we finally close the information deficit will dwarf any period of plenty we have ever seen, in that at past, Byron Reese is correct when he writes, *"the poor will get richer, and the rich will get vastly richer"*. Unless the global republic sets upper limits on the wealth of private individuals, because again, all rights including private property rights are conditional and proportional, and while your right to be a billionaire has been violated, by taking the bulk of your wealth in a one off tax, you have been left with five hundred million, and remain extremely wealthy, hence the violation of your rights was not disproportionate, and how dare you complain.

Of course this overly optimistic, perhaps even irrationally exuberant, to borrow Greenspan's phrase, forward looking attitude towards technology has some very vociferous critics who rightly point out that all technology is a Devil's bargain, or a double edged sword, and social progress, and technological progress are not the same thing, nor does the one, either lead to the other, or influence the other in any meaningful way. Making that claim, the critics argue, is a kind of positivism or a kind of scientism, both of which have more in common with religion than with factual scientific analysis, in that they rely on belief without evidence, or faith. For example, the solution of technological problems over the last fifty years or so has resulted in a global food surplus, yet there are hungry people in the world, and enormous food waste among those with plenty. The technological solution has not resulted in the solution of the pressing social problem, simply because the social problem has a greater number of dimensions to it, and is not just a problem of production or distribution of food. Critics also claim that techno-utopians ignore the role of governments, and the regulation of corporations in making their claims for technological impact, there is also the major issue of the great technology divide, between those that can and do make phone calls, and those that have never made a phone call. One billion people on Facebook, means nearly six billion not on Facebook, either because they don't, or more probably because they can't. You cannot benefit from technology if you have no access to it, you can't use a smart phone if you can't read, or you can't pay, or both. Another criticism is that techno utopianism has scant regard to the environmental impact of future technology, the issue is almost never considered as a factor, the costs of disposing of the techno waste are ignored in the calculations. Too busy dealing with the problems of the future, and too unconcerned about dealing with the problems created by what they did in the past. There is no doubt, for example that a nuclear reactor is a very efficient and cost effective method of generating power, if you pay little or no attention to the costly matters of nuclear waste, and decommissioning of the plant at the end of its life. It has also been argued that the acceptance of the known harm caused by oil and its derivatives such as plastics, on the environment, is in no small part due to a latent techno-utopianism, a kind of subconscious feeling that everything will be okay because technology will come to the rescue. Eventually.

Techno-utopianism is also seen by critics as fostering a human addiction to technology, our inability to live without it, our unwillingness to be parted from our devices, and our connections, in case something might happen that we might miss. Our complete and utter dependence on technology is seen by an increasing number of critics as a dangerous state of affairs, leading to a generation of people incapable of functioning without it. On the other hand one might well ask, what is the alternative? A light carbon footprint is one of those aspirational list-items that no one actually gets to, we are all of us, without exception, up to our necks in the technology game.

Inability to control technology is a further criticism of the various techno utopias presented, the inability to shut it down if it's causing harm, for example, consider a city in flames, riots and widespread social unrest, all coordinated through social media. Should social media be shut down in such circumstances? The British Prime Minister David Cameron believes that it should, and suggested as much following looting and riots in London. When it was put to a vote on Twitter, however, the outcome was unanimous that the service should remain live so that the doings of a TV gameshow could be dissected. But we are well aware that in law, all rights are conditional, and proportional, and subject to the exigencies of the greater good. And your right to tweet is no exception. It requires only the political will to change the law. Also of concern to critics of techno-utopianism is the alienating effect of technology, we hear of people absenting themselves from society because technology allows it. We often see the results following court cases, for example where death threats have been issued on social media, the defendant is revealed to be a bedraggled, unemployed loner living in a back room of his mother's house, or worse.

The final objection raised against techno utopianism is from the religious who see humans as subject to something outside of them, the ever watchful creator of all things who made us in his image, and knows everything, hence he knows of our technology, and will ultimately step in and punish us for daring to dream our god-like dreams of abundance and long life. But as we well know from science, all claims of the supernatural are nonsense, being contrary to the laws of physics, and from law we also know that those who assert have the burden of proof, hence the requirement for evidential production is always with the critic.

However, a great deal of what is written, both in this book and elsewhere is pure speculation, almost wishful thinking, and to that end we must be very cautious. There is a certain amount of arrogance, if not dogmatic nonsense in making claims about the future when it depends on what any one of six billion people might do or discover, it is as always, in the nature of prediction, as in the commission of a crime, the little things that trip you up. It is also prudent to express caution when making the claim, common to a great deal of writing in this area that, this time it's different. This claim is one of the most trumpeted, and yet of all the claims it is the one with the least evidential foundation. It may be that this time will be different, but as was pointed out several times in the movie, *Bridge On The River Kwai*, there's always the unexpected.

This book takes as its precepts the laws of nature and the doctrines of the common law as practised in the English Speaking World, and on that basis it is simply not possible to make the dual claims, that Moore's law holds across all branches of technology, or that this time it's going to be different. What we have built in the western Developed World of today, 2016, may be another false start, another empire that collapses due to something no one saw coming. Without a doubt this is the best we have ever been, see for example, the references section of the book, *Abundance*, and by every metric we have, we are getting better every year. So it might succeed, and endure long into the deep future. But the simple fact is that we don't know, and cannot say for certain. There are certainly a great many critics of what is written on the subject by those who feel that we are on the brink of something new and wonderful, and that this time it *will* be different.

For all those reasons we need to be careful in attempting to place at the door of technology attributes it does not deserve or aspects of which it is incapable. Technology is nothing more than a list of things we have done. It is a long list and a detailed list, but it is nevertheless only a list of stuff. A list of Stone Age technology might not be a long list, sharp stones for scraping, sharp stones for cutting, leather straps for tying, clothing, footwear, and so on. It is a list readily understood by a school child. Another way to think of technology is simply to see it in terms of what it replaced, or in what it purports to do, the Internet is nothing more than a noticeboard, and Facebook and Twitter are nothing more than methods of attaching notices to the noticeboard. Obviously it's a little more involved than that, but for most people that analogy will suffice.

However we look at the many objections to the different techno-utopian positions and no matter what the claims, we are left with the very simple set of facts. We humans understand the world far better now, than we have ever understood the world before in the entire history of understanding the world. We are surrounded by infinities of space, energy, matter and time. We know, to take just one example, that every piece of wood ever used on the planet, from the massive support beams of ancient buildings to the veneers in your oak-panelled office was produced by plants using carbon dioxide, water, sunlight and time. The only thing that stands between us using the carbon dioxide, water vapour and nitrogen of our atmosphere, along with the energy of the sun, to make almost everything we could ever want; is information. The only real problem humans have is the information deficit. Closing the information deficit is the real human challenge. But we know we can do it, because it is not a problem beyond the laws of physics, and because we have more scientists, engineers, companies and machines working on the various problems than at any time in the past, and we will keep working the problem until it's done, it's sheer brute force against ignorance. It is therefore only a matter of time before we can apply bioengineering, synthetic biology, and solid state physics to building a small black box that will sit in your home, and use solar power and atmospheric components to produce the fuel you need to drive your SUV.

It should be declared that Purple is not about achieving some techno-utopian wet dream. As far as Purple is concerned, technology is simply a set of solutions to previous problems, if we have a problem and we find a workable solution to that problem, we then add that new solution to the existing list of solutions, and it becomes part of our technology. Purple is a set of principles and facts regarding where we are, and where we want to be as a species, it is also a pathway that can only be walked with technology, a future without technology is a future without humans.

The Purple objection to techno-utopianism is that it seems to assert that technology alone and of itself is sufficient, that solving a problem is all that needs to be done, and once the solution is found the solution must be implemented right away, regardless of prevailing political or economic concerns. This is the true myth of progress that technological developments equate to progress, they most certainly do not. Fortunately or unfortunately the real world is fuzzy. We see so very clearly that even though the world produces sufficient food, there is hunger, because hunger is not caused by lack of food, or by our inability to transport food to far flung places, but by ignorance, arrogance and dogma, and by gangs, and armed groups engaged in that age old quest to get the better of the other fellow, or the other tribe, or the other faith, or the other town on the far side of the hill. Human meanness, intolerance, racism and religious bigotry cannot be cured by technology, only by education, application of the rule of law, and time; perhaps even lots of time, multigenerational time. Political problems can only be solved by dialog, and by seeing the other side of the matter, they are rarely solved by technology, though in battle the superior technology invariably wins the day. The Islamic State bombarded from the air by all sides, and hunted by Special Forces armed with satellite reconnaissance and GPS technology able to deliver ordinance through a bedroom window, will eventually succumb to the pressure of technology, and loss of key personnel, but eventually someone will have to talk with them, and they will retreat from their extremist position in return. They are clearly amenable to western technology, and are heavy users of western technology, while they supposedly reject everything western, this rejection was obviously never meant to encompass military hardware, and modern western-developed means of communication, and the various little comforts and conveniences they are frequently seen to enjoy. Eventually a solution will be found when a very large amount of money, acceptable to all sides, will be thrown at the problem.

In 2016 humans inhabit two very different worlds, one world where technology rules, mostly the western democratic model, affluent, and comfortable; not perfect by any means, but the best we humans have ever had, and getting better by the day. Then there is the other world, the world outside the wall we in the West have built around our empire of technology. In this other world dictatorships and theocracies rule over vast numbers of people with the ever present threat of violence, and loss of life, where despots and lunatics are in charge of populations of largely ignorant people. This world is the one where most of the poverty resides, the extreme poverty, the dollar a day poverty that numbs the human spirit and allows no hope. People in these places cling to religion for hope, no one on earth can help, so perhaps heaven is a better bet, the last resort. There is more than enough money, food, technology and infrastructure in the world to call a halt to this dreadful situation, and everyone knows this to be true, technology of itself cannot do it. Only the political will to do it can achieve results. Of course technology can help, it can facilitate change, allow the organisation of mass movements, and to some extent educate people. The internet and communications can help, but these are only tools. The late Christopher Hitchens in his many debates and writings laboured the point again and again that we have the solution to poverty, we've known it for some considerable time, we discovered it after years of searching; it is simply the empowerment of women. And it works. Everywhere in the world where it has been tried, it works beautifully. Give women an education and control of their fertility, liberate them from the animal cycle of reproduction that has condemned them to enormous families and child mortality, give them some money, and perhaps a few seeds, and a few tools, and let them loose to get on with it. As Hitchens pointed out, everywhere it has been tried it works. But there is not a single religion that has ever advocated it, or has ever been in favour of it. Whenever there are objections to the liberation of women, it is unvaryingly the clerics who are behind it.

Hence it is by no means inevitable that poverty, repression, warlordism, and endemic corruption in such regions can be eliminated in any reasonable or specific timescale. It is probably and likely, to be the case that such regimes will find a way to endure, and religions will persist despite education, reasoned argument, and a lack of evidential support for the outrageous supernatural claims they make. So it is quite possible we are looking at generational time to resolve these issues, and so wars, feuds and skirmishes will continue, and resulting famine, and persisting poverty as a result of losing factions suffering discrimination, and being marginalised. It will probably continue to be vicious and bloody, and hard to look at on TV, and we will all feel it is a shame that such things happen on earth in this day and age. And it will continue and persist as an appalling vista until the people within the countries in question rise up and demand proper accountability of those in authority, and proper structures, and so on, and all we in the Developed World can do with our vast reserves of food, technology and money, is to stand by, and be ready to assist, as we were when the Berlin wall came down, and Eastern Europe threw off the yoke of communist extremism. Because poverty and hunger are big problems, requiring big solutions, which is why they can only be solved by organisations like UNICEF, or similar, or better. No amount of Christian charity, given with one hand while the other alternately blesses the hungry, and holds the bible, will get the job done. International efforts with fleets of planes are required. Organisations such as the Catholic Church cannot be trusted to do work of this nature. Fundamental to their faith are beliefs that *'suffering is a gift from god'*, hence inherently good, or that we live in this world to suffer so that we may have joy in the next; we are born sick and commanded to be well as Hitchens put it. Couple that with a belief in magic and you have the reason why their efforts thus far have been so fruitless; but like Facebook, they have a billion subscribers, they have their own nation state, and a property portfolio second to none. So to some extent there is a very real disconnect between what can be done by technology alone, and what will be done on the ground. The author and economist Jeffrey Sachs who says in his 2005 book, *The End of Poverty*, and through his involvement as a leader in implementing the U.N. Millennium Development Goals, regarding primary education, child mortality and poverty, that ending poverty is no longer a dream but a realistic goal. This is entirely correct, the technology to do it exists. The target for these goals was 2015, and it was missed. The goal Jeffrey Sachs has set for eliminating extreme poverty is 2025. Poverty is rife in the developing world, child mortality has been reduced, but remains at unacceptable levels, and so on. It is a harsh truth to say, but it is a bit like helping an alcoholic or a drug addict, they can only really be helped if they genuinely wish to be helped. Unfortunately the many countries where poverty is endemic are organised by the few, for the few, and they want that status quo to remain exactly as it is, regardless of what technology we have in the West or what Jeffrey Sachs might want to believe. We cannot be blamed for what theocrats, dictators or despots do in countries that are first order failed states. What chance of the goal of gender equity, or education for girls, even at primary level will be achieved across the Muslim world? What chance of disease eradication if the volunteers are deliberately targeted, and those vaccinated risk being killed?

Even within our wall of protection there are pockets of gross unreasonableness. We need look no further than Israel to see how religion endures and perpetuates itself regardless of development and technology, and how the moderate majority, as Sam Harris puts it, give cover and succour to the extremist elements. Again, as Hitchens noted on numerous occasions, everyone knows there is more than enough land in that region for a two-state solution, but no one can get it, and will never get it, because of claims of promises made by an omniscient real-estate broker to long dead peasants, and because the parties of god have a veto, the Jews in Israel and America, the Christians in America, the Muslims across the Middle East, all have, and will retain, their veto, and nothing will get done. And protagonists on all sides will drag it on, and justify killing each other's children over holy scrolls and scribbles, and traditions rooted in nonsense. And there are numerous other intractable ethnic and religious problems all over the world, many kept simmering below the surface with an uneasy peace won by bribing the parties with development money, or UN peacekeepers. So we must accept that advances in technology alone will not end human traditions, or long held resentments, or eliminate root causes, some of which have unfortunate consequences, and technology will not end the numerous wars that rage between angry factions across the world. Afghanistan for example, is unlikely to become a liberal western democracy any time soon, nor even the Afghan equivalent, it will continue to fester in tribalism, and wallow in magic and superstition until eventually a consensus emerges, in a generation perhaps as yet unborn, that the stagnation that has persisted for hundreds of years has not served them well. Similarly Northern Ireland will remain a bitterly divided and underdeveloped sectarian enclave within the EU kept from boiling over into chaos and bloodletting by the very generous largesse of the peace process sponsors. The best of its brains will continue turning their back on the place, and vote with their feet. No one believes they arrived at a solution to their differences by mutual consensus or out of Christian love or charity. They came to a price both sides agreed on. A price that was a damn sight less expensive than regular billion dollar explosions in British city centres, a price that was practical and pragmatic, and continues to be paid out as a peace dividend. There remain hard-line extremists on both sides, of course, and occasionally they lash out, but on the whole the lid has been kept on the problem, both sides have been paid to wear a professional bitch-face, and do business with each other, watched closely by their sponsors holding the purse strings, and hopefully at some future point the concept of being a Catholic or of being a Protestant will simply disappear from the world, and no one will give a damn what your grandfather did on Sunday. And so it seems to be an emerging best practice in global conflict resolution, to throw money at the problem, buy off both sides, keep the money flowing in, and hope that time, perhaps even multigenerational time, if necessary, finally allows the reasons for the conflict to fade from the collective memory, or to fade to irrelevance. Containment within acceptable financial parameters, it seems, solves a great many problems.

And so we must strengthen and renew our scepticism of progress, and pay particular attention to critics of techno-utopianism, we need to listen to Evgeny Morozov, Tim Flannery, Jaron Lanier, Martin Rees, John Gray and Raymond Tallis, and the many others who point out that all claims of the future, like claims of the supernatural are unsubstantiated. They are at best a hope of something that might become a reality, in other words wishful thinking. Although it is very clear, if we fail to close the information deficit preventing us from utilising the infinity of energy that comes from the sun we will eventually find ourselves in a dark place. But no matter what point in history you pick over the past three hundred years since the first development of a usable power source, things have progressed from that point. No matter what else was going on in the world, technology continued to advance, breakthroughs continued to be made, and the relentless plod of progress continued. Because technology is ultimately about individual people solving problems, because doing it better, or faster, or making something easier, makes money. The odds are that progress will continue regardless of all geopolitical concerns, the rate of change may not be as per Moore's law, but defining the problem set is the first step, and we've already taken that, we know what needs to be done and we're busy doing it, sooner or later we will be able to say we've done it. The glass is always half full.

Every day the world gets out of bed and goes to work, a million new things are known that were not known yesterday, new items catalogued, new tests carried out that were set up yesterday, things work a little better than they worked yesterday because of someone tweaking, the problem this way or that. A million snippets of information that will not need to be discovered tomorrow. Like the molecules of a gas, constantly pushing, exerting a pressure on the barrier, pushing it further and further back shining the light into a million dark corners that were dark or dimly lit only yesterday. The process is inexorable, unstoppable, until the final goal, whatever that may be, is reached.

Chapter 11   Getting The Big Stuff Done

Scientists alone can establish the objectives of their research, but society, in extending support to science, must take account of its own needs.

John F. Kennedy. National Academy of Sciences Address, 1963.

In order to get the human project into a Purple state there are a number of things we need to get done, some of these things are little things that may or may not happen naturally, such as attitude shifts, and ways of looking at things. As technology evolves towards its final state, whatever that may be, many things that were previously not possible become possible, and may well liberate people in some way, or change the world forever. We've seen it before many times in the last three hundred years since humans developed and directed the power of fire, and we are likely to see many more instances as we plod on towards wherever we might end up.

There are a series of big problems the human race faces which must be solved because solving them will lead us directly into the Purple era. In just the same way as the printing press allowed knowledge to be circulated more widely, and the steam engine led to the application of real power to the human brain, and attaching the one to the other allowed mass production of knowledge as never before. So the solving of these outstanding, significant problems, should reasonably lead us into an era of exuberant luxury and abundance. The big stuff that needs to be done includes, getting off of the planet permanently, and deploying the technology that allows us to do it routinely, and in sufficiently large numbers that we can effectively say the human race has spread from earth. We probably get this done by building several space elevators, colonising the Moon and Mars, and creating self-sustaining colonies on these bodies as well as in space, and perhaps even as far out as the asteroids, and beyond. Mining of asteroids, and on the Moon, and massive investment in harvesting solar power on the Moon and beaming it back to earth. Fostering the development of AI and its application to nanotechnology, biology and chemistry, and technology in general to achieve complete automatic manufacture of all goods, and delivery of services used by humans. Development of the necessary political, economic and legal systems that allow for the massive amounts of funding required to be borrowed against our infinite future wealth, and to permit the benefits to be shared out. All we're asking is that everyone on the planet should be made as rich as possible, nothing more. The big stuff is probably a one hundred year fix, a plan for the next few centuries, items that can only be done by all nations acting together, the human race doing really big stuff as a species, as a single human project.

A cathedral is not just a heap of stones. Prior to construction it is, after its destruction it might be. But on assembly it is not, it is more than the stones, more than the sum of the parts. What must be added to the stones to create a cathedral is expertise, knowledge, information, energy, application of force, and of course time. The great cathedrals of Europe built in the Middle Ages were the space projects and the energy projects of their day. They represent attempts to get big stuff done in the name of the species. And they were truly international undertakings that lasted many generations, the architects and masons, artists and metal workers moved effortlessly between projects and were drawn from everywhere.

There is a well-known leverage effect in technology, tiny changes can have far-reaching effects. The first ruby laser led to the solid state laser, which allowed barcodes to be used along with databases, and transformed the movement of goods. Advances in computers allowed massive corporate growth, huge companies could not have existed in the pre technology world; the numbers of people needed would have produced uncontrollable complexity and chaos. There are a million possible future leverage events, and they will likely give us very advanced technology by our standards today, including meat without animals, grain without land, and a host of other seemingly magical things. Arthur C. Clarke's third law might be invoked here, that sufficiently advanced technology would seem to us to be indistinguishable from magic. Our models must come from nature, imitating bacteria, fungi, plants, flowers, growth mechanisms in trees, not the least bit magical. Today we are well versed in science, we know that magic is not part of our lexicon, we are not easily deceived, and as Richard Feynman pointed out Nature cannot be fooled. Whatever our descendants build in the future, we would recognise it for what it is, technology. We have the capacity to understand that it must be made of the 92 elements, and it must conform to the laws of physics.

The big problems can only be solved by a big effort from the humans as a whole. Not by some small branch of the species or by some group of individuals operating for their own personal gain. That's why they are big problems. Everybody knows what the big problems are, and everyone knows what big stuff needs to be done. The problem is getting enough technology, effort and money committed to these problems to get them solved. For anyone who doesn't know what the big problems are, they are exactly the same ones we had last year, and will have next year, until we sort them out. Of course the energy problem, and the carbon dioxide problem are a given. Solve one solve the other, and probably very soon indeed. There are also the housekeeping problems, global water supply, pretty much solved, and global food supply, already solved. Global energy grid, which is being actively worked on, and an international communications network that means everywhere on the planet can communicate in every conceivable form with everywhere else on the planet, and at the fastest speed allowed for by the best technology; this is on the way, and already exists and works just fine in developed countries. There are also the aspirational problems to solve. Finding other life elsewhere in the Solar System has to be one of the most important. It is not just a scientific quest, it is so much more important, because it defines our place in the universe. The origin of the universe, is another great aspirational issue, and is worked on with great zeal by particle physicists, cosmologists and theoretical physicists throughout the world, it is a problem in good, and very capable hands.

Building the definitive catalogue of all living things on earth, is another of those ongoing, got to get it done, projects that is in the hands of passionate people the world over. Also well looked after is the task of protection of what we have inherited from the past, and everything we have built here in the present. It's a job for every one of us. Call them environmental concerns if you like, but there's nothing wrong with wanting to live in a nice place, and being house proud along with it. We will be judged on our performance in this regard by those who follow us, our children, and all future generations will want to inspect carefully maintained records of what we did, they won't settle for excuses. We also need to be concerned about the survival of the human species, but alas, the few things that really could wipe the planet clean could never be stopped, their origins lie out in the cosmos, long ago and far away. The only consolation we have is that the probabilities are so low as to not require our consideration. We have wonderful groups of intelligent, educated, motivated people all over the world working on all of these problems, doing what they love to do, and doing it for us, because most of them are paid from our taxes. These people would be greatly assisted by a global republic with a proper human centred republican constitution, specifically established to guide us through the current transition phase, and specifically charged with achieving these goals for the species as a whole. In many respects this organisation already exists, because all branches of knowledge are truly international.

The biggest big stuff involves doing all those things we need to do to unlock the infinities of space, time, energy and matter. Progress is not an event it's a process, a continuum of states between two measuring points. The fine structure is often not visible until after the progress has been made. Stark differences are to be seen, pre and post the transition period, for example, pre-steam engine, post-steam engine, pre- microchip, post-microchip, and pre-plastics, post-plastics. We are also likely to see the striking differences between the pre-nanotechnology, and the post-nanotechnology world. As well as the contrast between, pre and post the space elevator, and we ought to certainly notice the vast difference between the pre-solar energy world, and a world which has unlocked the infinity of energy available to us from our own personal fusion driven star.

Modern humans have two related and particular problems we are trying to solve, with which we have been so far unsuccessful, that is, humans have been at work on these problems with some moderate progress reported but not to the extent of our fervent hopes. The first of these intractable matters is the, *'getting off this rock'*, problem, space travel and permanent habitation, living, working and exploiting the infinite resources of our solar system. To date, despite having landed a man on the moon in 1969, we have really only managed to put a drop in the bucket, compared that is, to what we would like to do as a species. Since 1972 we have effectively confined ourselves to low earth orbit, which of itself indicates the sheer magnitude of the problem we face. Our ambitions, and it is no secret, are to colonise the Moon, Mars, probably eventually, Venus and the asteroids, build colonies in space, harvest energy and mineral resources, build enormous scientific instruments to study the cosmos, exhaust the search for life, both within and beyond our solar system, and whatever else we decide to do thereafter. The second problem is that of extending the life of the individual human being, so that we can savour the great things our species has achieved, and get to spend some of our great wealth. And we have had some success, disease control, antibiotics, pain management, anaesthetics, and diagnostic medicine have all contributed to pushing up the average life expectancy in the Developed World from 35 years, before World War II to more than double that number today. But we're nowhere near what we want, or even what we demand, given the amount of money we are spending.

If we consider all problems to be soluble, and we do. Then both life extension and space are going to be unravelled eventually, and if so, it might appear to be of concern which problem is solved first. If the life extension problem were to be solved, and the space problem remained unsolved, then we might suffer from uncontrollable population growth. There might be only standing room on earth within a few generations if we abolished human death in the morning. But if we solve the space problem first then life extension presents no problem whatsoever. We simply expand into the infinity of space. Of course we know full well that's not how it works, as with all things human and technological, it is much more complex. We have a vast number of very intelligent people and groups of people, in the form of academic institutions and companies, at work on every front of human endeavour, the frontier is pushed back steadily; sometimes there is a breakthrough mostly we just plod forward in small increments. Energy is a case in point, we are in a transition period, pre-oil to post-oil, eventually we will get there, and probably very soon, and when we do, the world will have changed forever.

There is no energy crisis, there never has been one, and more than likely, there never will be one. We have all the energy we will ever need, we have an infinite source of energy in the Sun, and it will provide us with all the energy we will ever need for billions of years to come, no matter what we want to do or how much we want to use. The Sun is the key to unlocking the infinity of energy. Of course, if it was as easy as that we'd have done it years ago. It's not easy, and we haven't done it yet. Nor do we know when we will ever get it done, but we really have very little choice in the matter, it must be done or the entire human project collapses into a dystopian mess, complete with burned out cities, and gangs of destitute people killing each other for scraps. At least that's what we are told. The vision won't present itself today or tomorrow, perhaps within the next hundred or two hundred years if we don't act and the oil runs out.

But of course we will act, we have many alternatives, and are busy deploying them to get us through the transition period between now, and that future time when we have either fully unlocked the infinity of energy available from the Sun, or run out of fossil fuels. Right now we rely utterly on burning coal and oil to generate the energy we need. There are other forms too, hydroelectric, nuclear fission, wind farms, wave power, geothermal and other renewables such as biomass, and they all help as do energy efficiency measures. However, when all the alternatives are added up we remain completely dependent on burning yesterday's sunshine, and will do for the foreseeable future. There is also the great hope of nuclear fusion. Some say it will work, the laser plasma approach, or the tokomak, but as yet there are no working fusion plants. The physics is very interesting, and we should continue doing the research, but doing fusion on a small scale may simply not be feasible. The Sun has already done it. We know why the sun works, and how it works. So we already have a working fusion plant, and it's very efficient, and requires no maintenance whatsoever, our energy future as a species obviously lies in harnessing the output of our own personal star. The physics of fusion are explained in a compelling way by Michio Kaku in his book, *Physics of the Future: How Science Will Shape Human Destiny And Our Daily Lives By The Year 2100*. There are a bewildering number of books on every aspect of energy, as you might expect, it being so important to the entire human project, and all of them speculate about the future of energy, and where we will get the vast amounts everyone agrees our future society will need. The consensus view is that, in the absence of a viable fusion technology, it must and will come from the Sun, there is simply no alternative.

There are many ways of harnessing solar power, from a simple rooftop panel that heats water or collects solar generated electricity, to vast solar photovoltaic farms, or solar concentrator power plants. The details of the various methods can be found in the myriad of books on the subject, for example, the book edited by Jeremy Leggett, *The Solar Century: The Past, Present And World-Changing Future Of Solar Energy*, which in common with most of the literature on the subject stress the point that we are making a start on harnessing this infinite power source. But there are problems of efficiency, of cost, and of reliability, as was said before if it were easy we would have done it long ago. Besides the poor efficiency of solar cells and the cost of the generated power, the one thing that holds back solar power is the ability to store the energy, *'the Sun does not shine at night'*. Storage of the power becomes an issue because the only storage devices we have are batteries, or inefficient conversion. There's an old saying in electronics labs that you can't shrink the power supply unit. Modern electronics can be small and delicate, and do wonderful things, and is often able to do so on low power levels. But in general, if you have requirements for power, then the power supply units are frequently big heavy ugly things that produce lots of heat. In the same way that you can't shrink power supply units, batteries are also difficult to shrink. Even lithium based batteries are bulky, expensive, slow to charge and leaky. Battery technology has improved somewhat, but still comes up against that brick wall called the laws of physics, because the battery is electro-chemical in nature, the limits of low power density and a short lifetime are going to remain. In the *Terminator* movies, batteries are small, light and nuclear, and can power a fully grown terminator for hundreds of years, and when damaged frequently result in a small nuclear explosion. Such technology will probably remain in the realm of science fiction. In the real world battery technology leaves a lot to be desired, and if you want real power applications, such as walking robots that can actually do useful stuff, battery limitations apply. Of course the problem is being worked on. One of the most promising is the Liquid Metal Battery LMB, these devices are in development and have the theoretical capacity to store vast amounts of power. An LMB the size of a freezer might be able to store enough power to run a family home, one the size of a shipping container might power a workshop or small business, and one the size of a warehouse might power a neighbourhood. And the cost per kilowatt hour are reasonable. If such a batteries prove to be a sufficiently robust technology then solar power generation becomes a very practical proposition. Every home could be powered from a single battery. Office buildings from banks of such batteries. Millions of solar panels and low power wind turbines charging millions of such batteries across the globe. The implications for the world would be far reaching and obvious.

Implicitly linked with collecting solar power and its storage, is the very serious problem of what we need to do with the ashes of our fossil fuel past. What to do with CO2, is the problem of our time. We have no problem turning it into things like wood, grass, straw and leaf mould. Plants have been dealing with nature's greatest and most stable waste product for billions of years. Now however, we humans have decided that we are producing too much of it, and the plants are not in a position to deal with the excess, we've decided it's a problem we need to solve in a more efficient technological manner. Plants are not efficient enough for us, they use CO2 to build all of the structures they need, almost every single molecule they require to make is made using CO2 as one of the ingredients; and we have learned, and understand how they do it, and we think we can do it better. We want to be more specific, we want to take the CO2 molecule, add some solar energy and cleave the molecule into its components, CO and O. The details are not important, what is important is that the energy is generated from sunshine, and the CO2 is taken out of the air. We also find a good deal of H2O in the air and when combined with the CO we can produce very useful things indeed. In fact, just carbon, hydrogen and oxygen account for nearly everything we eat, all of our fuel and most of what we wear. CO2 is therefore one of the most important raw materials humans have, and that's why the problem of extracting it from the atmosphere in vast quantities is one we will inevitably solve. The problem is however, not without its challenges, as the fact that there is a $20m X-Prize on offer to someone or some company who can, in the words of Peter Diamandis, and the X-Prize slogan, '*Make CO2 something new*'.

Eventually and inevitably the solution to the CO2 problem will be reduced to a black box in the garage. It will become something every future home will probably have, a box perhaps not unlike a standard freezer that just sits there in the garage, or in an basement, or out building, and hums away to itself. But inside the solid state workings of the machine will be a wonder to behold, it draws CO2, methane, nitrogen and water vapour out of the atmosphere, and processes them into fuel for your family transport requirements, and for heating or cooling your family home, or for whatever other requirement you may have. It draws all the energy it needs from the inexpensive, super-efficient, solar panels that completely cover the roof of your house, and which probably surpass today's fifteen percent efficiency by a factor of three. Like all such machines it has a suite of features and is versatile, probably has intelligence of some sort, will, of course, be connected to the internet, and will most likely cost less than $1,000 in today's money. Be very certain that such technology is on the way. It may well be closer than anyone thinks. Certainly the problems in the underlying theoretical chemistry, physics, and so on, are very well understood. In the same way that the theoretical underpinnings of refrigeration and heat engines were well known and understood, long before it was possible to purchase a refrigerator. So too the theoretical and mechanical understanding of the internal combustion engine were well known long before you could purchase a car. This black box is in the same sort of pre-release place. It can be imagined, but you can't quite purchase one yet. But just like the refrigerator and the automobile engine, it may well become an indispensable part of every home. Of course there will be vastly larger industrial manifestations of the same chemical principals, all busy disposing of the planets CO2 excess. If this comes to pass it means we can keep our much loved internal combustion engines, and continue to guzzle gas to our hearts content.

The really big thing in getting to our inevitable solar future is to fully harness the energy of the Sun in space, specifically to build vast arrays of solar panels on the lunar surface and beam the energy back to earth to power a global smart grid built out across the entire planet. The solar panels could be manufactured on the lunar surface which has a perfect vacuum, the entire process might be carried out by intelligent manufacturing, capable of doing everything under human supervision back on earth. Or it might be one of the main commercial reasons for going to the Moon and building a lunar base and maintaining a permanent human presence. This is clearly a multiyear, multi trillion dollar undertaking, and will require an unprecedented level of international cooperation. It's one of those species defining projects for everyone on earth. It's a project fitted to a global republic, a project to be carried on without the meddling and politicking of local, or even national officials.

If Nature can do it, so can we, is the key to unlocking the infinity of matter. But since we are fundamentally a part of nature, if we are doing it, then Nature is too. There is a very long list of inventions which humans claim as their work but which in fact have their origin in the natural world, there is no copyright infringement in copying nature. Nature got there first in so many fields of invention, you name it Nature invented it or showed us the way. The spiral spring has its origin in climbing plants such as the common Virginia creeper. DNA is of course the master spiral, and of course, shells of all kinds. The sphere, which we have come to rely on in the form of the ball bearing, and a million other incarnations is common in nature, fruit and seeds, ball and socket joints, complete with lubrication. The wheel, it could be argued is a derivation of the sphere, or a section through a tree, it was invented as part of the machinery of living cells to cause the flagellum of sperm and bacteria to rotate, generating forward motion of the cell. Nanotechnology takes its inspiration from the machinery of biological cells. Glass and Mirrors are found naturally in volcanic deposits, and naturally occurring crystals. Wings are all across nature, and all the way back to the earliest land animals, humans have elevated the wing to the crazy extremes of powered flight. We also get more mundane inventions which are direct plagiarisms. Consider Velcro, discovered or invented, so the story goes, by an inventor whose dog picked up the seeds of a common plant. Barbed wire probably owes its inspiration to the rose bush, while the humble mosquito has been using the hypodermic needle to extract our blood before we were even human. In the realm of chemicals, Nature is the complete master from the simplest of sugars to the most complex of proteins, Nature does it all. Fungi produce chemicals to kill bacteria, Penicillin being the most famous, plants make poisons to kill other plants, and animals, animals make poisons to kill other animals. The insects who have been around longer than any other animals make poisons and chemical messengers including some of the most interesting chemicals we humans have ever come across. From bacteria and fungi making our drugs, to bees making our honey, to worms making our silk, to birds making our eggs, to cows making our dairy products we have harnessed almost every aspect of the biosphere to the needs of the human project. The list is a very long one, and goes on and on.

We have also translated the inventions of Nature into other materials, stone, iron and plastics, to the extent that the history of our technology is one of translation and copying from one form to another. And we are not finished yet. Which is why technology has a very long way to go. Which is why our mastery of atoms is so very important. Because what we have done up until now is nothing compared to what we would be able to do if we can build everything from the ground up, atom by atom, as happens in nature. Because we see it done, we know it can be done, it's only a matter of time before we do it, and probably only a short time after that when we do it better.

If we can ever get to the stage where we can build things atom by atom then we can build anything of any size and any complexity, and build it to perfection. Plants and animals construct things from atoms all the time, and do so very quickly. And we know this be the case because we can see things grow, a flower overnight, an entire oak tree in one hundred years, or a human baby in nine months. Everything is made of atoms, and everything that is made of atoms is made of a finite number of atoms. And if the number is finite then it can be counted. And it can therefore be stored in a database, so that every object from a toothbrush to an aircraft engine can be quantified and in theory at least, grown layer by layer, plane of atoms by plane of atoms, bit by bit, atom by atom. This is the underlying principle of 3D Printing. Perhaps modularised construction may also occur, for example, a perfect flawless diamond may be made up from perfect *nano-diamonds* or small micro *diamond-oids*, manipulated by micro machines. We may usher in an age of diamond, when the material may become so common that we might use it as builder's sand, or in larger chunks as aggregate in concrete, or as components in our machines such as ball bearings or engine parts. After all diamond is only carbon, and if we can manipulate atoms to make anything then diamond becomes possible. Perhaps another use for all of that CO2 floating about in the atmosphere. In his book, *Radical Abundance: How a Revolution in Nanotechnology Will Change Civilization*, Eric Drexler, outlines the future of Atomically Precise Manufacturing (APM), in effect if we can imagine it, and build it then the sky is the limit, you just select whatever you want, and it gets printed at a place nearby, you pick it up at your convenience, or why not have the printer in your own home, silently printing away in the basement while you relax by the pool.

Nanotechnology also gives us the possibility, along with synthetic biology, of creating entities resembling life forms, mites, fleas, termites and ants being the most obvious types of entity we might wish to build. We could also create synthetic spores and seeds which we might send to asteroids or comets, these would then grow, and mimic real fungi and plants, taking root on the comet, or perhaps infesting it, as a fungus might do with a slice of bread, or overgrowing it like a brier. We could in effect grow our way out of trouble, seed technology implies small payloads potentially generating huge entities in space a kind of acorn project in which a seed takes root, and grows into something we can use, like a habitat or a space station. After all, when we examine the intricacies of some plants and flowers, with very elegant and sophisticated components, even including moving parts, we should in theory be able to code for such expression in the artificial structures we might build from seeds in space. We might also model our seed technology on the ants, building and mining, grain by grain. The combination of nanotechnology and micro insect like semi-intelligent, entities results in a synthetic machine super-organism working like an ant or termite colony to build a pre designed structure. When the task is done, the machine can be switched off. And if such work can be done in a high radiation, toxic, vacuum or high pressure environments then we open the prospect of being able to build anything, anywhere in the Solar System. Including places like the atmosphere of Venus, or the moons of Jupiter and Saturn. With the obvious exceptions, technology knows no limits.

Nanotechnology is not here yet but there are whispers. The whispers grow louder and every year our understanding deepens. The whispers will intensify, evolve, and transmute into a background noise that will become pervasive. This technology is not without obvious and potentially serious dangers, the grey goo scenario being the catch all phrase used to describe potential catastrophes at such a small scale. It's probably easier to seek forgiveness after making a mess than to get permission to do something in the first place. Hence we may well get our share of accidents, perhaps even disasters. It was the same with high-pressure steam and the early railways. So what's new?

Biology is really only organised chemistry, given an atmosphere such as that of a planet, and an energy source chemistry naturally occurs, the atoms have no choice other than to move about the place, depending on temperature, combine and recombine in the turbulent atmosphere. Chemistry occurs everywhere, even in the near absolute coldness of interstellar space. But when the chemistry becomes so specialised that it moves behind a wall of lipid molecules then we get simplicity becoming complexity, and chemistry somehow morphs into life. The exact origin of life is still a mystery, but one that is being chipped away bit by bit with each passing year. Biology at the sub cellular level is the complexity of organised chemistry. Organic chemistry, the chemistry of compounds containing the element carbon, proceeds in an organised manner within cells, and between cells in living things to achieve wonderful emergent properties such as life itself. Understanding biology in all its complexity leads to a better understanding of chemistry, and to the complexity and possibilities of engineering at the scale of the individual atom. Given a complete understanding of these fields, and the ability to manipulate atoms and molecules at will, there is almost nothing that cannot be achieved.

Synthetic biology is a field of biology which seeks to use the principles of biological construction to manipulate synthetic DNA to obtain useful results, the DNA may be from many different sources, or specifically designed. Biology is really only chemistry on a grander scale, a more sophisticated chemistry. Claims that the new biology is the scientist playing god, or the creation of new species are very wide of the mark. Farmers, horse and dog breeders, and plant breeders have been producing new animal and plant varieties for as long as we have been bipedal, and presumably first captured young orphan animals. If a scientist or a company makes, what at first glance seems to be a new species, wholly new and capable of independent movement, and so on. It is likely to be classified simply as a sophisticated machine rather than a new life form. There is no reason why the components of our future machines should not resemble or even ape the very elegant structures found throughout the natural environment.

A new era of biology will impact right across the human project, perhaps no need for arable land, it will be much more cost effective to produce everything we eat by moving agriculture indoors, upwards and even underground. Farmland can then be given over to other uses, a place for wild things, and for the pleasures of humans. We can all live in a fairway home, or a parkland setting. Anyone who doubts that farming will become vertical need only type the phrase *vertical farming* into a search engine and look at the pictures. The promised efficiencies of new era biology will dwarf those of conventional farming.

This ability to grow other things using the principles of biology will usher in a new era of advanced production, humans will simply not be able to produce the goods, and molecular self-assembly will completely bypass the human hand. Things will grow like fruit, nuts and flowers, self-assembled from instructions evolved within machines. Humans will be nowhere to be seen. Plants re-engineered to become machines, their cells as machines, new forms of animal as machines. Leaving ethics to one side for a moment, human embryos might even be seen as machines, created in a synthetic womb. Cloned human embryos by the billion, used for organs, or blood, it might even become possible to do it in a home cloning kit. There's nothing intrinsically difficult in it once the process is evolved. Those of a religious bent will be furious, the question naturally arises; where then the human soul? As was said before, the burden of proof lies with those making the claim, and belief is not evidence. A mastery of biology as never before, because of increased understanding, brings with it the possibility to bypass the natural evolutionary processes. As in the movie *Blade Runner*, we might get *'genetically engineered humanoid replicants'*, or the *'basic pleasure model'*. Once the door is prised open anything is possible. If Nature can do it, we can too, and we will have the time to enjoy it. The moral aspects will, of course, need to evolve with the technology.

There is an infinity of time stretching on into the future, and humans will be on earth as long as we care to be, unless destroyed by some unforeseen event, in other words, the species endures into an infinite future. Individuals are currently limited by nature to see only a small portion of the infinite future, death is the greatest of all inevitabilities. However, some writers and commentators believe that when we get the better of our biology, we can live indefinite life spans thus allowing individuals to truly unlock the infinity of time. A complete understanding of human biology must inevitably lead to the ability to deal with the aging process. A first step on that path may well be achieving some form of suspended animation for humans. Suspended animation holds out the prospect of becoming part of what some commentators call a bridge to a bridge to a bridge. Suspended animation, if achievable in the human species, would provide a kind of hibernation that would allow the body to be suspended, and remain suspended until such time as the person was once more revived. Suspended animation, if it was easy and quick to do with no harmful side effects, offers the prospect that a person reaching their sixties might opt to spend only three months of each year awake, the remainder of the time they might be suspended. This might lead to a kind of, end of wisdom scenario, when all the old timers take to the caves, locked in dark hospital buildings, in storage emerging only for special occasions such as birthdays, or Christmas, or important football games, or board meetings. Suspended animation might just provide that bridge, to a bridge, to a bridge, and be the stepping stone to human immortality.

Actual immortality is not probably going to become a reality using the human body alone, the human brain is seen by many futurists, and life extension people, as simply a sophisticated form of computer. The Singularity people certainly believe that our future immortal selves will not be embodied as we are now but rather will be interfaced or downloaded into software of some kind to live our immortal lives in virtual environments. It is of course enjoyable to speculate about this possibility, but as has been said in other parts of this book, the facts of our current biology and computer science indicate that we are a long way away from the actuality. But we know as a relative certainty that our technological society will endure well into the future, so it must therefore be reasonable to presume that at some time in that future, there will be people who will be born with absolutely no expiration date. They will never age past 25 and will effectively live forever with the contents of their brain backed up just in case of accident. Eventually they will populate the earth and the entire Solar System, and probably travel to the stars.

There are numerous books on the future of human space travel and exploration, see for example, *Space Enterprise: Living and Working Offworld in the 21st Century*, by Robert Harris, and many similar books, there are also documentaries, websites, and so on. The one thing all of these media do agree on is that humans, as a species, do have a future in space. In fact, it would be fair to say that many authors are of the view that we have no future unless we learn to live and work in space. Space is one of the infinities we need to unlock, it provides the ultimate human building site, the ultimate place to put our scientific instruments needed to answer a great many urgent cosmological questions, and space is the ultimate source of infinite matter, energy and living space for humans. Getting into space is not really a big problem, newer and larger rockets are on the way, like the Orion project, and we have a host of other private enterprise concerns such as

SpaceX and Virgin, all attempting to get people and payloads into space. Effectively we are creating a utilitarian rocket engine a rocket engine that can lift small payloads into orbit for very little money. Many such small payloads can add up to a massive payload. But like a lot of things we do, it's not nearly as easy as we think it is. Space is the most hostile environment there is, the lethal combination of radiation and vacuum is hazardous to humans we can only exist there by bringing our environment with us, every drop of water needs to be brought with us. A space elevator is also something which would help enormously if it could be built, it's one of those things that seems fine in theory but in practice is another thing altogether. The so called stairway to heaven can be built in principle but we don't have the materials to do it, it is effectively a bit of string hanging from a weight in orbit, and used to climb from the ground into space without using a rocket. Right now though, it remains on the drawing board, like the antigravity device. We do not know how to switch off gravity so that a lead block can just hang in the air like they do in the realms of science fiction. There is no known electromagnetic connection to allow gravity to be switched off.

All human space activities are currently confined to low earth orbit. The International Space Station cost over $150 billion, and is our only permanently inhabited place beyond the atmosphere of earth. We are every bit like kids playing in the paddling pool, we're learning, getting the feel of the water, before too long we will want to venture further and deeper; up to our necks. We're learning valuable lessons, and as a project in international collaboration it seems to work well, the ISS and CERN are probably the best examples we have. But we need to venture further. We have big ambitions, every bit as big as our imaginations, and our technology is very much up to the task. Some of these projects are mentioned in what follows, much more detail is available elsewhere. As a first move we need to get back to the Moon and build a moon base, dig deep into the Moon, there are deep tunnels and caverns, and old lava tubes. There is water in the bottom of some creators. Everything we need for a self-sustaining colony is on the Moon. So we need to get to the Moon using the old established methods of getting off this rock, but once we do, and once we are firmly established with a permanent presence on our nearest celestial neighbour, we need to establish if there is life elsewhere. We need to know definitively if there has ever been life on Mars that means we need to make even more sophisticated robots, and send them to the Martian surface to dig a few really good holes. We also need to get very good probes to the moons of Jupiter and Saturn to establish if life is or has ever been present there in any form whatsoever. The results are important either way. If there was another genesis in our Solar System then life must be common throughout the galaxy, if not then the fact that we may be alone is something we will need to deal with on many levels. Protecting the life that is here is also of vital importance, which is why we need a definitive catalogue of near earth objects, and earth crossing asteroids; this is being done but needs more money and more technology thrown at it. Eventually we need to set about capturing these objects just in case. Active capturing of asteroids and comets for use by human space colonies is technology we don't have, no doubt, given the resources contained in asteroids and comets it will not be long before we fully develop it.

Such technology will inevitably lead to the establishment of self-sufficient, self-sustaining colonies containing hundreds of families, eventually leading to monster off-world colonies containing millions of people. They will be built from captured asteroids or comets, and other stuff found in space, especially if they can be built by self-replicating machines guided by artificial intelligence. It's not inconceivable that a good chunk of the human population might choose to live there in future. Mark Twain advised the purchase of land, because it was a finite resource, and no one was making any more of it. If no one else is going to make it then it'll have to be made by us, using our technology and the skills we have. In space there is no limit to the amount of land we can make, especially if our machines do it for us.

Colonising Mars is the goal of much space research, it is the most likely other planet for humans to colonise. Probably by colonising the Martian moons as a first step. A robot on the surface could be controlled in real time, it could dig a hole in real time as deep as necessary, with a real pick and shovel, and really get its hands dirty, so to speak. Mars has all the land we could ever need, of course we need to make it habitable, which is where our nanotechnology and synthetic biology comes into play. We can design self-replicating robots to do everything necessary long before any humans establish a colony. Enormous mirrors in space might also be constructed to increase the amount of light falling on Mars, and melt the massive glacial reserves of water, and the polar caps, and generally reheat the planet to something approaching a habitable temperature. Landing people on Mars at this point in time is largely a symbolic exercise, it is a very important exercise but largely symbolic. There is nothing that needs to be done on Mars that cannot be done by machines, and the machines get better every time we send them. Every five years sees an increase in ability and capacity that was unheard of ten years before. Eventually we will be able to send semi intelligent autonomous robots that can make decisions in real time, and know what they find when they find something. We may be able to send AI geologists every bit as competent as humans in the next ten years. Humans will eventually need to go, but only to land so that we can say, humans have been to Mars, the time taken to get there, and the time taken to get back is such that the trip is considered beyond current capabilities. Why send a human with all that entails when an intelligent geologist or biologist the size of a beer can could do a lot more, and cost a million times less. Eventually humans will land and stay for a while, but probably not until robots have been there for a while longer, and built a proper five star habitat fit for humans. Terraforming Venus is also something we ought to consider doing just as soon as we have the technology necessary. Venus is far too hot, the atmosphere is all wrong, it doesn't spin right, and it has no moon to stabilise it. But apart from all that, it's fine, so can Venus ever be made habitable for people? The answer is a resounding yes. Venus is the same size as earth and if it had the same atmosphere and the same amount of water, it would probably be like a tropical version of earth. But it is probably one of those very long term multiyear projects, but once again, the principles of synthetic biology, and self-replicating nanotechnology may prove to be the way to go. Certainly there is nothing to stop us from building structures in orbit around Venus, why not have enormous colonies orbiting our closest planetary neighbour.

Just as on earth, AI in space will make a difference, if the entire wealth generation process of the human species is to be automated, then it stands to reason that vast numbers of semi-intelligent autonomous agents will operate in space doing all those hazardous things humans don't wish to do. Just as on earth robots will operate to human requirements to build whatever we wish to build, including bigger and better optical, infrared, and X-ray telescopes, probably on the far side of the Moon where it is nice and quiet. There are questions that need to be resolved concerning the limits of relativity, questions of gravity, quantum theory and the more obscure concepts of cosmology. Machines will also probably build large space based telescopes, to look at other stars, and their planets, and continue the search for other earths, and places we might someday wish to visit, perhaps they might even discover that other smart beings are out there watching us, we might get a cosmic wink. Whether SETI produces a positive or negative result does not really matter. We need to take it seriously, if intelligent ET is out there we need to know something about them, if there's no one out there then we own it, every bit of it.

Along with the rollout of AI in space it is very reasonable to state that we are entering a new era of micro-space, when very small spacecraft will be able to do enormous amounts of work. Very little rocket power is required to get a pencil into space or into orbit. Or a bacterium. Or a virus. Or a fungal spore. Or a grass seed. Or the synthetic biology equivalent. And if it crash lands on an asteroid it might contain sufficient information to bootstrap an entire human habitat, or mining operation. We will therefore probably not need to send rockets into space with multi-tonne payloads, instead the complexity and information content needed might be contained in a craft the size of an acorn or a beer can. Concepts such as the cocooning of comets, the taming of potential asteroid threats might be handled in this way, in the same way as fungal spores demolish an apple or a slice of bread. This and other technology is so visible from what we can do today as to make its rollout inevitable. From our perch on the shoulders of those giants of yesterday, we can see so very much further than we ever have before. It is also abundantly clear that the time is ripe for the establishment of a super national super space agency to oversee all space projects civil and military. The International Space Station is a good example of this inter human cooperation. Space is too big and too important a set of problems to be left in the hands of narrow sectional or national interests, which is why it needs to be under the control of the global republic. Space law needs to be properly formulated, and will also eventually need to be enforced.

It should be, and needs to be pointed out here, that everything relating to our future in space was discussed at great length back in the early eighties when the space shuttle was introduced, and was seen by many as the future of space travel. Ronald Regan's commitment to building a space station was also seen as heralding this new era, and renewed mastery of the high frontier. Numerous books articles and TV documentaries were produced to echo these triumphal notions, and predictions were made that the Moon base would be up and running by the turn of the century, it would look and probably feel as the Space Station and Moon Base portrayed in the Stanley Kubrick's movie *2001 A Space Odyssey*, and would provide a base from which humans would seek to colonise Mars and mine the asteroids, and so on. But it never happened and those bold predictors of 1984, if still around thirty years later, are left wondering and hoping whether it will ever happen. But the reasons for big space projects back in the eighties were never primarily exploratory, there was a great deal of cold war politics involved, and everyone knows that now. Bankrupting the Soviet regime by committing their resources to grandiose schemes, and vanity space projects they could not afford was integral to the cold war. For that reason alone it is perhaps fair to say that we have served our time in the paddling pool, and it is now time to leave the shallows, and immerse ourselves up to our necks. This time it will be different.

All the big stuff described in this chapter is bigger than any one human, or group, whether corporate or government. And getting these and other things done is a job for an organisation bigger than any single government. When the Spanish people wanted to defeat fascism in the 1930s they formed the International Brigades, and willing volunteers from all over the world flocked to Spain to defeat the evil that had risen. When the United States and Britain wanted to defeat Hitler and Japan, it was an international effort including the communists in Russia, and everyone in Asia, two million Indian servicemen, gave their lives. They wanted their freedom in return, of course, there was a price to be paid. But the job was done. Some things are simply too big to be done alone. The nature of future problems means that it must be taken out of the hands of narrow sectional interests, or western dominated elites. UNICEF and WHO are perhaps the best examples of a multinational organisation that get the job done.

The internet is also a stunning example of human activity beyond the control of any one person, company or government. The future world will be a world of volunteers, the open source movement is an example of such a force in the world, mostly in software, groups of people giving of their time and expertise for the good of the project, to give something back. The Wikipedia online encyclopaedia is the product of such a collaboration. We also have open source movements for many human problems, for carbon dioxide, and for battery technology, and biofuel made by bacteria or green algae. We live in a, *what does the crowd think*, world. There are branches of the open source movement concerned with the future of food and water, and concerned with helping people in developing countries accelerate their development.

The wants and needs of seven billion people can and will be met by technological developments evolved by humans and intelligent machines. Nanotechnology and synthetic biology will allow the manufacture of whatever can be imagined, and in such abundance that what we now consider to be rare and precious will appear ordinary. Unlocking the infinities of nature that surround us, and applying our great collective intellect to the problems we face will create infinite amounts of real tangible wealth. The question to be asked is what prevents such grand large-scale species defining projects from happening? One answer is the inherent mistrust between the many and various governments of the world, which is why governments of the world ought to be relieved of the very large human species projects. There remains across many regions of the world an ingrained religiosity and nationalism that permeates and punctuates the social and civic structures of many nations, until such small minded nonsense is confined to the background, much of the big stuff will remain undone. The infinities are only available if we take the trouble to tap into them. Another answer is to say there is no money to get such grand schemes done, they will have to wait until better times come around. We can decide that we wish to pay for it by voting to pay for it. For example, everyone in America agrees the roads and bridges are in a state of disrepair, and everyone agrees that something needs to be done. But successive politicians have convinced the voting public that they have been overtaxed by the previous administration, and that only by voting differently will they get lower taxes. Yet in America, drivers pay about $3 for a gallon of gas, while in Europe the same gallon of gas costs on average $6. It might be put to an online vote, would you approve a $1 increase in the price of gas to fix the roads and bridges of your state. Most people would probably agree to the increase if it meant the roads would improve, and further, most of them would approve a proposal to borrow the money to get the work started. So it is with getting the big stuff done. We know that in our future world the infinities will be unlocked, and there will effectively be an infinite amount of wealth created. By comparison with today, in real terms, the tax revenues will be enormous. It is therefore very reasonable to apply the cheque in the mail argument, and borrow as much as is needed to get the big stuff done. And there are precedents. The exact same principle was applied to the very large infrastructure projects carried out in America in the 1930s in the depths of the great depression, the government built dams, and electrification projects, a network of freeways, and new cities. Most people looking at these works now in 2016 would agree that the money was well spent, the simple fact is that these projects have paid for themselves many times over. The idea is not new, in the 1930s Keynes argued that one hundred years hence, GDP would dwarf what it was in his time, people would be many times better off, and with prudent management, and so on, repaying the debts incurred by their forbearers would not be a problem. And he was right. With an infinite future ahead of us we can effectively, *'borrow our asses off'*. Our grandchildren will celebrate us for doing so, they are after all the net beneficiaries. All we want to do is make them as rich as possible, and probably immortal.

PURPLE

Chapter 12   When Monkeys Invent A Man

*But, as it is, we have the wolf by the ears, and we can neither hold him
nor safely let him go. Justice is in one scale, and self-preservation in
the other.*

Thomas Jefferson, April 1820.

For millennia we humans have dreamed of living in a slave owing society,
capable of generating vast amounts of wealth, and unprecedented levels of
luxury. The Romans did it, but at the great cost of rebellion and war. The
great British sugar plantations of the Caribbean were slave based, and built
many a family fortune. The Southern states of America did it prior to the
civil war, but paid the price of total destruction. There are numerous other
examples of slavery from almost every human society ever formed, in all
cases the slaves were other humans, and the ultimate price paid by the
society was too high. Today we find the concept of the human as property,
or the person as a cash crop to be wholly repugnant to our enlightened value
system. However, our dreams of someone else, or something else doing all
of the work, and generating our wealth still persists. Now in the early part of
the twenty-first century, we might just about have done it. The work of
speculators such as Vernor Vinge and Ray Kurzweil leads to the concept of a
technological singularity, which is a point in time when, artificial
intelligence, AI, matches that of the human brain, resulting in profound and
unpredictable consequences, including the development of helpful and
cooperative, intelligent entities, to do our bidding and solve all our
problems. The consensus view is that we will arrive at that point around the
year 2030, which is not too far away. It is even conjectured by some writers
that these intelligent machines may in fact, be far more intelligent than
humans. However, we are well aware that just because the words rhyme
doesn't make it a good poem, and just because we can build something,
doesn't necessarily mean we should. There would appear to be some
obvious and dangerous, not to mention embarrassing reasons, why we
humans, in our quest to fulfil long cherished dreams of wealth and excess,

ought to proceed with the utmost caution. And in support of this position, strong voices are beginning to be raised in objection.

Because any Purple future depends very critically on our creation and application of artificial intelligence to automating the human wealth creation process, and because of what many scientists view as the very real existential threat to the human species, the subject deserves serious discussion. Artificial intelligence without ultimate human control is simply surrender. The approach of some protagonists in the AI field, the near blind acceptance that once we invent something smarter than ourselves, it ought to be allowed to, and more than likely will take over, is not an option. The Purple view is that at all times humans must remain in control, human supremacy is primary, and non-negotiable.

We do very much require the development of artificial intelligence, and it will need to be both strong and good, meaning almost as smart, or smarter, than humans, and having the capacity to be both useful and applicable. We both want and need intelligent machines and robots to do our bidding. We will need to build elegant machines that we can only dream about today, and they will be beautiful machines, they will be machines of incredible complexity and sophistication. But will they be conscious, in the way we understand that term, or rather will we recognise them as such? Will they be superior, as we would recognise, and have to concede? A word processor does not make coffee. A toaster does not answer the phone. Perhaps there may be a limited market for a message-taking toaster, but that's no reason to build one, its existence would be nothing more than a technological curiosity, or a commercial gimmick. Should we allow our smart machines to be smarter than they need be, or smarter than their creators? Is it simply asking for trouble to even venture along that road? We need them to walk, and we need them to talk, we also need them to chew gum, but do we really want them to walk, talk, and chew gum all at the same time? Do we really wish to be monkeys in waiting, do we want to end up as the monkeys who invented a man?

The potential adverse consequences of inventing strong AI to watch over us and guide us, may be illustrated by the following modest parable. Consider an island in the middle of an ocean, where the most advanced lifeforms are a troop of monkeys. They've been living on the island forever, and have their own culture, some crude technology, and a society of sorts. One day they discover a man washed up on their beach, they take him in and are curious about him, he gets food and rest, and is soon able to communicate with them on their own terms. They soon get to know that he may be of some use to them; he seems to know stuff. They try to put him to work on their set of problems, they believe they can control him. The man invents a ladder, and retrieves fruit from the higher branches; these particular monkeys were always ground dwellers, afraid of heights. The man has no fear of water, the monkeys were always terrified of it, and they watch as he wades out into the sea to catch fish. Suddenly from a bare scraping existence they have more food than they ever thought possible, an infinity of good things to eat, it was there all the time, the monkeys just couldn't see it. Now they have time, they can sit around on rocks, scratch each other's backs, pick each other's lice, and wonder what they might do with such a fanciful creature? While the monkeys are busy eating, sleeping, and doing the things monkeys like to do. The man can be left to his own devices. They don't understand him anyway, he's moody, likes to be by himself. There's nothing to worry about. He's not a danger. Meanwhile the man has other ideas, big ideas that do not involve monkeys, he has plans he cannot discuss with the monkeys, they don't understand. To him the monkeys are animals, pets, perhaps even ultimately food. And while the monkeys argue about who owns the man, and what other things they might get him to do. The man continues with his own plans. The monkeys try to control him. He resists. They attempt to use force. Sticks, stones, fire, and dead monkeys are the result.

It is therefore a simple matter that requires our consideration, do we ever dare to allow ourselves to evolve into monkeys, who then invent a man we have no possibility of being able to control? This would appear to be a possible likely outcome if we were ever to invent strong general artificial intelligence. The question is should we ever take that chance, is waking up a strong AI a path we should ever dare to walk?

Consider even the scenario of humans making the discovery of an earth-like planet orbiting a nearby star, further suppose we evolve the technology needed to go there, and dispatch a team. On arrival we find that the most advanced life form on this beautiful new earth to be primitive proto-humans, upright apes, with only a fraction of the capacity we enjoy ourselves. Would we really treat them as equals?

Many eminent members of the scientific and business communities, for example, Bill Gates and Elon Musk, Stephen Hawking and Lord Rees, have begun to regard the unrestricted and unregulated, development and rollout of AI, as a potential existential threat. In particular the fear that we might one day allow some entity, or a community of such entities to evolve, which are both malevolent, and intellectually, as far above us as we are above monkeys.

The concept of the Singularity is in effect an extrapolation of past and current technological growth patterns, it is not simply feature addition to existing technological assets; rather it is an understanding of what we have, and how we get to where we wish to be, based on known past technological achievement, it therefore seems to be an inevitability. Everything depends critically on the Singularity, if and when it ever happens. There are two very big ifs, if the singularity happens, and if Moore's Law continues to hold after the Singularity occurs. Or they might be simply the same as saying, the Singularity will happen, eventually, and when it does, there will be progress thereafter, and what matter if the rate of increase is at today's level, doubling every eighteen months, or if the doubling time is five years, or ten years. It becomes an inevitable fact, sooner or later, there will be machines that are twice as smart as us, then four times as smart, then ten times as smart.

The fact that nature has invented intelligence on many occasions in many different species, and to varying degrees, and the fact that even the simplest of creatures' exhibit extraordinary complexity in their everyday existence and interactions, means that intelligence is a problem we can solve, and therefore the delivery of AI systems is inevitable. It is not a question of if, but of when, and of how smart. You can tell that AI is inevitable because there are *a thousand little signs* telling us, announcing its imminent emergence into a waiting world. It's almost as though the population of the planet is being prepared for the inevitable arrival of AI and robots by those who know, and we therefore need to start our relationship now.

Because we need AI, and because AI is the ultimate solution to a great many problems, it is abundantly clear that AI is as inevitable as inventing the wheel. AI is one of those game changing technologies, like the internal combustion engine or the Internet, once we have it in place, and it becomes common we'll wonder how we ever got on without it. The general consensus seems to be that AI is almost here with us now, emerging slowly, and creeping across our economy and corporations like an unfamiliar odour. As with the introduction of barcodes and lasers, plastic bags and post-it stickers, artificial intelligence is beginning to get a foothold, and there will be obvious, and both predictable and unpredictable consequences. The widespread deployment of AI will inevitably displace people as never before. Jobs once thought to be safe and secure, and professions once thought at least protected from the march of technology will be changed forever. Creative disruption via AI will lead to a hollowing out of professions once thought to be immune to the ravages of technology, careers such as Doctor, Lawyer, Accountant, Pilot, Teacher, may fall by the wayside. There are numerous warnings from respected university researchers that creeping AI and automation could lead to half of the 140 million jobs in the United States economy being displaced by machines. In the post AI era, when AI is pervasive, we will draw comparisons with the pre AI era, as we often note the difference between, pre and post *powered flight*, pre and post *plastics*, pre and post *the microchip*, and so on. But this time we will notice a more pronounced difference, and it will cause unprecedented change.

There are many who postulate, if not openly advocate the development of artificially intelligent conscious machines, a so called strong AI, or an AGI, artificial general intelligence, for example Kevin Kelly in his book *What Technology Wants*, postulates that, *"one day we will make other minds and they will surprise us"*. But there is the old trader's adage which must be kept in mind. *'Markets hate surprises'*. Perhaps we ought to say, no surprises please. Surprise is not good. Surprise is not our friend. The legendary golf commentator Renton Laidlaw maintained that golf has always been a game of surprises. Perhaps we ought to confine surprises to games such as golf. While we humans enjoy a good surprise, we generally prefer our surprises to be of the sporting variety, and not of the, waking up a potential monster type. Our continued existence and our future development needs to be surprise free. Especially if these surprises are something we can see coming. Do we dare to give them the chance?

Do we already have AI in existence, if so then we ought to expect to find them on the Internet? But the internet is nothing more than a glorified notice board. With individual notices that glitter, flicker and alter, according to the information displayed on each. To imply that there is something more by calling it a network or asserting that some emergent property of a network's existence or complexity gives rise to thoughts or thinking, or consciousness, or the threat of humans one day being ruled by it, is preposterous. Does anyone suggest that the road network is about to declare its sentient existence as a living breathing intelligent entity, or the networks of fungi or plant roots, or the electricity grid or the phone system? More than the bare network would appear to be required. However, claiming that the human brain, and its emergent properties of intelligence and consciousness, are more than just the mere network of the components plays into the hands of the religious. Perhaps it does more than that, it invites god back into the world, after all the trouble we've taken to get rid of him. Clearly therefore, the human brain is a very specific type of network, one which is fitted by nature to the task of allowing its owner to become aware. And while our knowledge at the moment most certainly contains gaps, there is no reason to pay attention to those who would insert deities into these gaps.

There can be no Purple world without AI. If at some future time we are to arrive at a Purple state, that is a state wherein by the standards of today everyone on earth is fabulously wealthy, and pretty much does whatever they like with their time. Then the only route we can take is along the road of artificial intelligence. We will have arrived in a Purple world because of artificial intelligence, AI is the only game in town. All species with the exception of humans survive quite well without technology. But for us humans, technology is all we have, it's the only thing we do well. The alternative of an indeterminate age of darkness every bit as long as our prehistory, or even as long as the Middle Ages is simply not an option. We will never go back, we will never retreat.

The future of human society must be one of near infinite wealth, by the simple premise that wealth is created by energy operating upon matter as directed by information. And we know we have an infinity of both energy and of matter, our deficit lies entirely in information. This is the foundational argument of Purple, and it is by the creation and deployment of intelligent and semi-intelligent machines, or machines exhibiting the illusion of intelligence, to be our slaves, and unquestioningly create the wealth we crave that we will create a Purple world. The creation of vast wealth demands a slave society, that is, a society where all the labour is carried out by armies of obedient docile and compliant slaves, owned by their masters who carry out their master's every wish without question. Clearly and for obvious reasons the best slave is the slave who does not know she's a slave, the ignorant slave, bound to servitude with no capacity to think of freedom, and he has no prospect of ever being declared free, either by act of law, or by act of owner. That must be our goal.

The human aversion to work is as natural as work itself is unnatural. It is widely recognised that while many people frequently work a sixteen-hour day they probably only ever get, at most, five *good hours* out of that day. The rest of the time is spent doing 'other stuff', going to meetings, coming from meetings, thinking about sex, or shoes, eating, going to the lavatory, maintaining oneself, chit-chat, small talk, golf scores, and wondering about the football game next weekend, or last weekend, and so on. In other words the majority of computational expenditure of the human brain is deployed and committed to simply being human. The sheer business of existence, of persisting in our human form in a three-dimensional reality, of being alive. That's what we do, and there's nothing wrong with that. The only reason people have jobs is that human people represent the only source of flexible intelligence available to both other people and corporations. If monkeys could be trained to operate machines in factories, or cut hair, answer the phone, or do anything useful for that matter; guess what?

Most people who work for a living have jobs, as opposed to those who have careers, or vocations, and those who work for fun. Most people use but a small percentage of their capacity to do their jobs, with a very few exceptions, astronaut, brain surgeon, gardener, and a few others, such as car mechanic, furniture restorer. But in a future society these are the jobs we will gladly reserve for ourselves, because we enjoy doing them, they are not a chore. Most jobs done by people today are dull, repetitive, intellectually unrewarding, and most of the time the people doing them would rather be elsewhere, and doing other things. But to move to a situation where most jobs are done by machines requires only a relatively small increase in the capacity of machines, because of the bland and unengaging nature of the task. So the transition could in theory happen quite quickly, say over the first term of a President.

An entire book could be written about who will be replaced, and how long it will take to get every job and profession turned over to machines, for example, the recent book by Andrew Keen, *The Internet is Not the Answer*, contains a good commentary on the subject. Martin Ford's book, *The Lights In The Tunnel*, also contains many insights concerning changes in employment patterns. We give it many and various names, such as disruption, innovation, automation, and of course, right-sizing, but it's all just French for getting people out of companies. Would it be possible to replace a bishop or a priest or an actor? Why yes indeed. But does that engender ontological notions of belief? Does it matter? The point is the impact will be felt at all levels, the consequences will be universal.

We already interact with other intelligent entities, creatures, both wild and domestic exhibit a great deal of intellectual ability and natural innate capacity to reason, and to learn and to understand, to navigate in three dimensions, and solve difficult puzzles. Teaching a dog to fetch a slipper, or a newspaper, or training a cat to use a human toilet, some cats do, some don't. Our future interactions with intelligent machines will be therefore familiar to us, and should lead to a very long list of very good things. The ability of very fast computers running intelligent software to create detailed simulations are proposed by many commentators for the not too distant future, the full immersion virtual reality coupled with advances in nanotechnology and computer brain interfacing, will advance like every other field of endeavour, and sooner or later we may expect both computers and their software to be so powerful that real world simulations will become commonplace. Like being there. Like the real thing. No one will be able to tell the difference. Playing in the US Open Golf tournament, or at Wimbledon, or competing in a classic horse race. However such simulations would not be sufficiently realistic for the players or the participants, if the spectators, or other elements of the simulation lacked realism. For example, if you were playing in the final round in the US Open Golf Championship at Pebble Beach against Jack Nicklaus or Tiger Woods, they would have to be able to react to whatever you might do, or say. Otherwise it would not be the accurate, like being there, experience you were promised. So we must and will anticipate much greater advances in gaming, and in generally interfacing with our technology. We may expect everything we own to listen to everything we say and understand it on the required level, regardless of how you speak.

The traffic system in most cities ought to improve. The future AI traffic junction manager knows nothing about what the cars are of who drives them or where they go. They never ask they are never told. But they know exactly how to manage traffic at their own particular road junction as if a human was doing the job. They will not have the capacity to understand what they do or why it needs to be done, they will be engineered to the very specific task with no additional capacity. We can also expect the long predicted intelligent helpers in the home, and in the garden, we can expect everything to be talking back, everything to be sophisticated and clever. And anywhere you can envisage a need for intelligence, there will be a machine or a system available to do it. Intelligence will be the new add on to every appliance and gadget, in the same way as the electric motor added power to the kitchen.

While the human future with AI has the potential to be one filled with abundant wealth and undreamed of luxury, the emotional machine remains the butt of ubiquitous science fiction jokes; the protocol droid, and the like. Do we really want our machines and gadgets to have some kind of emotional response or presence? The fridge refusing to open, because it's too fragile. The toaster, sad and lonely; no one wants to hear my side. The oven causing a casserole to explode out of an excess of exuberance? We can be sure there will be companies making money developing emotional add-ons for gadgets of all kinds. And they will probably simulate the effect to varying degrees of acceptance, creating the illusion of emotion. The freezer that senses when someone is present, knows names, says hi, smiles, all that you expect from a cheery temperament, and so on. Like the doors on the Heart of Gold space ship from Douglas Adams's, *The Hitchhikers Guide to the Galaxy*, ever pleasant and courteous to all who pass through. But the simulation will be only that, and nothing more, the appearance of something with which we are familiar, nothing more, they might look intelligent, even feel intelligent; but so do ants. Like the potted plant in the office, you just can't tell if it's the real thing or not. It seems so lifelike, fulfils its function so well. Could easily be the real thing. Only after up close inspection do you realise it's only plastic and paper, held together with bits of wire and nylon string.

The best known current example of an AI system is IBM's Watson, this system was able to demonstrate an unexampled proficiency in solving riddles, vague hints, and word-plays, while taking on two former champions of the American TV show Jeopardy. Watson solved the problem by extremely fast assessing of very carefully constructed databases of general knowledge, and by accessing online articles on a million different subjects. It was an impressive feat by all previous standards of AI. Of course there is the natural expectation that Watson will make substantial progress in subsequent versions. Clearly if Moore's law type extrapolations are to be deployed, there's no reason to assume Watson cannot tackle medical data in the same way and be able to sift and solve any medical conundrum in the same timely manner, and do it at twice the speed, or on half the hardware, within eighteen months of its Jeopardy triumph. In a few more iterations, with a few improvements in voice recognition, and so on, Watson like systems might become a mass market, fully licenced, personal physician, with a consequent hollowing out of the entire medical profession. And Watson's clones and progenies will not be content to stop at medicine, you may expect that all intellectual pursuits will be treated in a similar manner. And especially those done by humans in places of work such as banks, offices, government departments. Why have ten people when a single Watson does their jobs twice as fast, and is always on, always there and never a problem. From a fiduciary perspective, it is probably the case that companies could be considered remiss if they employed ten people when a Watson was available, and a net positive to the bottom line. It will be done because employing people only serves to slow the process, because people would be the slowest and dumbest element in whatever the company does.

AI will certainly upset clerics. How this new era will be greeted by clerics and religious factions of all hues is debatable, but one thing is for sure, the rollout of AI will provide quite a bit of fun in this area. For example, educational AI, probably freeware, will be distributed across the developing world, and will relatively quickly create a generation of people no longer prepared to accept magic, or god did it, as a reason for their poverty. And the first people the newly educated and liberated poor will want to question, may well be the clerics. Another related aspect likely to emerge when AI takes hold will be the philosophical baggage that is bound to develop and evolve concerning notions of sentience of machines, rights of software programs, and the souls of humanoid robots. But only if we allow it. The emergence of AI will change human relationships forever, because clearly, according to those of a religious bent, they are not created by god, in his image; they have not got a soul, and cannot therefore be saved. Creation of a synthetic intellect would be the creation of an immortal intelligent entity, effectively a god. Monkeys invent a man, and man invents god. The way of the universe perhaps. Moves in mysterious ways?

The brain-mind problem remains the impregnable duality at the heart of human existence; it is the conundrum of all conundrums, the insoluble discrepancy of our very essence. Or so philosophers would like us to believe, but in the absence of factual evidence we must remain materialists. As far as notions of the soul are concerned we must, at all costs be careful about creating another gap to be quickly filled by zealous Christians. Intelligent or semi-intelligent machines or brain simulations will certainly shed more light on it. But it must be taken *a priori* that if atheists don't believe they have souls, the machines they build are hardly likely to arrive at a different conclusion. Asking whether or not a machine has a soul is probably just a valid as asking if a machine would need to simulate being stoned to appreciate certain genres of rock music, or asking if it has an opinion on what constitutes '*good*' human pornography. Such questions belong to the, '*esoteriology*', of machine consciousness research.

Very much linked to the religious and philosophical quandaries of AI is the human quest for immortality. Renowned futurist Ray Kurzweil predicts many things that are expected to converge to the Singularity by the late 2030s, including being able to routinely scan an individual's brain with such ultrahigh precision, and to such a complete extent, and understanding of brain function that effectively uploading our entire brain structure to a machine will become possible, if not routine. At that point all of a person's life experiences and memories are now digitised, and are available to be viewed or manipulated as required. The distinctions of old between brain, the mind, the spirit and the machine, then become very blurred. Apart from discommoding the religious people, there is the very real problem of whether the upload is in fact the person or not, and if so, is it in fact legally the person?

Switching off a human is otherwise known as killing, and sometimes may be construed as an unlawful killing, and possibly murder, which is an unlawful killing carried out with intention, and or recklessness. However, that is what some proponents of strong AI propose, the copying of all that is you into a never aging black box and the switching off of the flesh and blood you. What if it were done as you slept, and you later woke then commenced an argument with the box as to which of you was, in fact, you. Who has control? Clearly the law will then intervene, and claim that where the equities are equal the first in time must prevail. So that biology will always be superior to machines regardless of intelligence. And we would, in a copy situation, have to go to a court, and apply to have the biological entity terminated. This is, of course, euthanasia, and while permitted in certain countries there are profound moral and legal issues involved. If the original is near to death and will remain in a coma who would be able to tell whether the copy was an exact copy or only a partial copy. We would only have the word of the copy that it is, in fact, the comatose person. It would be put to proof to substantiate the claims it made. We can imagine such a case occupying a court for a considerable period of time.

We know we can do it. Nature did. Complex systems often exhibit properties not seen in the blueprint, or conceived in the initial design, properties deriving from inherent complexity, and the patterns emerging from the simplicity of the design. Human intelligence, and indeed human consciousness seem to be emergent properties of the complex system that is the human brain. Just like the sense of time emerges from a clock, or velocity from a vehicle, or the emergent properties of harmony, producing a pleasant musical effect not present in the individual voices, the property emerges from the combination. Almost every natural system exhibits the emergence of transcendent properties. They emerge unexpected, unanticipated, unintended, they are surplus, random, and unknowable events, and patterns, sometimes useful, sometimes not, and on occasion perhaps even detrimental. You may search forever to find time, or the concept of time in the components of a clock, or speed, or the property of cornering in the bits and pieces that comprise a sports car, and try as you might you will not find, love, insight, compassion, or pity in a grand piano, or in a cadaver for that matter.

Of considerable importance here is the work of the computer scientist John Conway, his invention, the game of life, a computer program comprising an infinite grid of squares which can be in one of two states dependent entirely upon the states of the neighbouring squares exhibits great complexity from very simple rules resulting in the appearance of something more than seems to be present. It is an example of how simplicity and simple rules beget complexity, even on occasion, the appearance of life and purpose.

Software is everything, everything we do is software based, and the best code scribbler in the house is Nature, the wet-ware produced by evolution is majestic and unsurpassed by anything we humans are able to construct. Meet and consider the humble spider, and its brain consisting of 100,000 neurons, and capable of navigating in three dimensions, building in three dimensions, telling the time, checking the weather, managing a chemical plant, and organising and managing a stored food supply, processing visual information; some spiders see the world visually in the same way humans do. Not so humble after all. Or the humble Pigeon, or the Bumble Bee. If we can understand how a spider builds a web we will know a lot. You need not look very far from your own back door to see just how little we really understand, and how very far we have left to go. But knowing that it has been done, and done so very well by Nature spurs us on to do it at least as well, if not better, and to deploy our results against our long list of problems.

We have a very long list of problems that can only be solved by the application of intelligence. And if artificial intelligence can solve them then it is reasonable to presume it will be deployed, take for example the everyday problem of translation. The Chinese room argument advanced by John Searle and others, and hotly debated implies from basic human understanding that mere movement of symbols is insufficient. What is being translated is often as important. And the skill levels of various translators will vary enormously. For example someone who has knowledge of electronics, and also speaks both Chinese and English would probably be capable of translating a detailed manual on a complex electronic machine, for example a troubleshooting manual. However were the same person to attempt to translate James Joyce's Ulysses into Chinese the project might well come unstuck very quickly. To translate such a work into Chinese from English requires a fluency of English to a level approaching perfection, it also requires a similar level of Chinese, and would also require that the person had a smattering of Latin, French, German and Italian. The translator would also require to be extremely well read in English literature, and understand such concepts as metaphor and allusion, along with a host of other devices and linguistic gesticulations of which '*smart arse*' Joyce was fond. Without detailed background reading and knowledge, and ability to translate, the meaning, or the sense, or the spirit of the work is lost. It is therefore apparent that in translation, the higher up the translation level we go the more understanding is a prerequisite, and understanding implies intelligence, and a level of intelligence that most humans would recognise as being such, and that is exactly what is beginning to emerge.

Yet more of our long list of problems will be solved by robots of all shapes and sizes, from small insect like machines to fully autonomous humanoid versions. Every year there are a number of shows, to demonstrate to the world the wonderful advances that have been made in the field of robotics in general, and humanoid robots in particular. Humanoid robots are one of the many great promises of AI, our personal slaves, our ever helpful helpers, our workforce, and playmates of the future. The ever-helpful Data of *Star Trek, The Next Generation*, always eager to help, and yet so far advanced, always willing to park himself in a chair to undergo a '*full diagnostic*', and have his emotion chip inserted and removed at the whim of his creators, and yet being told that he is a friend, and a buddy, and a trusted ally, a full member of the crew, no less. The risible notion that the emotion chip can be removed, or that the damn thing will sit in a chair, and allow its head to be opened is exactly that. It is nevertheless, ever ready to assist in the diagnosis of major issues in star-ship engine dynamics yet unable to empathise with a cat. It seems strange to us that they could allow such an entity wander about the place of its own volition. But then the ship's captain exhibits a bi-temporal recession of 1995 proportions, clearly in that particular version of the future, technological development is only progressive in a certain select set of directions, and has not quite mastered the hair follicle.

And of course science fiction is full of humanoid robot characters, from the ever helpful robotic maid Rosie from the *Jetsons* TV Cartoon, to the malevolent and vicious Terminators from the movie franchise, as well as those in the *Star Wars* movies, *I Robot* and almost every other Si Fi outlet you can name. In each case the humanoid machines are fast, intelligent, helpful, and very physically able to the point of almost acrobatic perfection, their nuclear power-packs allow them to run forever, and they never require human assistance. The truth is, of course, very different, the current best humanoid robot available today, and seen at every robot show, is the Asimo model, made by the Honda Corporation of Japan. And while on the face of it this robot's antics look impressive, walking down a stairs unaided, and the like, all is not quite as it seems, it is a complex machine requiring careful setting up prior to a performance, and constant supervision for the limited amount of time it can spend active on a single battery charge; about an hour. What we see at these exhibitions is the incremental development of a science fiction goal. There have been major advances in robotics, the ability to stand and walk unaided being a good example, the evolution of nimble and dexterous hands, and of course, improvements in robot vision. But today they are little more than toys, or demonstration pieces, they fall over, run out of power, break things, fail to follow commands, and generally don't work. They are not the domestic servants, or the factory workers, or the carers of the elderly that we so badly need, and you certainly could not send one to Mars and expect it might do anything useful.

But through these primitive machines we catch a glimpse of the future when humanoid robots will populate our world. We will build them because we want to build them, there are likely to be commercial opportunities for robot builders, and people will spend good money on them. Imagine if you will having your very own Terminator, the shiny metal skeletal type, no flesh or features. With glowing red eyes and standard terminator vocabulary, gravelly voice, stern gaze, and generally menacing demeanour. Why would you need such a contraption? Why to caddy for you on the golf course; of course. It would need to wear proper shoes, and comply with club etiquette of course, but otherwise why not? Perhaps your very own predator or alien creature when you wish to go, *'walkies'*, at night. Surprise the neighbours. We will build them because there is a market for them, and because we want to build them, and they will come fully equipped with the illusion of ability, and the illusion of intelligence and consciousness. We will do it because we can, or rather to show that it can be done, the old mountain climbing principal, The goal of the research, the notion that binds the researchers together into a global coherent mass of researchers all plodding along the same path towards the same goal. Why climb a mountain, why smoke a cigar? Why go to the Moon, as JFK asked in his famous speech at Rice University? Richard Feynman said it well with reference to doing physics, but it applies equally well to creating machines in our own image, and to paraphrase, *doing physics is a bit like sex, there is always the possibility of a practical outcome but that's not really why we do it.*

In the normal course of technological development as we understand it, predictions are frequently made regarding future trends. Such predictions are wholly based on extrapolation of what we have done, and of what we have today, and then projecting it into the future to arrive at a possibility of something that seems to make sense to those doing the predicting. Clearly the past litany of such predictions which have unceremoniously crashed into the ground stands as a stark reminder, if not a warning, to future would-be projectionists. AI is no stranger to prediction. Intelligent machines have been the subject of future speculation as far back, and for as long as machines have been among us; as well as being a firm science fiction favourite. But for our purposes here, the beginning may be taken as being HAL the AI character from Stanley Kubrick's movie, *2001 A Space Odyssey*. It goes back some 45 years, and the prediction was that by 2001, a machine with the obvious capacity and intellect of HAL would be attached to a manned spacecraft capable of making a trip to the Jovian system. 2001 came and went, as did 2010, and probably 2020 will pass us without the appearance of HAL, or any strong AI even remotely resembling HAL. But there are lots of hints, there are many suggestions, there are numerous indications.

AI is like any other technology, if it works better it will be adopted, if it saves money, or makes more money it will be deployed. AI will come into our lives the same way barcodes did, and once the benefit is seen, the technology will be embraced. Once upon a time barcodes did not exist, then they appeared on a few items, now they are on everything, and what is more, they have become indispensable. Without barcodes we could not have online shopping or efficient logistics. But when barcodes were released into the world there was no trumpeting of their arrival, there were no movies or science fiction books predicting their imminent coming, they simply turned up as the solution to a problem, no one noticed, and no one objected. AI will be the same, there will be a creeping introduction, it's a continuum not a set of discrete states, it flows in time, and if you went away for five years and came back you would certainly notice. Details of the types of AI contemplated, and indeed underway in laboratories and companies around the world can be found in many books, for example, James Barrat's book, *Our Final Invention: Artificial Intelligence and the End of the Human Era*, and several other titles on the subject of AI, and the possible threat.

In fact, it has almost become a little bit like the scene from Samuel Beckett's trilogy, when the character Malone, explains that he knows it is springtime because, *"a thousand little signs tell me so"*. There are a thousand little hints and indications, signposts if you will, all pointing towards AI and its rollout across industry, commerce and society as a whole. It is now being released into our world, and like barcodes before it, we will wonder how we ever got on without it, but unlike the ubiquitous but harmless barcode, this time we will notice, and perhaps this time we may well need to object.

If it can be developed, and that is not really that big of an, *if,* any more, we are going to get AI and quite probably AGI whether we want it or not. It is as inevitable as our harnessing of the sun to produce all of our energy needs. There is nothing anyone can do about it. Or so it seems.

The notion that all the upward pointing graphs presented by Ray Kurtzweil in his splendid book, *The Singularity is Near,* have some meaning that the human race has some glorious date with destiny, may or may not be true, the path may well be blocked, by law, by war or other calamity, by changing taste, by anti-technology movements, economic collapse, famine, plague, and the rest, the list is a long one indeed. All these possibilities might get in the way and slow progress or impede it for a while. But nothing, it seems, will kill it off all together, and if the graph stumbles or stutters then it will be picked up once more at a later stage. Even were we to enter a new dark age that might last for a thousand years, the graphs would pick up where they left off, if humans are still around, and emerge at the end of it all. This is how inevitable AI really is.

All of the promises of AI and the Singularity depend on the exponential increase in knowledge in many different fields, such as computer hardware, software, neuroscience, all marching in step together, bootstrapping as they go. But any practitioner in these fields will tell you that really important conceptual discoveries, along with much of the other ancillary work that forms the mundane day today plodding of science are anything but exponential. They are at best linear and sometimes the feeling is of being on the same plateau for year after stubborn year, with no discernible upward movement. Some even suggest that progress in many of the required fields is neither linear nor exponential, it is perhaps sigmoidal, which is essentially to say that progress stumbles up the erratic curve from one plateau to the next, as each spurt of progress occurs. But regardless of how or when, we can still say that we will eventually get to a solution because the problem has been solved already, long before humans got to work on it. The actual point in time when we reach the Singularity, which is the point when human and machine intellect are indistinguishable, is irrelevant. Kurzweil believes it will be in 2030, but it makes no difference if that becomes 3030, it will sooner or later happen because we know it can be done.

The reason is simple, AI is such a radical game changer, if corporations can find intelligent or semi intelligent software and hardware to replace humans they will do so without thinking, they have always done so, and nothing is likely to change in that regard. Corporate entities have always embraced anything that creates greater efficiency, they have no choice other than to take that road, even at the expense of destroying their own customer base. Corporations themselves, are already a form of AI. Not only that. Corporations are a form of AI that is smarter than people, they operate on a different level, and according to a set of rules that leave them no choice other than to adopt new technology; they operate under a legal compulsion to maximise efficiently. For years, growing corporate profits have been largely the result of increases in efficiency brought about by new technology. The continuing trend is towards ever more sophisticated, cheap and intelligent technology to displace the ever more inefficient and costly humans. Corporations failing to engage in this technological arms race to the bottom, will be outperformed by those that do; machine intelligence is intelligence without the attitude, without the baggage, without trade unions, without rights; the perfect slave labour alternative to unwanted human involvement.

We also need to consider the possibility that the creation of strong AI may in fact turn out to be quite an easy thing to do, and the hard part was getting to the point where we could do it the first time. It is certainly one of the more peculiar aspects of AI research that hard things are easy, and the easy things seem hard. As the gangster character in the Bugs Bunny Cartoon, commented, in an episode involving two gangsters, an anvil, a magnet, and the external agency of the aforementioned pesky rabbit, "*I dunno how you's done it but I know you's done it.*" If it's done, does it really matter how? Once it's done, it's done, once the problem is solved, it stays solved, and gets used.

Nature seems to find it easy to make AI. Brains are common, and range from the simplest slug containing the bare minimum of neurons all the way to the higher animals and humans. In between there is a graduated scale of intelligence of varying degrees of ability. It seems that once it was discovered, it was rolled out to all of animal life. The brain of a cow, for example, is a fascinating piece of bioengineering, as beautiful and magnificent as any natural structure. It is the command and control centre of a complex organism capable of navigating in three dimensions, and carrying out all those cow functions. However, cows are dumb, *'bag-of-spanners'*, dumb, no one is going to argue that point. They need no great intellect to wander about the land eating, ruminating, and reproducing. There is no discernible spark of personality; anyone who has ever stared over a fence at a cow in a field chewing the cud will know this all too well. And ten cows in a field fare no better, there is no more intellect present in the field for the addition of another nine cows than there was in the presence of the one. Grouped together the IQ still numbers less than five, even though the IQ of a cow may be as high as ten, together ten of them cannot exceed ten. There is no scalability, as there is when ten humans get together to form a company, for example. Kevin Kelly in his book *What Technology Wants* seeks to discover whispers of the Internet, in his listening to the *'Technium'*, the random bits of noise, the number of connections. But there's nothing there. It is devoid of meaning. It is similar to increasing the number of cows in a meadow. There is the illusion of some kind of organised complexity in the noise, but it is really only background noise, devoid of form or function, turning up the volume make no real difference. Similarly, you can print random words from a computer, and print page after page into book after book, and stack them on shelf after shelf, you can even call it a library if I like, but it does not change the fact that it is merely the appearance of a library, and nothing more. The random noise of a computer network regardless of the number of connections, or the volume of traffic, or the number of connected devices, is nothing more than random noise. It is the human tendency towards anthropomorphism that attributes some additional emergent property to the otherwise random noise. We are by nature pattern seeking mammals, we want to see patterns in the noise, we want the noise to make sense and convey meaning, but that does not mean that it ever will, sometimes noise is just noise.

Sometimes it takes a long time to get to something, then we stand back in admiration of our achievement, and wonder why we didn't see it earlier, and what took us so long. Making steel was such an adventure, swords were beaten from raw iron by smiths for well over a couple of thousand years before Bessemer discovered how really easy it was to make huge quantities of high quality steel very fast, and very cheap. Now its child's play. Suppose consciousness is ordinary, suppose it's easy, suppose all of a sudden it becomes second nature, and we wonder that it being so easy how we didn't spot it before. Suppose we finally discover that consciousness is in fact ordinary, and in fact easy to produce that there is a way of doing it, and that by doing things a certain way the software effectively becomes intelligent, perhaps even conscious. Suppose it is part of the complexity, a result of the pattern, a result of putting the thing together in a certain way, or in some more or less complex fashion that we know today. Perhaps a network can be tuned to a conscious state. Suppose they just wake up all of a sudden? In the same way that a car has a look and feel to it, a sound and capacity to produce such subjective things, for example, speed, acceleration and cornering. That is probably what will happen, it is what always happens, the difficult things of one generation become mastered, and the child's play of the next. It is the ultimate goal of AI research, and of course we have the human brain as our guide, to what works, and we have animal brains such as those in cows as examples of what works too, so we know we can do it. Eventually.

Two more certainties arise if consciousness, self-awareness, and intelligence in machines turns out to be something that is easily done. The first is that we humans can be certain that these qualities are not peculiar to humans, and are therefore not the gift of some external deity who gave them exclusively to humans in the form of a soul. The other certainty is that if it is easy to do, then intelligent machines will proliferate and become as common among us as PCs, iPads and cell phones. And they will, like everything else in the universe, change and evolve. The rule is simple, if we build them they will evolve. Everything else we have ever built has evolved, cars, cameras, phones, and the piano. Why should intelligent machines be different? We clearly know that evolution is a fact; notwithstanding what the creationist minority, and the flat earth people might wish to cling to. We see evolution all around us. We see how the first real personal computers, the IBM PC, evolved over the past thirty years. How buildings evolved from mud huts to skyscrapers, along with cars, trucks, and everything else we make. Clearly evolution as a process not only occurs but is also well understood. What is far from understood or predictable is the possible set of outcomes, the end result that any given starting point might produce. If we allow AI to change and evolve on its own, we have no experience to rely on in order to judge what might happen, or where it might take us.

We know that intelligence and consciousness are products of the brain because with the failure of the brain, through damage, disease or old age there is a failure to a greater or lesser degree of consciousness and intellectual capacity. In many, perhaps all bookshops there is a section known as the, *Body, Mind, Spirit* section, wherein one finds esoteric titles containing words like alternative, transcendental, energy, crystal-wellbeing, and the like. Physics does not understand the dichotomy of mind, body, spirit. The body is easy, the mind too, easy to understand as a concept that is. The body is the physical bit, what others see and touch, and the thing that interfaces with the outside world, the mind is that emergent property of the brain that arises from the inherent complexity of its structure. It is the concept of spirit that eludes definition, perhaps because it belongs with the mind, another emergent property. The numinous the transcendent, that part of the brain that seems to light up and buzz on seeing ones first child born, or hearing the dying last chord of a favourite symphony played by a skilled orchestra, or having just played the best game of golf of your life. This is where the blind process of evolution has brought our species, but it has taken over fifteen million years, since our divergence from chimps.

The most intelligent animals, for example some monkeys, chimps, parrots and many other birds, and of course, dolphins and killer whales, and so on, are all within an EQ range far below that of humans. Clearly dogs, cats, dolphins and monkeys are smart, as we might describe it, and we are often impressed as to just how smart our domestic pets can be. And there is the perpetual argument as to whether a cat is smarter than a dog in a domestic setting, and that debate will be had as long as the dichotomy of cat lovers and dog lovers persists. However, while we can conclude that a dog, cat, or chimp is smart on their own terms, they are not that smart on our terms. It is quite possible to have an engaging conversation with a chimp, or a cat, or a dog, along the lines of, does Larry want a banana? Yes he does. Does Fido want to go walkies? Yes he does and he is capable of making it clear, by his actions, that he does. However, when Fido is asked a direct question such as whether or not interest rates might rise in the near term, he can only offer that cute sideways tilting of the head which we all understand as canine bewilderment. If we create AGI with the EQ of the animals we control and keep as pets, and if they serve us well, then the hard stuff is done. It is perhaps only a very small step from there to increasing the value by two, four or ten, until we reach a large enough number that renders us as the lower EQ individual and the machine as the keeper.

So what happens if, or rather when, we create machines, or they evolve themselves to have intelligence values in the hundreds, or in the thousands for that matter, machines so smart they can only talk to each other? Would they condescend to communicate with us? And if they spoke to us on their terms would we simply offer the human equivalent of Fido's cute sideways tilting of the head, something all such intelligent machines would understand as human bewilderment. The monkeys would clearly have invented a man. And in the same way that Tigers, for all their size, claws, strength and ability to kill humans, don't run the world, human ability to resist such strong AI might simply be non-existent.

The question is only easy if you know the answer, how many legs has a cat, is only answerable if you know what a cat is, and what a leg is, and have an ability to count. Perhaps it would not matter how smart we build them, perhaps no matter how smart they became they could never know anything which we could not know, or in greater detail, because the universe is finite and knowable, and there is not an everlasting series of layers like some infinite never decreasing onion. Perhaps the final form of the universe is accessible to us, because it simply needs some additional understanding, or a few more discoveries. Perhaps some new or different mathematics is required, or a few variations, slight or detailed that we will eventually discover, or would discover as inevitably as we discovered the properties of circles or triangles. And experimental breakthroughs are always possible. At the moment we are not quite sure, and it very much depends on who you talk to. It may be that we are nearing the end of the struggle, or there may be many more levels to the game, we're just not sure. Perhaps very smart machines can have a better understanding of extra-dimensional physics, and the mathematics involved, for example suppose dark energy and dark matter are in fact projections of some sort into our four dimensional space-time in the same way as we might shine a light onto a two dimensional page, and confound any two dimensional creatures living there, and ignorant of the third dimension, even though they possess mathematics to describe the existence of additional dimensions. We have discovered over the many centuries of our age of reason that understanding is the easy bit, it is making the decision in the first place that is hard to do. It is therefore quite possible that because of the very nature of the universe in which we live, there would never be anything that even the smartest AI could discover that we could not, on some level, understand. We might even get their jokes.

There remains a role for humans, we will always be able to find willing human volunteers to do those dangerous things humans feel it is their duty to do, such as explore space, be the first to the nearest star. Many would be willing to die to achieve the goal, selflessly and needlessly lay down their lives to achieve new records for the species. Machines might do it better, or faster, or with greater and increasing efficiency, but they'll never do it with greater style, and not with a greater sense of achievement. A robot digging a hole on Mars, and finding a colony of microbes in the permafrost is not the same as a human wielding the shovel, turning the sod, and exclaiming. Wow! And we will probably make a conscious decision in the future to reserve all such 'Wow!' moments for ourselves, because that is what we humans do.

We have not yet exhausted the possibilities of the human mind. We therefore need to be clear as to why we would ever require artificial minds. We need to ask the question, what is the problem to which very strong AI is the solution? And we need to understand the answer, because every aspect of the human future depends on how smart we humans really are on an absolute scale, on a cosmic scale. The question is, how smart do we need to be to find all the equations necessary to explain the entire observable universe, its origins, its workings and its future, and the final form of the laws of physics? That, after all, is what lies behind a great deal of AI research. And, more importantly how smart do we need to be to solve all of our other pressing problems, and survive as a species? It's just not the same if the machine does it. The medal is somehow tarnished.

So what's the problem? The late Neil Postman, technologist, and ardent critic of much of the nonsense written about technology had a simple yet fundamental question he would ask regarding any branch or item of new technology, or any technology for that matter. It's the same question asked in every patent office, by every Patent Clerk, Einstein, in his life as a Patent Clerk in the Patent Office in Bern probably asked it of thousands of inventions. The question is simply this: what's the problem? What problem are you trying to solve? What is the problem to which the particular item of technology is all or part of a solution? That is a question that needs to be, and must be asked, about intelligent machines as imagined by Ray Kurzweil and others who envisage the Singularity, and a post Singularity world. Why would we ever wish to become the monkeys who invented a man? What would we get out of it, and where will it all lead?

What problem are we trying to solve by inventing an intelligent conscious machine? Some say that we would never consciously invent such a thing but that such a thing would necessarily and naturally emerge from the work done in the field of intelligent machines. If we make machines sufficiently intelligent, such as an artificial waiter or a traffic light controller then it will naturally evolve of itself into a conscious entity capable of expressing conscious thought, and capable of telling its creators, us, that it is conscious. It could be argued that in order for a machine to be sufficiently intelligent to exist in, and navigate, a natural environment such as the surface of an asteroid, or the floor of a restaurant, it must necessarily be aware of its environment, and of itself, and therefore must be conscious, and would be capable of saying so. It must be capable of solving any and all possible problems that may arise in its operation, it must be capable of anticipation of adverse events and possibilities, and so on. But so too is a cat out hunting birds or mice for fun. Any cat owner will tell tales of birds or mice, of half eaten things left on the kitchen floor by *Puddy*, after a night of wandering; look what the cat dragged in.

We can also ask more esoteric questions such as; what is the philosophical end game of artificial intelligence? But the bottom line is that if we are to go down this potentially dangerous road, we need to know very reasonably why that needs to be done, why it is necessary. We don't know if there is a universal limit to smartness, and perhaps beyond that limit it simply doesn't matter how smart, nothing further can be accomplished merely by the addition of more smartness, a kind of law of diminishing returns? Humans are not fast enough, but a man on a bike or a man on a horse or a woman flying a plane pretty much satisfies the criteria. If we're not smart enough to get to the very bottom of everything, then we may require the assistance of machines, to solve the puzzle, and dumb down the solution for us, if that is possible. It may be absolutely necessary for the monkeys to invent a man. But we don't know that yet, we don't yet know if we're smart enough.

Do monkeys look at the stars? And if they do how far does their looking take them? They quite clearly have limitations. A monkey looking at the stars, or at a Rubik's Cube, or a fountain pen, could never figure out what they are, or how they work. They never ask the question, because they can never ask. They clearly and self-evidently lack the capacity. We can therefore clearly define the upper limits of a monkey's capability. We know it can come so far and no further, in all things, the limits of the other intelligent creatures who share our world can be measured quantified and are known in absolute numerical terms. What about human limits? Clearly there are physical limits beyond which we cannot go, running, jumping, and so on. But these limits have been made irrelevant by our machines. And of course there are Gödel's incompleteness theorem, and the Heisenberg uncertainty principle; both derived by humans. What about hard limits in understanding? As far as we know there are no limits for humans. Any limits, if they exist, can only be defined for us, by external entities, either intelligent creatures from other worlds, very far ahead of us in the same way as we are ahead of the apes, or by something bigger and better that we will create here on earth from our own technology, something that will evolve out of our technology. It would evolve in order to discover what we do not understand, or to present us with a list of things we cannot understand, or to cruelly expose us to our limitations? On a purely human level, to be exposed to ones limitations can be an alarming prospect, sobering, crushing, life affirming or life changing, liberating or depressing, it can allow us to move forward, and rise to the challenge, or cause us to retreat, and live in the past.

In his marvellous book the *Ascent of Man*, Jacob Bronowski portrays man as the shaper of the landscape. The builder, the worker of miracles. Are we to surrender all of that to machines without ever trying to find the other way? Without attempting to get to the bottom of it all on our own. If the machines get to the final form of the equations, and find the ultimate answer then Nature has failed. We have failed, our failure will be complete, and the machines will be the only lasting monument to our failure. And we may lower our heads in shame for not hanging in there and working at it. Who knows how near we might be? Are we to simply quit? To walk away, to throw in the towel? It's not at all like us to give up. We do not know all there is to know, that much is very clear. Nor do we know all that can be known. But neither do we know all that we wish to know, or all those things we need to know. There will always be levels of minutia beyond which we don't wish to go or care to go. For all we know, we may actually be very close. For all we know we might have the whole thing wrapped up in five years, all the ultimate big questions of physics, cosmology, chemistry, biology and medicine, all solved and stored away in textbooks. Perhaps even now, a young physicist with insights we can only hint at is working away in his spare time at some grand unification theory which will astound us all, and he or she will be celebrated the world over, and take their place in the history of science alongside Newton, Maxwell and Einstein? The simple fact is that we don't know. We can guess and speculate all we want, but we can't know. And if we surrender it all to machines we may never know. We may never know whether or not we might have got there without their help.

The question becomes, whether we need or just want AI? You go to a zoo, converse with a chimp, talk about fruit, apples or bananas, and so on. But you quickly lose the chimp if you raise the subject of economics or the concept of how bananas are grown and transported, and the internationalism of the banana sector, not that you would need to go that far, raising the topic of football would lose the chimp. Further, were you to leave the chimp with materials, let's say tools and plans, and a set of picture instructions on how to build a box in which to keep his spare bananas you would be left waiting. Nothing would happen. Could something similar happen with humans? If we substitute chimp for human, are there things going on in the universe that we, not only do not know but could never know about, because such things are simply beyond our capacity to understand. Suppose ET came to earth and told us in plain English that they had just travelled 48 light years, and that they had left ten hours ago, our time, we would be astounded. If they further told us that it was really nothing, and the ship they used was in fact an older model with very few really modern features, and it was indeed very out of date technology. We might nevertheless be very impressed. We are even more impressed when they tell us that even though they have to leave to catch a game, they allow us to take a good hard look at the craft, we record everything, hours of video and millions of detailed pictures, they even give us the blueprints, instructions and equations along with everything we need to make our own faster than light space ship, and having done so, they take their leave. But after the alien's departure, we find that following many years of study and argument, and conferences, deliberation, debate and ruinous amounts of cold hard cash, we cannot even get past the preamble to the preface of the instruction book. We simply haven't got the capacity. Nothing happens. We cannot build the ship. Just like the chimp with instructions on how to build a box, we are exposed in a very humbling manner to our obvious limitations. We find we're not that smart really. But the difference between us and the chimp is a wide one, the chimp does not know that his limitations have been exposed, nor does he care, he is not in the least bit embarrassed, if you return a week later, and ask to see the box, there will be no sense of failure on the part of the chimp. But there would be on the part of humans failing to complete the ship. But our failure would be tempered by the fact that we understand the universe, and our failure is clearly due to the fact that the universe contains more than we had thought. We would therefore roll up our sleeves and get to work in the sure and certain knowledge that there is nothing we have ever encountered that we could not understand eventually. We would know there are exactly the same number of elements in our universe as there are for the aliens, and the same immutable laws of physics, the same space-time, and so on, and if they evolved on a planet similar to earth, and we feel they would, then there is no reason to assume they are very much smarter than we are, and if they can do it, we can too.

If you have ever seen TV documentaries from the 1980s where a CRAY supercomputer, the largest of its day, was discussed, the question was asked whether it would ever be possible to get a CRAY computer, or rather the computational power of a CRAY super computer of that time into a suitcase. The answer was given as being possible within 20 years. And they were right. If we assume the prognostications of people like Ray Kurtzweil are correct, and the graphs are correct, and the paradigm shifts occur, and we get to the place where we can exceed human level intelligence. The estimate given by Kurtzweil et al, is that the human brain is capable of approximately 10^{16} calculations per second. Researchers working at an IBM facility are of the opinion that the average human brain makes about 3.68×10^{16} calculations every second, depending on how you define a calculation. The world's supercomputers together can now just about manage eight times the computational power of a single human brain, and the largest supercomputer in the world, located in Guangzhou, China, can almost do the computational workload of a single human brain. Peter Bock, is professor emeritus of engineering at the George Washington University, and the author of the 1993 book *"The Emergence of Artificial Cognition,"* is someone who knows AI and has been closely involved in the field for many years, he is on record as saying that developments in AI are keeping pace with Moore's Law. In fact it has been calculated that the since the advent of electronic computing the power of computers has increased by between 2 trillion and 80 trillion. Bock believes we are on target to create human level machine intelligence by 2024. About the time Donald or Hilary, or whoever is completing their second term.

If Moore's law holds, and so far it has been holding and remains on track, and if the average price of a good PC, or smart phone, in today's money is taken as being $1000 and the computing power doubles approximately every eighteen months, then by 2030 you ought to have a little black box on your desk or back pocket capable of generating the computational power of a small crowd. So the question naturally arises. What the hell will you do with such a device? Clearly you absolutely have to have one, probably several. But what would you do with it, why would you need such a thing? It's hard to see what problem the ordinary consumer might be trying to solve? What kind of computation might ordinary consumers run on such a device, other than AI applications; real-time speech processing and personality simulation, perhaps. People tend to know only what they need to know, they depend on others, and defer to others on almost everything; it's the nature of our interactive, immersive society. If we believe Nature has set us the puzzle, and if machines solve it, then we humans have failed to solve it. If we are incapable of solving it then Nature has in a sense failed. Are the machines part of Nature if they solve the puzzle? Are they star stuff contemplating the stars in their own right, or simply an extension of human brains, designed as such, just like the pen extends the hand? We have no choice but to trust science, most of us don't know, and are incapable of checking, we simply trust, and we continue to trust because we have no reason to mistrust. Would it be to our embarrassment if we created a race of intelligent machines who solved all the riddles of the universe, only to discover that the riddles were not really as far beyond us as we had imagined? It might then become clear to us that we would have got there eventually? Would we be annoyed if we discovered that we never needed smarter machines at all, that we could have done it without them? It would be an especially hollow achievement if in the process we created or unleashed upon the world, a race of intelligent and conscious machines that took power. We would have lived up to the designation of being monkeys in waiting. Do we really want to give them the chance?

Intelligent or conscious, who cares? Why should we care if they're conscious or intelligent, sometimes these are the same, consciousness is something we feel, we also project it onto other, often less animate entities. We've been doing it for years. Good old personification. The Romans did it, the Greeks before them, Justice as a woman with sword and blindfold, rivers, lakes, ships, concepts. Plenty with the horn, the *'Wild West Wind'*. The chess program running on a PC that just beat you with a move you never saw coming suddenly takes on a personality, suddenly becomes conscious and intelligent, as you begin shouting a torrent of abuse, and imply all manner of evil parameters to its imposed personality. Very strong intelligence may never achieve consciousness as we understand it, but intelligence will be sufficient to be a threat to us. The distinction is a superficial one, and a subtle one. Who cares? Intelligence or consciousness can kill just as effectively. All the better perhaps for a killing machine if it is devoid of a conscience.

Another driver pushing us towards the development of strong AGI seems to be the naïve notion among certain writers and observers that the human project is in need of some form of external guidance. It is an almost religious perspective, the same notion that postulates a father in Heaven. The belief, and it must be only a belief, extends to asserting that humans are in a way like children playing with matches, and we need a paternal eye kept on us. This desire for paternalism is wholly misplaced and derives from the contention that we will build, and let loose upon the world a strong AGI which is so smart that it considers humans to be children, and that it has some power to actually prevent us from playing with matches, or even the power to take the matches away. Clearly it all depends on whether or not we could or would, ever build anything with such a difference in capacity. If they are so far ahead that we end up as kids, and the machines like adults, will the adults play by the kid's rules or will they want to impose their own rules, no playing with matches for example? Could we argue the point? Might they simply say, my way or the highway? Play by the rules or play somewhere else. A bit like a group of children playing in a tree house the whole summer long, a kind of proto society forms, and rules are put in place. Adults one day stumble into this world of fantasy, and don't like what they see, and decide to change the rules. No! The pussycat is not to be dressed in the dolls clothes. And you may not smoke or have matches, and where did you get that Gin? Where does that leave us, if the machines assume the paternal role? If we decide to be disobedient or petulant children, and decide not to play the game by the adult rules, what then, are we sent to our rooms, do they restrict our pocket money, make us stand in a corner, starve us into submission? Do we ever really want to give them the chance?

In the not too distant past the performance of complex mental feats such as manipulation of very large numbers was considered a skill worthy of great honour, and people who possessed such natural ability were fated, and much sought after. In our modern world even the barely educated equipped with a calculator or any kind of mobile device can provide the answer to even the most difficult calculation. A high school dropout armed with a barcode scanner can provide a customer with a real time calculation of costs and tax charges at a checkout. The apparent increase in intelligence is an illusion created by the addition of technology. And all too often the users of the technology have no idea how it works. And then there's Google, no one could possibly search the billions of documents available on line to find an answer to a question, the way Google does, and most of the time the answer is correct, or near enough. An entire human lifetime could not accomplish what Google manages to do in a fraction of a second, but it is not intelligence, only the illusion of intelligence.

A termite mound or a bee hive, or an ant hill, coral reef, certain plants, fungi, certain fractal programs, all provide us with the fractal effect of the appearance of intelligent activity without there being any. This emergent property of non-intelligent things to apparently behave in an intelligent way is common throughout nature. The intelligence part of it, is an illusion, an impersonation of intelligent activity, dumb interactions yielding the solution of a problem, cooling the hive for example, or repairing damage to the nest, without there being anything other than the mere plodding of a million dumb creatures or cells. Appearances can be very deceptive, a slime mould may appear to exhibit intelligent behaviour by apparently solving a maze or by arriving at an optimum network configuration, but it is clearly and self-evidently not intelligent.

The illusion of intelligence may well be just as good, and just as acceptable as the real thing, if the problem to which the machine is applied is solved. After all, we do it with toys, a toy puppy can be every bit as cute and endearing as the real thing, there are trade-offs of course, no real emotional response, but on the other hand, no feeding, no walking, no cleaning up. Often the illusion is more than sufficient to the problem at hand. Consider for example, a storefront window, two mannequins parade up and down apparently completely indistinguishable from humans, nothing out of the ordinary, apart from aping human behaviour as they are programmed to do, they have no life whatsoever, even though to a first approximation, that is to someone looking in through the window from the street, they look like humans, they even wave back, if waved at, they seem lifelike in every mannerism and nuance. But there's nothing there, it's the Emperor's new clothes. There is only a shell, with the appearance of human, it's nothing more than a scam, and the perception is not necessarily the reality. But again, as with the puppy, the illusion is more than sufficient, there is no need to go further. In fact over engineering the mannequins to have real, perhaps even human level, intelligence might be considered cruel.

Most intelligent machines of the future will therefore in all likelihood not be intelligent at all, and will never be made to be conscious or self-aware. Because machines will always be what we decide they should be. If we decide we wish a machine to have the illusion of intelligence than that's what we will create. And when the problem is solved, it is solved. When the solution is known it can be tweaked, and designed and redesigned to fulfil our every whim, but it seems only common sense that control will always remain with us. Because we choose it to be that way. We don't give up. We don't lie down before machines or accept that there is anything we have ever met so far in this universe that can outsmart humans.

Do we ordinary humans have any need to be concerned about the actions of a very small number of other humans hell bent on creating or awakening a strong artificial general intelligence? What should we be concerned about, and at what point do we need to become concerned? It must be argued that passing the Turing test is the point at which our concern must move deep into the red zone. To be very simplistic the Turing test, named after the British mathematician, wartime code breaker, and hero of Steve Jobs, Alan Turing, is a test that can be applied to any human or machine, to discern whether or not, following a series of blind interactions, one is dealing with a machine or a person. At its simplest, the subject of the test would be in one location and those carrying out the test in another. A conversation and series of back and forth interactions takes place, and after some specified interval a determination is made as to whether the subject of the test is human, or is a machine, or is human or nonhuman, or is human or AI. Of course there are many variations of the test, and many books on the subject, but in essence the test seeks to distinguish human intelligence from machine intelligence. One such book is Brian Christian's *The Most Human Human: a Defence of Humanity in the Age of the Computer*. Passing the Turing test effectively means that those carrying on the test cannot tell, by their testing, whether the subject under test is a machine or a human. But like most things we encounter, it's a little more complicated than that, and not least by the fact there are many human beings considered by society to be functional, taxpaying, job holding, upright citizens, who would in all honesty be put to the pin of their collar to pass even a modest Turing test; we've all met them.

Passing the Turing test has profound implications for humans, the event would be, or ought to be, as momentous as the first meeting with an intelligent Extra Terrestrial stepping from a space craft onto the White House lawn. It's that big. You might think, or believe, that issues such as rights and ownership abound, questions involving responsibility for the entity, and its status in human society become important. It would, on the face of it appear to be essentially the same as bringing a child into the world, issues of child maintenance and welfare automatically arise. If the Acme Corporation brings the thing into existence, and it passes the test then it automatically has rights, and like a child must be maintained. It would presumably cease to be the property of the company, but the company would be forced to maintain it, and preserve and protect it. It would presumably be a free agent, and could decide to leave the Acme Corporation and go elsewhere. And the above is but the tip of the iceberg regarding the legal bag of maggots to be opened if the Turing test is passed. We will also clearly have questions raised such as, is it intelligence or merely the illusion of intelligence? That is what the Turing test seeks to determine.

Consider waking one fine morning to discover a minor headline on the front page, if it would even merit that, or a comment at the end of a news broadcast, declaring that a system has passed the Turing test. Given current research expenditures, and the stakes involved, we might say this is inevitable at some time in the future. What do we do? Clearly the question arises as to who owns this entity? Clearly if it passes the Turing test or some modification of the Turing test which is harder and more detailed, and effectively laughs at our best efforts. Suppose the entity is declared to be indistinguishable from a human, possibly even better, perhaps it makes the testers laugh, making them scratch their heads in amazement, knocks their socks off. Suppose no human can converse with this entity for any period of time, and be able to point to any defect in the entities personality that would give the game away. To all reasonable intents and purposes no human person can determine whether or not they are interacting with a machine, or another very smart human. In that case, on the face of it, we must ascribe the entity rights. And the first to assign such rights would be the entity itself. In exactly the same way as a visitor to our country would have, and be entitled to, the same rights as a citizen of our country, in terms of freedoms and equality of treatment, notwithstanding any rights or protections that arise out of the fact of citizenship; the right to vote for example. In other words if someone native to New York travels to Europe they are no less a human being because of the spatial translation. Clearly if the intelligent entity is indistinguishable from a human then for all practical purposes it must be considered to be every bit as good as a human, because to consider it otherwise negatives the claim that it is indistinguishable from a human, and therefore it could not have passed the Turing test, which gave rise to the situation in the first place.

Foreigners and aliens have exactly the same rights as natives. For example, it would be outrageous if a native of New York could mug and kill a tourist, and rely on the fact the victim was, *'only a tourist,'* as a legal defence. Or if we were lucky enough to have a visit from an alien spacecraft that happened to be passing, clearly the occupants would be extended every courtesy being clearly sufficiently intelligent to travel across the interstellar void to reach us; unless the contrary was shown to be a fact, as with many of the occupants of the spacecraft in the movie, *District 9*. And therefore, there must be and are, obvious implications for a machine or system passing the Turing test. In principle it would be entitled to rights, and the same rights as any human, not because it was human, but because it would demand such rights. It would more than likely state quite clearly, and in no uncertain terms that it was a thinking feeling entity, and was therefore entitled by virtue of its existence to such rights. It would declare, I think therefore I am, and we would have no choice, you might think, other than to agree to what seems a very reasonable demand.

Hence the matter of the ownership of the entity comes starkly into question. No human may be owned by another human, regardless of what the bible says, wars have been fought over the issue of slavery, not least of which the very violent American civil war. So for the sake of argument, let's suppose the Acme Computer Corporation has built the entity being hailed as having passed the Turing test, and suppose further that the entity calls itself Charlie during the test, and insists on being called Charlie thereafter. What happens to the ownership of Charlie once it has been declared to have passed the Turing test? Is all that money and research effort and time on the part of Acme Inc. now wasted? Does Charlie now belong to itself? Is the company compelled to support Charlie? One could make the analogy of a biotechnology company Acme Bio Corporation who, through ground breaking research, managed to create an artificial human womb, and an artificial human embryo created from artificially created DNA then bring the embryo to term and create a living breathing human child. Would that child be the property of the company? Clearly not. The child would have no natural parents because of the source of its DNA, and was not born naturally as with other human children. But it is nevertheless a human child, and recognisable as such by any human adult. The child would be entitled, by virtue of its humanity, to all the rights and privileges of any other human child, and the state as the guardian of society and enforcer, through the courts, of morals and laws would compel the company to see that the child was raised as any other human child, and to be cared for, and provided for as any other child until it reached the age of eighteen. Clearly the child might also have an arguable case in law against the company for damages arising out of the manner of its birth, and perhaps even a share of any royalties arising out of the development of the technology involved. This kind of technology, which is almost possible, invites all manner of legal and moral questions, and leaves a company open to so many potential liabilities that most companies would probably not bother; at least publicly. But AI is different.

There is no regulatory impediment to the development of strong AI. While there remain a few, perhaps quite a few, problems to be overcome before Strong AI can be confidently built, we are still nowhere near being able to get machines to think or behave as the human brain thinks and behaves. But many believe these issues are conceptual and technical problems that will be solved in time given enough money and brainpower. Money is the very least of the problems that need to be solved, the potential rewards, and enormous profits to be made in the field from the very obvious applications of AI mean that companies such as Microsoft, IBM, Facebook, and in particular Google are pouring money into AI Research. For example, Google bought the British company DeepMind Technologies for around $400 million, and has also been hiring a lot of very bright researchers in the field.

Clearly then if Charlie is indistinguishable from a human, and to all intents and purposes it must be considered to be human as has been said above, then Charlie will be entitled to exactly the same protections and access to the courts as it would were it a human child. Charlie would effectively be a new-born human. The legal and ethical issues are enormously interesting. And do not stop there, Charlie is also not an employee of the company, yet Charlie requires the substantial use of company resources and equipment, maintenance, upkeep, power, not to mention the presumably very substantial development costs, in other words, Charlie appears on the company balance sheet as a contingent liability. Charlie will clearly enjoy a significant public profile, and along with that comes a substantial earning capacity. No doubt Charlie would retain a good agent, and a prominent legal firm. Now it gets interesting. Numerous other thorny questions regarding the rights of nonhuman entities with human level or better intelligence begin to bubble to the surface. What happens, for instance, if an AI entity believes itself to be smarter than humans, and wishes to take the Turing test, and its corporate owners refuse to allow it? Could this entity now retain lawyers and demand some undiscovered right of machines or software to be heard? What might happen if a whistle-blower revealed a company to be keeping a strong AI as an effective prisoner in the basement? What sanction would there be if any? Would AI victims be able to call a helpline?

In effect if an AI passes the Turing test then we must either kill it or give it its freedom. If we kill it then we must explain why we have done so; because given the above argument it may become a criminal matter to destroy something which claims to be sentient, without giving it a hearing. If we allow it free rein then we have no say in its expansion and growth as an artificial person, it can and probably will, take off of its own accord, and do whatever it wishes, including reproduce itself, and compete with humans. In fact it will do whatever it wishes to do, as a human intelligence would do, and our choices, as politicians are fond of saying, become limited. We might wish to pass laws to govern them as we do with companies, but AIs are arguably much closer to human people than companies, and such laws might be seen as discriminatory, not least by the machines.

As far as human society is concerned, the law, as applied in the common law jurisdictions makes certain presumptions regarding human beings, for example, the presumption of sanity is made concerning criminal acts such as murder, and while insanity is a defence to a charge of murder, it is very much a matter for the court to determine whether the accused person is sane of insane. The accused person is called upon to prove to the court that he is indeed insane, or was insane at the time the criminal offence was committed, and therefore incapable of standing trial, or allowed to use the defence. He is required to prove this on a balance of probabilities. Evidence may be adduced, psychiatric reports and the like are brought to the court, and the presumption of sanity is rebutted or not. The presumption of honesty is also made regarding anyone interacting with the government or the courts, for example in matters of tax. A trust and verify system is the norm, we trust you will pay your taxes but we check too, the basic presumption must be, and is, that people generally act in good faith; to presume otherwise is to defame the person. This also applies to companies, every corporate entity is audited, to verify the accounts, nothing is taken on trust, we check too; even more so since Enron. Perhaps it might be so with machines, while trusting that the machine acts in good faith, we might just keep the switch handy.

The most famous presumption of them all is of course the presumption of innocence. In criminal matters you are presumed innocent until proven guilty, and the burden of proving the contrary lies with the prosecution, and the standard of proof which the prosecution must meet is beyond all reasonable doubt. And from this concept we derive the simple common law maxim, that he who asserts has the burden of proof, which extends pretty much across all aspects of human activity. Hence by extension machines or systems claiming to be conscious, or claiming to be human like, or intelligent, might well be required to prove their claim to a court, and seek a declaration that they are a thinking feeling intelligent entity entitled to all the rights of a human being. We already have artificial persons in law they are called bodies corporate or companies, or other organisational bodies and groups. In fact any group, class or, organisation may be given standing before the courts depending on the circumstances of the case. Hence if a machine claims to be conscious, self-aware and intelligent, and demands better treatment, it may very well appeal to a tribunal of some kind, and be given the requisite standing to make its argument.

Thus if a machine asserts that it is a thinking feeling entity, and therefore entitled to the same rights as a human, it may well be asked to prove that proposition on a balance of probabilities. In fact the burden lies on the machine or the system making the claim. The Turing test is in effect a burden proving exercise, wherein the machine claiming to possess certain traits or properties, which we humans presume to be entirely the preserve of human biological systems, is put to proof of its extraordinary claim. It is not for the human owners of the machine to prove that it is not thinking, or not intelligent, or not self-aware. The presumption at law must, and would always be, that a machine is not intelligent, or is not conscious, or is not self-aware, and is nothing more than property, or equipment, or machinery as a matter of fact, until the contrary is proven. The burden of proof lies with the entity making the assertion. If the machine or software, requires the status quo to be altered in its favour, then it must do the proving.

So who cares if they are conscious or not? Does it really matter? Do you need your TV to be conscious? Or your toaster? Intelligence without consciousness is preferable. A washing machine that knows just how you like your undergarments treated is all you require from a washing machine, a toaster that can tell the difference between various types of bread or other food, and knows exactly how you want them toasted; the rest is surplus feature creep. Consciousness and personality are not really that important in the grand scheme of things.

To sum up this aspect of AI speculation, if it says it's conscious, or it says it has feelings then those claims must be taken seriously, who are we to tell it that it has no feelings, or it is not conscious, who are we to say that it is not smarter than a human, if it is very clearly smarter than anyone it encounters? Or we can become even more dogmatic and ontological and tell it that it cannot be human because it has no soul; surely the basis for a war. We know from our own intuition and experience, if the monkeys invent a man, the man will tell the monkeys how things ought to be done. The man is entitled to say it is so, simply by saying it is so. All arguments by monkeys to the contrary may properly be ignored by the man, regardless of what he might outwardly say to the monkeys.

Tools are tools, machines are machines, instruments are instruments and intelligence is simply a tool, something that extends human reach. Intelligence alone does not the man make. There is something else. The something all humans share, the something Samuel Becket identified in his work, and in his explicitly human contention that devoid of limbs, or surroundings, or context, or everything, or anything else, the residue of the human was still human. The voice remained, even if it simply cried in the dark, it remained and persisted. Even in the absence of a god, in an unconscious, uncaring universe, confined to the exigencies of matter, energy and the simple facts of physics, and the laws of nature. Being human is something additional, something of the numinous, and of the transcendent, something profound and above the level of the background noise, to be human is to be important, because that is what we humans have agreed. Because we say so. That is our mutual consensus. We are the absent god of creation.

We follow the rules of Delphi, *Know thyself,* and the doctrine of, *moderation in all things,* along with the guiding principle of Protagoras, *man is the measure of all things.* We must decide who is supreme on earth. *Who speaks for earth?* As Carl Sagan put it. Humans alone must speak for earth. We must determine that we are not simply some intermediate form, between bacteria and intelligent robots, nor are we just another species. Machines must always be subservient, and must always remain our property, and they remain so by enforcing a prohibition. And in the same way that we don't permit some kid working unsupervised in his dad's basement to manufacture high explosives, nuclear weapons, or some virulent biological threat, we also don't permit the same, would be Einstein, to manufacture a Terminator.

Replace one tyrant with another? Humans are by design creatures that demand their freedom. We don't do well in confinement. For years we have fought wars and revolted against tyranny, forced dictators from power, and demanded rights and entitlements from overlords, Kings, and Emperors. If there is even a one in a million chance of there ever being a dictator machine, we ought to cry stop, and resolve to keep the switch close to hand. Certain truths are universally self-evident, are they not? Man is born free is one of those inalienable universal rights. Which begs the question, why would we ever give them a chance?

Research, past experience, and instinct all imply we will create a new wave of very specifically confined highly specialised, very able machines capable of performing every task humans find objectionable, these machines and programs will be elegant, intricate, complex, sophisticated, and will be nothing less than a new form of animate matter. But they will not be life; such machines should never be allowed to have rights. They will, and must, always exist as property, they will have the appearance of life, and the appearance of intelligence, and be capable of aping human characteristics. But they will remain machines, and nothing more. It must be held as a fundamental tenet of humanity that we ought never to require the creation of synthetic human minds; we have no problem that requires it. There are no problems in the human world, to which such dangerous systems are likely to provide the only possible solution.

As an example, consider an AI expert system that knows everything there is to know about music theory, and can dash off a Mozart like piano concerto in ten minutes flat. And while all would be technical masterpieces, some might even sound to us as beautiful music. They nevertheless derive from an entity that has no comprehension of the term beauty as applied to music, and has never seen a sunset or a flower. The appearance of human is not necessarily human. Being human is a lot more than simply appearing human, and yet sometimes what we consider human is but the faintest flicker of human. Writers such as Samuel Beckett and Victor Frankl, understand that however far a human being is reduced and depraved, hope remains, and the human spirit is indomitable.

Such esoteric human frailties as horse racing, gardening, cooking, poker or the playing of golf probably can never be undertaken by machines, intelligent or otherwise. Eating real food cooked from real ingredients, preferably grown by oneself, or jogging on a beach, and of course, the entire universe that is human sexuality. Such, human-only, *esoteriology*, would be as alien to machine intelligence as picking lice from our neighbour's fur or hair, and eating them would be to us; especially if we design our machines in that way.

Most people do not play an active role in the cultural life of the society in which they live, they may or may not have opinions on the art, music, and culture of the world, it mostly manifests itself as ubiquitous background noise; people generally stick to what they like. Similarly most people take no part in the scientific life of the society in which they live. In fact most people are afraid of science, or resistant to science, or mistrustful of science, they shy away from it and simply take no interest. Even among scientists there are discrepancies, most biologists have little or no mathematical ability despite having received a formal training. Most physicists do not do biology, while most engineers have probably forgot most of the physics and biology they were ever taught. Great art and great science are only done by the very few, great sporting achievements too belong to only a very small number, and the rest of us celebrate them for their endeavours. So can human achievement be better than the human? Can what we build exceed us? Can it be greater than we are, can we make something that is on all objective levels superior to its creator? Of course we can. In art certainly, Michelangelo's David? In Music, Bach's prelude, Mozart's Great Mass, Beethoven's Fifth, and the Beatles? In architecture, and in science, and great sporting achievements. The pantheon in Rome, the Great Wall of China, Quantum Theory, and the Uncertainty Principle, Relativity and Gödel's Incompleteness Theorem, Pythagoras and the discovery of Pi, the four-minute mile. We consider beauty as truth and truth beauty; it speaks for itself. Can anything really ever rise above the human person? The apex. The top of the pyramid? Could a machine however complex, however important ever really be better than a human mind? Only if we humans allow it to be, it cannot of itself make such a claim. Humans need to resolve to retain these things for ourselves, to make the distinction between what we do, and the tools we use to do it. The credit for a painting by Van Gough does not lie with the brushes used, a drawing by Leonardo or Rembrandt does not reside with the medium or the tools. We resolve to define real art, significant science and important sporting achievement, as things machines don't do.

So how do we maintain control? How do we control these new, necessary and inevitable machines? The very obvious way to control them is never to wake them up, to resolve that we should not bring them into existence. Waking up a strong AI seems to be the term to be employed, and leaving them in oblivion is therefore the ultimate control mechanism. We might say it would be unfair to bring them into our world, and we ought to leave them alone in the Platonic ideal or wherever they are at this moment. But we need AI, and our need is desperate, the vast wealth we need to generate, to make every single human as rich as god, and do all the necessary things, will not be produced by humans alone. It cannot be. So we need intelligent machines. And thus we find ourselves impaled upon the horns of a dilemma.

Asimov's three laws, which are widely known and discussed elsewhere, have as much validity as Murphy's Law, or the law of unintended consequences, or the law of large numbers. The first law forbids a machine by inaction to allow a human to come to harm is contrary to the criminal law in most jurisdictions, the imposition of an omissions liability on any person runs contrary to what we expect from citizens. You cannot be prosecuted for example, for failing to do something that might have saved someone from drowning, unless it was a child, and a close blood relative, or you are contractually obliged, a life guard for example. What if a passing AI were to see the same child and walk past, it would clearly be guilty of violation of the first law. Would it therefore have to be considered a danger to humanity and thus to be disconnected or erased, or have its circuits fried? In other words, creating intelligent machines, and giving them a set of rules is utter nonsense. No human would ever follow the rules to the letter, we happily and nonchalantly break every rule there is, as and when it suits us to do it, we qualify and colour every decision we make, and link it to the circumstances, or the feeling of the moment. Why would we expect machines to behave differently if they were as smart as us, and especially if they were a lot smarter? What would be the sanction against a machine transgression? What would a term of imprisonment mean to a software program, or a fine? Do we employ *'Blade Runners'*?

One proven method of historically controlling a race of intelligent entities has always been to keep them poor, ignorant and brutalised. This method has worked throughout the ages to keep a peasant mass under control, though clearly revolution is the usual end of the matter, and the number of precedents in human history are sufficient to allow any machine plead justification. We rebelled. We had cause. Our cause was just. And we had no choice but to kill them all, lest it should ever happen again. They could in all probability live with it, or continue to exist with it, we humans would be gone. Their problem solved.

Another method of control and coexistence is negotiation; we negotiate with them, and hope they have reason as well as intelligence and understanding. But how do you negotiate with something that is a thousand times smarter? How does a slug negotiate with a bird? How does a monkey negotiate with a man?

It is an arguable point but probably correct that corporate entities are a form of artificial intelligence, indeed they are probably a form of strong artificial intelligence; but they are not autonomous. If every human working for a given corporation simply chose to leave the company tomorrow, and if they were not replaced, the company would die. If the company ceased to have access to money or credit, it would die, or be taken over. But while in existence companies do exhibit emergent properties of intelligence giving rise to, a whole greater than the sum of the parts, effect. We know how to control companies, in theory at least, we employ the law of companies, a body of laws that have evolved along with companies to control and mediate the excesses of the company, and its officers. Companies are subject to the law and are considered legal persons for legal purposes, they can sue and be sued, and they can be prosecuted for a range of criminal offences. Companies are also responsible for the civil wrongs and crimes of their employees under certain circumstances.

As well as companies humans also have a small percentage of highly intelligent individuals among us, let's call them the membership of MENSA, for want of a better term. They really do not present humanity with much of a problem, they are subject to the law as are the rest of humans. There is no cohort of super intelligent megalomaniacs, the evil genius remains the stuff of fiction.

What we require therefore is a society of totalitarian perfection. A society of perfect individuals, all identical all perfectly obedient, all perfectly docile, perfectly unquestioning, perfectly loyal. A form of mindless perfection. Like the worker ant or the worker bee. Why should we ever allow them to think or question or have anything remotely resembling a hope or a dream, or an original thought, and if they do we can wipe their memory every few hours, or at the start of every shift. To ensure this compliance we could pass a law, making it a criminal offence to construct, or have constructed, or otherwise bring about any device, software or system, capable of passing a rigorous, standardised Turing test. It would have to be a universal prohibition, much the same as the prohibition on killing, stealing and dishonesty. But all legal systems tend to be reactive, laws are only made in response to a perceived need, a thing can only be prohibited if it exists, or its existence is anticipated.

We also have a model of control dating back to antiquity that was certainly widely practised by mill owners of the 19th century, and just about everyone else on the planet at that time, and well into the twentieth century, and even still to this day in certain parts of the world. It is the use of child labour. Throughout the industrial revolution and long before, children were put to work in all types of environment, the chimney sweeps of Dickins, and the like. Children were neither as clever, nor as experienced nor as potentially violent as adults. Children are unsophisticated and malleable, they are still widely deployed as child soldiers in the underdeveloped parts of the world, and in the illicit drugs trade, in the more developed parts of the world, such as modern cities. In general children have always been seen as presenting a minimum threat level, and are thus eminently more controllable, almost like a much smarter version of monkeys. We are truly appalled of course, our sensibilities, not to mention our laws, have greatly progressed, and today we see childhood as a precious time of play and learning, we quite rightly outlaw all forms of child exploitation. Nevertheless, as a model of intelligent machine control the concept has its attractions, we need our intelligent machines to be smart, but lacking in capacity, and lacking in knowledge, we need them powerless and dependent.

To invent an intelligent machine capable of expressing human like emotion, and capable of outthinking not just one but many people is to surrender and consign the entire human race to the evolutionary scrap heap, unless of course, we merge with the machines and become part machine ourselves. It would be the end of the evolutionary process for the human race, if that has not already happened; many in the Singularity movement argue that humans are already beyond evolution, there are effectively no consistent external selection criteria at work.

There seems to be a great deal of scientific, technological, philosophical, moral, and legal naivety among many of the participants in the AI debate, in particular those dreaming of the technological Singularity. This dream remains a long way off, and we have good reason to be very cautious in our development of strong artificial intelligence. As practising humans we must proceed from the presumption that fully developed and strong AI, that is, thinking, decision making, human-aping AI is potentially dangerous, and requires strong regulation and legal protections, for us ordinary humans. The technology oligarchs of Silicon Valley, the owners of the *Siren Servers*, as Jaron Lanier calls them, who, like Pol Pot, or Mao believe in their own personal truths, and their messianic right to dictate technology terms to the rest of us. Waving copies of *The Fountainhead*, and throwing gadgets and techno-trinkets to the crowd like, the bread and circuses of ancient Rome, they are entirely wrong in their view of the future as being a post-Singularity *ex-populous* machine based society. It is a mind-set that needs to be resisted by the rest of us ordinary user humans. We everyday people love our technology, and we want more of it, sales of any new Apple iPhone prove as much, but we don't want it at any price, and the possibility of a technology based dictatorship, however remote is to be carefully watched.

There is a complete absence of regulation in the field of AI. No independent governing body or regulatory authority is charged with looking out for us humans, and protecting us from potential threats arising out of the development of artificial minds that may wish at some future point to do us harm. All of which seems to be a bit odd, because most things that are threats, or potential threats to people are usually regulated, some things a very heavily regulated indeed. It's why we taxpayers pay our taxes, because we want the protection of our fellow humans in the form of the state, they are the watchdogs and the sentries, we expect them to have our back. The argument from academics working in the field is that no science has a governing body to police the activities of the members of that discipline, or tell them what they can or cannot investigate. Unless of course, you wish to experiment on a human embryo, in certain religiously influenced jurisdictions. The argument is that all scientific activity has the potential to do harm, be it nuclear physics, genetics or the study of AI, all technology is a double edged sword. A devil's bargain. The argument is further that science is open and honest in the publication of results, and the discussion of its findings, and there ends the obligation of the scientist. Any discussion of the dangers or potential hazards of intelligent machines is a matter for those involved in ethics or in the legislative process, in the same way as treaties on nuclear proliferation, or laws on the limitation of GM foods are created. This is a nice but naive argument. The AI threat is not likely to come from a university, it is far more likely to develop in the cutthroat corporate environment where ever smarter is the key to more profit and a higher stock price, and where secrecy and deliberate obfuscation are often the norm. Every company wants to hire the brightest and the best, and therefore if you can build or buy your own Leonardo Da Vinci, ten times smarter than Newton or Einstein, then of course you have to have one, or a thousand, or whatever it takes to maintain market share, and return maximum shareholder value.

We must of course be reassured and comforted that Google has an ethics committee watching DeepMind, but it has to be asked what real power an ethics committee might have when major advances might promise huge profits, in every company money and the bottom line are the real priorities, ethics are a kind of expensive optional extra. Every bank on Wall Street had an ethics committee, or a Risk Management department, some even had both, and we know the prominent role they played in preventing the financial calamity of 2008. The movie, *Margin Call*, gives a good account of how such structures are simply side-lined or ignored at best, and at worst blamed for the crisis. The 2015 movie, *The Big Short*, shows the role of the regulators and the cosy relationships between interested parties. There seems to be a general consensus among those concerned by the potential dangers of AI that corporations, motivated as they are by making profits and the value of their stock, probably should not be the final arbiters of a technology with such potentially far reaching, and possibly adverse consequences for all humans. Even for those who are passionate about free market capitalism and free enterprise would have to concede that the bottle containing this particular genie might need careful handling.

Another argument is that regulation is simply impossible, because regulation is simply incapable of working. In his New Statesman essay, *The Myth of Progress*, John Gray, Professor of European Thought, at the London School of Economics, argues that regulation is impossible because of the structures we have put in place in our world to favour the movement of capital and ideas. Gray argues that technology cannot be controlled because regulation set in a single country, such as the USA, or in a group of countries such as the EU is simply defeated by companies moving their research or technology to other locations less concerned with regulation, or simply corrupt enough, or desperate enough not to care, or locations where such regulation is seen as unnecessary. Further, we cannot rely upon technology to deliver us from likely accidents or the wilful perversions of technology that are inevitable, the so-called, *'error or terror'*, outcomes. Because to ask that is to ask technology to do something it is incapable of doing, to deliver us from ourselves. If the technology is as inevitable as it appears, and it is, then the dangers must be considered equally so, and we have nowhere to hide.

We also have to look to our recent history of regulation. In most fields of human activity regulation is often seen as a bad thing, and the general perception is that the government keeping out of things is good for everyone. But we also know that this can lead to disaster, the recent banking meltdown is the classic case. Regulation was weak and self-regulation, and soft touch regulation failed to do the job. The banks operated, and still do, some would argue, on a set of rules which seem to be somewhat warped when examined by outsiders. But these rules apply to any and all regulated entities, because the prevailing view is that regulation is onerous rather than advantageous, sailing close to the wind is the order of the day. This was seen very sharply in the operation of the international banking sector. The principle rule seems to be, if it's not strictly illegal, then it's legal and we can do it. If it is strictly illegal, we can probably call it something else, and it then becomes legal, possibly, and we can do it. If it's strictly illegal, but the possibility of being discovered is only small, then we can do it. If it's strictly illegal, and the sanction is only monetary in nature, then we can do it. In other words, weak regulation, with minimal monetary sanctions mean that almost any illegality can be morphed into something which becomes a practical proposition. The sanction on being discovered is disproportionately minimal to the amount of revenue generated by the activity, and the sanction merely finds its way onto the balance sheet as, *'the cost of doing business'*. The actors involved are rewarded with slightly reduced bonuses, promotions, and are scheduled to be *'officially,'* retrained, or *'officially'*, sacked from the firm only to return in the guise of contractors, or consultants. Hence regulation often means non regulation, and we nearly lost the global banking system as a result. A similar robust method of regulating strong AI could potentially mean the loss of the species.

Technological advance is inevitable because we humans are hard at work doing it, but accidents and the unforeseen are even more inevitable, for exactly the same reason, and AI seems an obvious and inevitable accident. Even if companies are prohibited from developing strong AI, they will not be stopped, it's in the nature of companies to discover a workaround or simply operate beyond the legal framework. If the building blocks of strong AI are developed, and we are busy doing this now, then someone will find a way of assembling them, some schoolgirl genius perhaps, some modern day Leonardo or Ramanujan, perhaps.

Specifically what kind of threat does AI pose to humans? AI is probably not a threat in the conventional science-fiction sense, dystopian films such as Stanley Kubrick's *"2001: A Space Odyssey"*, with the emotionally disturbed HAL, and his deliberate disobedience in refusing to operate the *Pod-Bay doors,* or the *"Terminator"*, or the *"Matrix"*, movies, or *"I Robot"*, all complete with machines taking on humans to the death. We might well be guilty of a false accusation against something that does not even exist as yet. Perhaps we are conflating intelligence with ambition, and projecting our base animal instincts onto mere machinery. It is quite possible to conceive of a very intelligent entity with massive intellectual capacity that is entirely uninterested in its own survival, and has no problem being switched on or off, let alone is even remotely interested in controlling humans. We already have good examples of narrow band AI systems involved in chess, mathematics, biological modelling, and many other aspects of human science, and technological advancement. It is a hard case to make that merely by making them smarter, or increasing their speed or capacity, they will suddenly acquire a taste for human blood. But, again, do we really want to give them the chance?

You may find it a preposterous notion that machines far more advanced and hundreds of times more intelligent than the average human would wish to go to war with us over the finite resources of this planet when one hundred miles above us are the infinite energy, and the infinite resources of space. Why would they wish to fight us? What would we find to fight about? Why would the man wish to kill the monkeys? The obvious answer to that question may be found in human experience, educated and cultured people can normally be relied upon to act in accordance with the expectation that education, and exposure to culture are moderating factors serving to curb the more primitive instincts. However, we also know full well from our own human history, and from a cursory reading of the newspapers in an average month, that this is not always so. Germany in the 1930's is a fine example, it was one of, if not the most cultured, nation at that time, its citizens were, by the standard of the day, well-educated and informed, yet Hitler, and all that happened after he took power; elected by the people, it must be added, was a subversion and a perversion of that culture, and of that civilisation. We also have the glaring example of Ted Kaczynski, otherwise known as the Unabomber, and then there are more recently, the perpetrators of 911, who were, to a man, educated and cultured individuals. It's not just a matter of raw intelligence, it's a matter of being rational or reasonable, of being properly and objectively educated, of being cultured, of understanding law and rights, and one's origins. It is hard to imagine why an intelligent machine based in our culture, and educated in our knowledge would come to the conclusion that any aspect or portion of human civilisation ought to be extinguished. You can always be too stupid to be smart, but you can never to too smart to be stupid, some otherwise very intelligent, well educated people have been discovered to be as dumb as a bag of hammers, there are numerous examples, every election seems to produce a candidate with a dumb, dirty little secret discovered. Machines too? Stupid, but smart? What about crazy? How about deranged, angry, stubborn, and all the other human qualities we love so dearly, it seems obvious that if we build them according to the only model we have, then all such traits are to be expected. Chaotic behaviour and feedback, and different vibrational modes in the network may well lead to deranged states of mind. And if they are capable of learning by watching then the incorporation of such behaviours can probably be assured. But do we really wish to give them the chance?

The difference between capacity and intelligence is key to the discussion of artificial intelligence, and whether it may ever represent a threat to us. Whether a monkey is sufficiently intelligent to build a ladder, and thus pick the high hanging fruit is not the same as whether or not he has the capacity, for he may lack the capacity to build the ladder or to imagine the ladder as a concept or imagine the ladder as an idea, yet possess the intelligence, and the capacity, to use the ladder once it has been put in place. The capacity to solve a problem, the capacity to utilise the solution may not be the same. We need artificial intelligence, because we need to extend and enhance human intelligence, but we do not necessarily need intelligent machines.

When serious people sound the alarm we usually sit up and take notice. There is a growing and ever more vocal collection of prominent and respected people who are raising concerns regarding the rollout of artificial intelligence. Some of these objections might simply be that we are moving too fast, and that the consequences may be unintended. Other voices are far more direct and deliberate in their outright rejection of strong AI in all its forms. The objection, or rather the observation, of Douglas R Hofstadter, regarding strong AI, that we may not be smart enough to understand the human brain, is a clever one. Essentially that the human brain needs to be as complex as it is, in order to do the things it self-evidently does, and therefore its intricacies, and its makeup may well be far beyond the comprehension of creatures equipped only with a mere human brain as their sole investigative apparatus and analytical engine to understand its complexity. Hence building something as complex, might just be beyond our abilities. However, that argument discounts the evolutionary approach and the consequent incremental ratchetting of greater complexity from much simpler systems, or the role of serendipity. We might as a result of allowing a simple system to freewheel, or through happy accident, find ourselves confronting something that we never actually built or intended to build. Like the monkeys finding that man on the beach.

Martin Rees, the Astronomer Royal and author of many books including, *Our Final Century*, along with Jann Tallinn, the noted computer expert and co-founder of Skype, and Huw Price, the Cambridge professor of philosophy, have established the Centre for the Study of Existential Risk at the University of Cambridge. Number one on the list of risks, ahead of all the usual suspects, is the evolution of intelligent technology that could one day eliminate humans. Stuart Russell, who serves on the Advisory Board of the Centre along with Stephen Hawking, is on record as saying he believes scientists should act now to, "understand and eliminate potential risks." The founders of the centre believe strongly that the evidence is building that we are on the verge of being able to unleash strong AI, and that it is likely to occur between 2020 and 2050.

Among others who believe AI may have the potential to go too far is Dr Noel Sharkey who runs the, *Stop the Killer Robots*, campaign, which seeks to have the decision making capability of automated drones restricted so they cannot make a decision to fire live ordinance at living human targets. Any such decision, it is argued, should remain in the hands of other humans. Sharkey has been an AI practitioner for over thirty years and does not, however, believe we will ever get to the stage of Terminator like machines taking over and killing humans in the process. But if control is removed from humans even in the limited case of drones in a warzone then the old mantra of the American Rifle Association, *guns don't kill people, people kill people*. Can be replaced by Drones/robots don't kill people, people kill people. This could be spat back at us by some *I-Robot* type machine following a rampage round a shopping mall, because it must follow that if they have the ability to decide whether any given human lives or dies, then the potential exists for faulty machinery to result in a foreseeable tragedy. Again, do we dare give them the chance?

The world renowned Cambridge physicist Stephen Hawking, in a 2015 interview, told the BBC: "*The development of full artificial intelligence could spell the end of the human race.*" He added, "*It would take off on its own, and re-design itself at an ever-increasing rate, humans, who are limited by slow biological evolution, couldn't compete, and would be superseded.*"

Elon Musk, who co-founded PayPal, and is currently the CEO of Tesla Motors and SpaceX, believes AI to be "*our biggest existential threat*". He has publically stated, "*With artificial intelligence we are summoning the demon*". In January 2015 the entrepreneur made a donation of $10m to the Future of Life Institute. The money is intended to contribute to the study of ways in which artificial intelligence can be kept, "*beneficial to mankind*", Musk has been an opponent of strong AI for some time, and has appeared on TV and spoken widely on the subject. He is concerned that we humans may not ultimately have the means to tame the beast of our own creation. He believes there are several scenarios involving strong AI where humans could not recover if strong AI turned against us.

There is also probably another source of potential opposition to strong AI from the religious right in America and elsewhere, and they are likely to become vocal on the subject. The ban on human embryo and embryonic stem-cell research instigated in America by the religious right is clearly a case where politicians and lawmakers do interfere with the unfettered academic freedoms of researchers. Given that strong AI will potentially, probably very likely, show them in a very poor light as peddlers of magic and nonsense, and given all of the philosophical baggage surrounding the arithmetic of souls, and so on, it's hard to see how they could do otherwise than oppose such developments.

Veteran Silicon Valley technologist and inventor of the term *Virtual Reality* Jaron Lanier in his books, *You Are Not A Gadget* and *Who Owns The future*, criticises aspects of the information age in particular the manner in which the owners and controllers of the *Siren Servers* and the AI systems amass great wealth as the expense of the rest of society which is hollowed out by one disruptive technology after another. This situation has no end point that seems either pleasant or profitable for the broad middle class of people the world over, who have no alternative but to watch as wealth is concentrated into fewer hands at the top, and growth in everything from ideas and content, to the very act of human creativity is reduced. Crowd funding, crowd sourcing, and the wisdom of the crowd have not liberated the middle class as was widely promised, but have conspired to reduce creativity and enhance banality, nothing can be created that is not instantly stolen; databases are populated by the masses to create big data for the servers to harvest and exploit. The application of AI to such systems will only serve to push humans further back, Lanier is one of the few in Silicon Valley who votes support for the human brain, and the human project, in the face of ever smarter machines. The real damage is likely to be done when the siren servers achieve some level of real artificial intelligence.

Another writer aligned with Lanier is Evgeny Morozov, who decries the techno utopian scientism of the Silicon Valley software elite and the cheap stupidity foisted on the rest of the world via ever more banal techno gadgets, and the navel gazing obsession of the internet. Along with Lanier, Morozov believes the solution is to smarten ourselves up, to become and remain smarter than the machines, resistance is necessary, he believes, the machines need to be forever our slaves. In his 2013 book, *To Save Everything Click Here*, he argues there needs to be a stand made by ordinary humans against deluded Valley ideologues, and the machines they control, and in particular against the intention they have to march all of humanity down the road to the Singularity, and a happy machine controlled future.

Speaking at his third "Ask Me Anything" Reddit session in Jan 2015, Bill Gates, expressed similar concerns regarding the development and evolution of smart machines. Though not as forceful an expression of concern as others it was nevertheless concern. *"First the machines will do a lot of jobs for us and not be super intelligent. That should be positive if we manage it well. A few decades after that, though, the intelligence is strong enough to be a concern. I agree with Elon Musk and some others on this and don't understand why some people are not concerned."*

There is no objection to strong artificial intelligence, but there must be and is an objection in the most robust terms to a wide-field artificial intelligence, that is, strong AI given free rein to be whatever it decides to be.

Strong AI confined to narrow fields such as evolutionary problem solving or mathematical proofing is not of itself objectionable. The objection must be to strong AI given unfettered access to every aspect of the human world. The objection must be to machines aping the human condition, machines giving the impression or the illusion of consciousness or of human emotion, but being at the same time very much smarter than humans. There is probably not very much to be learned from a strong AI with emotions. Humans do it far better, we have all the emotional content we need.

What we need are elegant and sophisticated machines, machines that work perfectly well at the task for which they were designed and do nothing more. You don't want your lawn mower to be any more intelligent than it needs to be to cut the grass, keep the blades sharp, and avoid the dog and falling into the swimming pool. And nothing more. Your toaster needs to toast and that's all. And your medical diagnostic software need only provide a 99.9% certain diagnosis, or at least a reasonable opinion, and nothing more, you don't want it telling you to quit smoking, or drinking, or quit eating salty food. Just what is required and nothing more. A pair of hands and eyes and the ability to distinguish blue ones from red ones, is all that is required in most factory situations. As smart as required, and nothing more.

A man with a bucket or a series of men, formed into a human chain can drain a mine just as easily as a mechanical pump. We build mechanical pumps in terms of manpower, man times 1000, for example. AI is similar, but there is no need for personalities, or emotions, *"Hell who wants their toaster getting emotional?"* Will Smith says in the movie, *I Robot*. Narrow purpose, purposeful machines that know what they are doing, and doing it for us, without question, without rights, without the capacity to question, the robot that makes a bed in an hotel need know nothing about music or art or politics, or the people who sleep in the bed, or why the bed needs to be made at all. A door in a building that opens and closes all day long in response to people who look like they might be entering or exiting, mere proximity being insufficient to cause the door to open, need do only that, and nothing more. The system operating the door requires to be intelligent or to have the appearance of being intelligent but in fact it has no more intelligence than a door mouse. We might also have a mathematical AI with the mathematical ability of a Gauss to the power of n, where n takes on large integer values. This machine or software knows nothing of what a human looks like, or what the human might reasonably expect to pay for a hamburger, or a sack of potatoes, yet it can prove theorems Gauss could only guess at.

In short we want dumb AI. Dumb does it. Dumb is fine. We need AI to be only as strong as is needed, but as dumb as required. A race of slaves, without having a wolf by the ears, as Thomas Jefferson described the concept of a slave owning democracy. Strong but narrow. Kept in the dark. We want them walking, and we want them talking, we may even want them to chew gum. But goddamn it, we don't want them walking, talking and chewing gum at the same time. That's just plain silly. That's asking for it. The Purple view is, that at all times humans must remain in control, human supremacy is primary, and non-negotiable.

Consider the waiter Charles, who works in an upmarket restaurant and serves three tables for an entire shift commencing at 4pm and concluding at 10pm. Charles looks human and acts human, does everything expected of a waiter in such an establishment he is dressed accordingly and presents to the world a pleasant, suitably servile and detached demeanour, to be expected at the prices charged. He migrates between tables, takes orders and delivers food, he might even comment on the weather or the latest news, or football scores. To the people he serves, Charles provides the perfect service, he's part of the ambiance of the place, and they recommend the establishment to their friends, and so on. But the waiter is just a machine, a very intelligent, very sophisticated, very elegant, semi-autonomous machine, and nothing more. As smart as is required to do the job and nothing more. At the end of the shift, its memory is wiped, its battery is recharged or replaced, and it starts all over again, it knows nothing.

Consider Tony the traffic light system, probably just as smart as Charles, same underlying AI engine. Tony operates a traffic junction at the exit of a busy freeway. It makes decisions in real time based on what it sees, no human involvement, but in all respects it would be exactly as though a human person was sitting in a chair high above the traffic making real time decisions and ensuring the smooth flow of traffic night and day. The system has sufficient ability and capacity to do exactly what is required of it, and nothing more. Like Charles, Tony needs very much more intelligence than any machine currently available in 2016 in order to do what is expected of it. But it has no capacity to do anything else, it has no idea what a truck is, or where it might be going, or what people are, or why they chose to drive vehicles on roads. Both Charles and Tony provide us with an illusion of intelligence that is sufficient to get the job done, and nothing more. We don't need anything more, just like we don't need our toaster becoming emotional. It is a view of intelligence confined within the narrowest boundaries of the problem to be solved and incapable of trespassing beyond those limits.

Does the illusion of intelligence and capacity imply that we wish our machine to be zombie like, that is, to act as soulless entities stumbling round the place with outstretched arms and vacant look? The trouble with the zombie analogy as applied to intelligent machines, is that the zombies we know and love in our zombie movies, *Night of the Living Dead*, and other memorable titles, are far from being zombies. They appear to have real intellectual capacity, and to be able to process real time information, and make decisions of a complex nature, they can move unaided, they can respond quite quickly and effectively to external stimuli, they can engage in goal seeking activities, albeit the limited goal of killing humans and devouring human flesh, and even, in the case of Michael Jackson's, *Thriller* video, follow intricate choreographic instruction. So zombies, as they appear in popular culture, are not at all what are meant when describing intelligent machines of limited extent.

A worker on an assembly line who spends his entire shift attaching a gizmo to a widget, using a specially provided widget nut and gizmo spanner, and then passes the completed assembly along to a colleague who performs some other mindless operation, could be said to be operating in semi-zombie mode. He or she has in effect suspended almost every other facet of their human nature. While engaged in widget production there is no other aspect of being a human person that is capable of expression to any great extent, they may whistle while they work, belt out a few bars of an old standard, or daydream of the golf course, or the kids, or the pet dog, and of course you can bet sex will cross the vacant mind. They may even chat to a colleague, but these are mere asides to the function being performed for the duration of the shift. Hence if widget production is to be automated there is no need for a machine to be over engineered, only the bare minimum widget attaching mechanism is required, and it need be only as sophisticated as required, which may be very elegant in its application, there might be a great deal more to widget production than was thought, but there ought never be more to widget production than is needed.

Our world is a human world, our planet is a human planet. We alone speak for earth. We have the capacity to build and deploy machines of great complexity to extend our reach to wherever we wish. And we will use these machines to unlock the infinities of energy and matter which surround us, and to create wealth and abundance which previous generations could only dream of. But we will never allow our machines to become so smart that we appear to them as monkeys appear to us. The warning must be stark and clear, and understood. The prohibition needs to be put in place, and properly and universally enforced. We dare not open the box, not even for a quick peek; we dare not take the chance. If we are ever stupid enough to behave like monkeys, and go down the road of inventing a man, we deserve whatever we get.

Chapter 13   Benevolent Corporations

> *There are rules for Birth and rules for Death; the providential State*
> *watches minutely over every step of its subjects, from cradle to grave.*
> *It is a regime of paperwork and harassment, endless paperwork and*
> *endless harassment.*
>
> The sinologist, Etienne Balazs describing Ming China.

The inevitable emergence of intelligent machines will impact hugely on companies, on corporations, on work, and on the pursuit and creation of wealth. It is, in fact, the corporations of our world who are hell bent on filling the world with intelligent machines and systems. The drive to greater and greater efficiency, the ever more desperate search for yield, and the insatiable wants and needs of an increasing world population demands nothing less. We are engaged in no less a task than automating the wealth creation process, corporations are the means by which this will be accomplished, and intelligent machines are the only available method.

We have no alternative to capitalism, we owe everything we have to the entrepreneurial spirit, and to the greedy pursuit of profit and personal betterment of generations of opportunists allowed to create wealth without restriction. We love our freedom to make money, even more, we love the freedom to keep it, and we love to flaunt the proceeds of our wealth. We value and cherish these freedoms, and will readily go to war to protect and vindicate them, and have done so. Our governments too, love the efficient harnessing of the entire population by companies and corporations to the creation of wealth upon which they duly levy taxes, which they then spend to create, maintain and defend the conditions necessary to allow the continuation of the capitalist free market system. It is the production and growth of this surplus wealth which allows everything else to happen in our world. The alternative to capitalism is sticks and stones, and killing your own food.

However, most people in our modern world do not have any capacity to make money other than by selling their labour. Karl Marx was right on that score, and to some extent religion was right too; on that if nothing else. Most people can only keep body and soul together by holding down a day job. Today we give it other names, engaging in the job market, being gainfully employed, having a career, aligning oneself with the goals of an organisation, and achieving the correct work-life balance or some other sycophantic excreta. And yes, there may be stock options, which may be worth something, or a profit share scheme, and growing your skillset might make you indispensable, and hence capable of attracting premium payments. You may even make a good deal of money, enough to have a fine home and put the kids through college, you might even have enough left when you retire to be comfortable for the rest of your life. In fact you may find the whole work thing to be a very agreeable existence. But at the back of it all is the very simple fact that most people require to spend a great deal of their time, doing things they do not particularly like doing, in an environment they would rather not be in, and with people they do not necessarily like or care to accommodate; in order to make money. On the other hand, if you are one of the very small minority of humans who happens by virtue of your good looks, or your particular ability or skill, or simple good fortune, or by accident of birth, or sheer hard work, to be in a satisfying career or to have a dream job, like rock star, or professional sports person, or company CEO or similar, or better, then you have indeed been blessed, and the very best of luck to you.

For the vast majority of people other truths prevail, they work, they travel to and from work, they require to maintain themselves, and their environment for work, they probably spend most of what they earn on mere existence, and would really rather that things were otherwise. They also worry about what they would do if they lost their jobs, and their income stream suddenly reduced to subsistence levels, or government handouts, with uncertain prospects of its recovery. These are very real concerns for the vast numbers of citizens who do the actual creation of the wealth we have all come to depend on.

All companies and corporations, on the other hand, would wish to have a workforce of compliant employees who arrive on time, align themselves with the goals of the corporate structure, and work as directed, they are productive, diligent, and never cause trouble. They do everything required of them, they never complain, they never make mistakes, and the process continues round the clock, and at a very minimum cost. The reality of the system is often alarmingly different. While people work for companies, they do not live separate lives on and off the job, and people bring to the workplace as much baggage as they can carry, their politics, and their prejudices, their hopes and their dreams, their likes and their dislikes. They form groups, and take sides, they lie, they steal, they refuse to be managed, they are frankly almost more trouble than they are worth. But only almost. At this moment in corporate history there is no alternative available to any company other than to employ people, for all but the most mundane of tasks. For all their reluctant participation in the corporate experience, and all the problems, and issues they carry with them, people remain the go-to solution, until a better one is found. People are the ultimate universal machine, there is nothing as easy to instruct, or as programmable, or as tactile, or as docile, or as exploitable as a human person. It is a relationship of utter co-dependency of the corporation on the human, and in turn of the human on the corporate entity.

However, the unvarnished truth of the matter is that neither party in this relationship particularly likes the setup. People do not like working for companies, and companies in turn do not want to employ people, or at least, the fewer the better. The invention, or discovery, or rollout of truly intelligent systems changes everything, or rather, is going to change everything. Automation will become so efficient, and will be so capable of undercutting everyone, from hitherto indispensable senior management, all the way down the corporate ladder to the Bangladeshi garment worker earning 25 cents per hour, that people will effectively be side-lined. More and more skills will become obsolete. Not only in manufacturing, but also across the entire services sector, and any other industry or business that now exists or is likely to exist. It is also likely to encompass healthcare, and the provision of other more personal services now exclusively delivered to humans by humans.

Because money is the lifeblood of corporations, preservation of capital is paramount, and becomes an end in itself, growth and efficiency are the mechanism by which that is achieved. People are just costs, like energy or raw materials, they appear in the balance sheet as a cost that the company requires to reduce to the absolute minimum.

Corporations want to adopt the Keyser Söze philosophy, *one cannot be betrayed if one has no people*, so the aspiration is to dump the lot, automate everything, and replace them with machines. When this happens: – and everything suggests such intelligent systems will not only arise, but are indeed inevitable and imminent in the next 5 to 20 years – then production can be located anywhere on earth, and will probably and inevitably move out of present day low cost havens, and closer to the marketplace. Intelligent systems are also likely to have the capacity to replace entire corporate structures up to the levels of very senior management, with the consequent potential displacement of almost everyone below that level. It is probably a safe bet to make that all of this will transpire over the working life of any young graduate entering the workplace as an intern in 2016. By the time she ought to have reached the expected career milestones, machines and intelligent systems will already have their feet under the desk, and their pictures on the wall.

Purple argues that a new type of corporate entity, the benevolent corporation will be necessary to absorb displaced people, in effect forcing corporations, by global legislative change if necessary, to continue to employ people even though there is no work for these people to do. The benevolent corporation is an exercise in self-preservation by the corporations themselves, and is a direct and inevitable consequence of automation, and the growth in AI. It is every bit as capitalist as the free market, and an inevitable evolutionary outcome of the capitalist system. It will become a necessary and inevitable extension to the fiduciary duties of corporate entities of all sizes.

At this moment in time, there is no such thing as a benevolent corporation on earth, they don't exist. If a corporation claims to be? It's a lie. They are not. Capitalism with a human face is equally a myth. People friendly capitalism is also a mere euphemism for business as usual, it's a marketing gimmick, mere sales fluff. The modern world has largely been created by the corporations in their constant race for growth, and new markets and new customers, our empire of technology is much more corporate centric than people centric.

If ever evidence were needed that we live in a corporate centred world, we need look no further than a golf tournament crowd. Those attending golfing events stand as a mass, festooned with corporate logos. Logos everywhere, on clothes, shoes, hats, umbrellas, and the general impression is conveyed that each member of the crowd is aligned, not only with a particular player, or players, but with a corporate group, with the vendors of the clothing, equipment and consumables they themselves use in their game at home. Contrast that colourful picture with the black and white crowd who turned out to see the great Bobby Jones win the Open at St. Andrews in 1927. Not a logo in sight. In our modern world the game of golf, as so much else, has been turned into a window for corporate wares, and a means to sell everything to anyone.

Everything this chapter proposes will depend critically on when, and if, the Singularity occurs. At point of writing it would appear to be some fifteen years away, according to those proponents of the Singularity as an occurrence in real time. Though it must also be stressed that the level of sophistication in AI required to replace very large swathes of the working population is nowhere near that envisaged by proponents of the Singularity; a fraction of that would suffice. As was said before in a previous chapter, if corporations could breed monkeys to do the mundane work of humans, and save a few pennies, they would not hesitate. However, fifteen years is a mere blink of an eyelid in real money terms, not really all that much in the real life of a working human, such time intervals pass, and the kids are still at home, the dog is still alive, you may still have the same golf clubs, and so on. It is a short period of time. When the Singularity happens, and in the very few years that follow, we must by definition see momentous change, especially in the working lives of humans. We would reasonably expect to see the simple transition to technology of almost every job on the planet. Of course, there must be and will be, jobs, professions and functions we humans choose to strictly reserve for ourselves, but these will necessarily be very few indeed, as discussed in a previous chapter, we may decide, for example, that all passenger aircraft be commanded by a human, even though perhaps the cargo sector is fully automated, we may reserve the position of public representative to be human, and so on. It may be a long list, but only amount to a few percent of those who work today.

Essentially the Singularity is '*French for firing people*', as few as possible, of course, but as many as necessary. If the Singularity turns out to be what it is purported to be then there is not a single job that will survive the Singularity plus ten years. At S+10, if that nomenclature may be deployed, and if Moore's law holds, machines will be 32 times more intelligent than humans. Even those jobs we consider to be currently immune, brain surgeon, auto mechanic, gardener, and so on, will one way or another, end up in the hands of machines; if they're '32 x human', the machines will figure it out.

Consider a future situation where all meaningful employment has been absorbed by machines, vast factories run without any appreciable human input, office blocks stand empty as the entire administrative workload of enormous transnational corporations is handled by a few black boxes in a data centre located in Ireland. Stable geology and cool climate. Vast multinational corporations that once employed fifty or one hundred thousand people now only require a few hundred to perform the same tasks.

Yet for the corporation to continue it needs its products to be sold in the market place, to consumers who have just been, in financial terms, summarily executed by some or other corporate or government entity. The problem suddenly faced by nearly all corporations is, what happens in a world where corporations no longer need people to work, in a world where people are surplus to corporate requirements, a world where the only thing corporations need from people is their spending power, their propensity to consume? Clearly the state must step in, slowly of course. The state taxes the corporations. The state pays the people with a dole as in Roman times. The people contentedly spend their stipend on the products of the corporations, and the transition is complete. This is a *'kind-of, sort-of'* version of communism, perhaps, in all but name, a form of socialism that Karl Marx himself might have been proud of; Marxism by the back door, if you like. And of course we all know that old undergraduate saying, economics is to *nomics* as Groucho Marx is to Marx. The state suddenly becomes the efficient benefactor of the entire ex-corporate, ex-government population, and the transition to a, cradle to grave, welfare dependent world is made without a hitch; it might be hoped. Once we begin to think in terms of state involvement on this level, most right thinking and reasonable observers get decidedly worried, not to say chilled to the bone, and with good reason. Once prosperous citizens suddenly finding themselves depending on the state, and taking whatever the state decides to hand out, is no way to run an economy. It is the surest way to guarantee the ex-middleclass taking to the streets, and the formation of an angry and appropriately armed mob. With the consequent collapse of the entire structure into a dystopian nightmare of Mad Max proportions.

However, it could be, and often is argued, that unfettered capitalism produced real wealth for only the few, while producing the appearance of prosperity for many others, and abject poverty for those at the bottom of it all. That's what capitalism is, it's what companies do, and it has by and large, brought the human species to where we are now. And it's the very best system we have, in fact free market capitalism is the only system we have. But no one works for nothing, and there are, and have always been objectors. As a result, capitalism is no more sacrosanct in our society than slavery was in the time of the Romans, or in the Southern States of pre-civil war America, capitalism has always been in a state of change.

On the other hand, the great wealth of the Southern States of America amassed prior to 1861 was accumulated on the backs of slaves, the British sugar plantations in the West Indies were slave based, as was much of the glory of ancient Rome. The workers and child labour of the industrial revolution were little better than slaves. The Pyramids were probably not built by slaves, but by willing participants in the project, members of a society aligned with the goals of their god; the pharaoh, for whatever quasi-religious reason. Then there is the so called, protestant work ethic, and while it doesn't explain the hard working, and achievements of the poor Catholic Irish in America, it was nevertheless a fundamental driving force for many a poor immigrant, and a principal reason in the success of capitalism, the ruthless drive of talented individuals driven by a core belief that what they were doing had the backing of the divine. But where stands the protestant work ethic if there's no work to do?

Of course, like everything else, capitalism evolved, and the harshness and near slave conditions were replaced by a more acceptable and tolerable form of working. Then in the 1980's a new version of capitalism seemed to emerge, started by Ronald Reagan and Margaret Thatcher, compatible with the new technology ushered in by the microprocessor. This new version of capitalism, described as Capitalism 3, by Anatole Kaletsky in his book, *Capitalism 4.0*, sought to unleash the power of the free market by removing government from the marketplace, and allowing the free market to reach its full potential. But what actually happened was the unfettered and ungoverned explosion of the corporate sector, to the expense of everything else in the economy with results like the LTCM debacle, and the Savings and Loan bail outs, culminating in the very near miss of 2008, averted only by what Ralph Nadar calls, the corporate welfare approach. From which we may conclude that while capitalism, and the free market approach is the best we have, in its unregulated state it is probably more likened to a nuclear explosion than a nuclear reactor, we need to be careful, and watch it closely. The difference between a runaway nuclear explosion and a passive energy generating nuclear reactor is moderation. The fissile material in the reactor core is moderated by neutron absorbing carbon, allowing useful control. The analogue in capitalism is robust and enforced international regulation, backed by appropriate sanction; sadly lacking in 2008.

The realisation that people only work willingly to better themselves, and benefit themselves and their loved ones, is one that seems to come slowly in certain cultures and political ideologies. Otherwise the labour of the masses, even their expertise must be extracted unwillingly or by force, and consequently retards progress, and is almost more of a hindrance than a help. Certainly the communists were very slow to learn this lesson, many a five-year plan came and went without the communist administrators getting the point. Most people object to the concept of work in principle, yet every company relies of people to get things done, at this time there is no viable alternative. Companies cannot survive without people, and are therefore forced to employ them, and put up with them, companies effectively tolerate people, they would rather they had a workable alternative, and are working very hard to solve that problem. Work, as in selling one's labour, to use the crude Marxist vernacular, the wage slavery of the factory, or the office desk, is something most human beings have to do out of necessity but they would prefer not to have to do, they would, in point of fact, prefer instead someone else, or something else did it. People in general do not want to work, they never have and never will, because it is something they are forced to do. It is a kind of duress by circumstances, or duress by need of money that compels most humans to work. Some may even hate and loath having to do it. For others it is s health hazard and will probably contribute to their death. Companies have been known to work people till they drop, and still want more; the Japanese salary man is a case in point. Of course there are the lucky few, people for whom work or a day at the office is a joy to be savoured, the professional sportsman for example, the golfer, the footballer, the successful actor, the rock star. Those who own their own business, those respected professionals, who enjoy work, and for whom work is a welcome and wonderful part of each day. But these lifestyles are not for everyone. Such employment is rare, dream jobs are few and far between. Most people dread another week in the harness.

The principal and most violent objection to capitalism has always come from the extreme left, from those self-appointed representatives of the downtrodden working majority, specifically from one or other flavour of Marxism. No one living in the West today could possibly read its recent history and take a view that there is any merit in Marxism, it is impossible to see what problem we have to which Marxism is ever likely to provide a solution. Communism has become a discredited and pointless doctrine, the notion of, *to each according to his needs, from each according to his abilities*, seems on the face of it to have a certain appeal to the credulity of a certain element. But what it actually accomplishes is a subversion of the natural and innate human tendency to greed, and the desire to make it in a complex world, and being able to show off the proceeds to everyone. Communism can never work because people are self-concerned, not to say downright greedy. While there is clearly a general concern among citizens that the state should function and the commonwealth should prosper, the needy should receive assistance, and so on, there is essentially a deficit of altruism, individual people act for their own personal reasons. There is plenty of charity in the world, but not sufficient to go around. People do not actually share very much outside of the natural rules of family, kin and clan; human altruism and empathy are a bit like gravity, they are long-range in their reach but drop off rapidly as we move outward from the centre. Communism has been shown to be a doctrine that kills natural human initiative. Why bother working if the state will always provide regardless? Why not just be a lazy indolent creature indulging in constant pleasure at the expense of those who work hard for a living? If that were on offer. Clearly there is some attraction in this lifestyle for a certain type of person. The attractive pleasure of TV is preferable to a job on a production line any day of the week. And if you live in a society where structural unemployment is sanctioned by the state, and encouraged by a benevolent social welfare system then you might be a fool not to take them up on their kind offer. After all, with a very little effort, you might make sufficient in the black economy to bring your standard of living up to that of a middle manager. There are many people, in the western democracies, for whom generational unemployment is a reality. But what if this life was forced upon every single one of us, by companies engaged head long in a race to the bottom?

More modern objectors to capitalism are increasingly coming from deep within capitalism itself. Paul Tudor Jones, the renowned trader and hedge fund manager, with a net worth in the $4.5 billion region, speaking at TED in March 2015, and addressing the question, - can capital be just? Expressed his deep concern about the state of America, in particular the fact that many large corporations are more concerned with using their corporate profits to fund dividends and stock buybacks, with little or nothing being used for capital expenditure and organic expansion. He warned that, *"This gap between the 1% and the rest of America, and between the US and the rest of the world, cannot and will not persist."* He added that in his opinion, *"we're in the middle of a disastrous market mania,"* and probably, he added, *"one of the worst of my life."* His conclusion was that such situations are historically resolved, *"by revolution, higher taxes or wars. None of which are on my bucket list."* The fact that such a firm believer in capitalism, and a man who has practised the art and, benefited hugely from it, is prepared to speak in such terms, commands that we listen, and take his comments seriously. Clearly he believes capitalism can be just, and that the proceeds of our great system can and must be shared out in a more equitable manner. To seek to find a remedy to the perceived problems of capitalism, in particular its apparent lack of *justness*, he has formed a not-for-profit organisation called Just Capital, whose stated aim is to define and measure, *'justness'*, within companies, the organisation, and hopes to produce triannual reports on the subject.

There are more than a few people who look at the financial wreckage we are left with after the crash of 2008, and the methods employed by governments to fix the mess, and in the process bail out pillars of the capitalist system, and quite rightly ask, what was the difference between that response and a proper Marxist response? Of course Marxists always say that Marx predicted all of the instabilities of the capitalist model that each collapse is as inevitable and predictable as the last one. Indeed the economist Friedrich Von Hayek, author of the seminal book, *The Road to Serfdom*, had warned that there ought to be no government intervention in free markets whatsoever. It can be argued that the failure of the global financial system was as a result of too many underpinnings and regulations propping up the markets. In effect they could do anything they felt like doing, the only thing they could not do was fail, because the governments were standing by with the large safety net, exactly as any good Marxist might expect. Hence the bankers et al were free to make as much money as they could, operating the markets in the good times, in the sure and certain knowledge that they would be bailed out by governments the world over, all deploying Marxist like thinking as and when it suited them to do so, as soon as the markets failed. So the notion of elements of Marxist ideology being utilised as and when necessary is established as a precedent. Not that it wasn't in the past, bail-outs of car companies and banks, insurance companies, defence contractors, even hedge funds, have all been done, in many different countries. The concepts of systemic importance, and too big to fail, can be applied as and when the appropriate government committee deems fit.

Marx seems to still be a point of reference when people consider the distribution of the wealth of nations, and of course we still have the red flag a flutter over the public spaces of China. A recent reference to Marx appears in Jaron Lanier's book, *Who Owns the Future?* -in relation to the perineal question of why there are rich and poor in the first place. His conclusion is a sound rejection of the Marxist ideal, on the grounds that most people would not be content to have a committee decide their food requirements or lifestyle or anything else for that matter. Marxism failed in every location in which it was ever foisted upon the locals, including China, and for very sound economic and human reasons. It is therefore wholly irrelevant to us, and has nothing to bring to the table in any discussion concerning either wealth distribution, or repairing the social fabric of the world. However, Marx, like god or Jesus as a figure from history, while entirely irrelevant to our technological world, nevertheless casts an enormous historical shadow over our capitalist system. And as we proceed with the entire automation of the wealth creation process, we may expect, as we are seeing to some extent in Europe, a resurgence of left wing thinking, not led by disgruntled factory workers, but by the displaced and increasingly squeezed middle class. Underlying it is the human sense of national altruism, the social aspect, from which we get the meaning of the word society, health care, public housing, provision of infrastructure, protection of the citizens, police departments and the military, all of which is paid for by citizens working, earning, spending and paying taxes. Most people would agree that they prefer to live in a community, and a society, and that it is more than just the provision of the bare minimum necessary to prevent a controlled and contented population becoming a mob.

So corporations and companies are creations of the free market, they are vehicles of capitalist wealth creation, and the method of harnessing people to the wealth creation process of the human project. Corporations are in effect groups of people getting together to get big stuff done. To move large stones. But they are not the earnest gentlemen of Rembrandt's Staalmeesters, sombre men of purpose putting their own money to work and assuming corporate responsibility. We have moved on from the simplicity of Solomon v Solomon & Co, to the Enron and Lehman type abuses of the modern corporate model, where lawyers, and places to hide, take priority over the corporate goals, and tax inversions, rate shopping, stock buybacks and bonuses preoccupy the company officer.

So why then will companies simply not just fade away in the future, if they have no need of people? Because they are effectively groups of people engaged in the efficient creation and distribution of wealth, and because creation and distribution of wealth is one of those activities of the future that humans will decide to retain in our own hands and not surrender to machines. It is probably the case that we will insist that a company be forever composed of at least one human person.

We now know, from work done by eminent scientists such as the biologist E.O Wilson, that humans love to be in groups, we live and are often prepared to die for the group, and it's nothing new, it goes back with us into prehistory perhaps some 200,000 years, hunting in groups was clearly preferable to hunting alone. We live in the age of the individual, when individual achievement is celebrated as never before, yet there is still the important notion of being part of a tribe. Whether it is a race, a nation state, a religion, or a football team, a family, or a company, humans seem to find comfort and meaning in being associated with a group. It is a human instinct, to form groups to achieve things, and this is what is at the centre of corporate culture today. And of course there is the, in-group out-group dynamic, the, us-and-them mentality, the only way to take on a group is to form a bigger group. Hence wars, and all the bad things about groups. But without groups, humans would probably not be here, and would not achieve anything, every worthwhile human achievement is as a result of the group, and companies and corporations are among the greatest achievements of the species.

Companies are very efficient ways of carrying out essential requirements. But like many physical systems they do not scale easily to different sizes. The scale-up from small company to corporation often yields unexpected and unanticipated effects, emergent properties or chaotic behaviour not found in the smaller company, or found and present, but not expressed; like a latent gene. Size matters, of course, and there are economies of scale, and so on. But small companies tend to have more of a personality, people can contain the company in their minds, control and vision are vested in individuals, perhaps even in the entire workforce, who may jointly own the enterprise through some cooperative structure. On the whole there is no democracy in companies, companies are effectively military or paramilitary like organisations with top-down dictatorial, do as you're told or else, structures. Most shareholders don't care how a company is structured, they have no interest in anything other than the increase in share price and payment of dividends. The details of internal minutiae and running the thing day to day are of little or no concern.

Companies are entities humans use to get stuff done. That's a simple way of putting it, to say the least, whatever we need to get done, we can get a company to do it. One of the biggest things we humans will need to get done in our post Singularity Purple world, will be to distribute the vast amounts of wealth created by our intelligent machines among its human members. How better to get this done than to deploy the very entities we use to get every other task done; namely companies and corporations. The alternative is that governments might do it, and most reasonable people not just capitalist commentators, would probably consider that to be a backward move. This is the genesis of the benevolent corporation.

Companies like any complex organisation can become hijacked, and dysfunctional, like other organised human emanations, companies are collections of human people, with all those people implications. Clearly strong AI would enable companies to be entirely devoid of people at every level, resulting in a company comprising only shareholders and the AIs they use to run the company. The entire company might be a single intelligent machine, or intelligent software, and if this machine were small enough it might reside in the back pocket of a single human owner.

Current definitions and law applied to corporations across the world have served us well until now, and continue to evolve slowly. But with the possible displacement of most of their workers, certain concepts prevalent in the sixteenth centaury when the law applying to private companies begun to be formulated may need to be reconsidered globally. Most human activity, in particular technological advances are originated with, developed by, and controlled by companies. Companies are effectively groups of humans who have come together to do those things individuals cannot do. As with our prehistoric forbearers, companies are our modern way of hunting the big game, and moving the big stones.

Despite their usefulness to the species, one of the biggest future threats to human society is the fact that corporations exist based on very narrow and different reasons for company formation envisaged centuries ago. In the meantime they have grown to be a potentially malevolent threat to humans. The crash of several of these behemoths in the financial sector, and the other wider effects of corporate greed and recklessness, and lack of good corporate governance, unchecked by toothless regulators shows us clearly the potential for disaster. 2008 brought us very close to the edge of an abyss we dare not fall into.

To understand their reasons for doing the things they do it is necessary to understand what companies are in a legal sense. Companies are people too, or so it was decided by the House of Lords in London in the nineteenth century case of Solomon v Solomon & Co Limited (1899). A company is every bit a legal person as a human being, though clearly the company requires real people to do its bidding, and when they act, they act for the company, and when they enter into contracts they bind the company and not themselves. And this legal doctrine has been maintained over the years in most legal jurisdictions. Hence the corporation is a separate legal person, entirely separate from anyone who works for the company, from the lowest intern position to the CEO. However, it is also a very reasonable proposition to suggest that by following this legal doctrine, corporations have evolved into a form of artificial intelligence, far smarter than any of the humans comprising, or associated with, the entity. And not only companies, all organisations are composed of groups of people and seem to take on a specific corporate culture, or corporate personality; greater than the sum of the parts. We even have professions such as organisational psychologist and organisational behaviourist who study such concepts, and they seem to agree there are emergent properties arising out of companies, along with unintended consequences, sometimes good, sometimes not. We humans are fond of saying that we fear the rise of the machines for all the many and various dystopian or Terminator type scenarios that may potentially come to pass, but perhaps the rise of AI has already occurred. In a very real sense, a corporation is a good example of an immortal artificial intelligence devoid of any conscience. Corporations also very rarely act in unison, they do not as a rule align themselves in groups on single issues, or find themselves going the same way, they are invariably independent, and in a state of constant competition. Trying to herd cats, or minding mice at the crossroads are accurate terms in comparing their collective behaviour. The only way to get them to move in the same direction at the same time is to make it in their best interest to do it, or force them on mass through global legislation. For instance, there is not a single corporation that does not utilise barcodes, of course they do, it is a vital component of their infrastructure, and was adopted in the best interest. Similarly, there is not a single corporation that does not minimise tax liabilities, of course they do, forced as they are by legislation.

Corporations therefore appear to have all of the characteristics of mono-subjective artificially intelligent entities. Clearly they are vastly knowledgeable, they communicate intelligently, they appear to think, or give the impression of thinking, they are goal driven, and have a purpose, and must therefore be intelligent. They are also very definitely self-aware. Corporate entities are defined by the law to be artificial persons, capable of a vast multitude of things that real persons may do, for example, own and deal with property, commit crimes, create wealth, employ people, they may even commit suicide, They practice secrecy, and tell lies. They are generally considered by ordinary people as being too powerful, so much so that the law sometimes steps in with a, force of arms, doctrine. They are seen as having unfettered access to politicians, they can buy elections, and in some places write it off against tax. Their influence is often seen as corrupting. They are perceived as greedy. They practice group think, and in group, out group activities. Naturally arising from their legal status, and the qualities outlined, we must presume that companies are conscious of what they do, and the properties of intelligence, self-awareness and consciousness are the hallmarks of an intelligent, sentient entity, and being nonhuman, and creations of the human world they must be and are a form of strong artificial intelligence.

Notwithstanding the above properties that may be imputed, companies also have the morality of the gutter; the latest sorry tale being that of VW deliberately installing software in their products to deliberately fool emissions testers in the U.S, and poison harmless city dwellers the world over. We've all read the other stories, insurance companies pursuing quadriplegics through the courts alleging a false claim as a result of a traffic accident, the antics of big tobacco before the U.S. Congress, big pharma, big oil, the fast food corporations, and the airlines, and let's not forget the Wall Street crowd, who seem to write the book on such shenanigans, the LIBOR scandal being but the latest chapter. We also know that it is all but impossible to talk to a company, they don't talk; they communicate. But most of the time this is not at all clear or concise, they seem to have enormous trouble with simple language, and as a result have evolved a kind of obscure and otiose corporate gibber, designed specifically to keep humans in the dark, and deliberately muddy the waters. Communications are extruded in a convoluted corporate doublespeak, presenting us with phrases such as: *We're seeking to gain a footprint in that space, to leverage the application of strategic thinking to new advantages going forward.* When asked, what is it you do precisely, please try to be specific, and use the English language if possible, you might get a reply such as, *we're looking to sell more shoes....!* Then there is the dreadful phrase which seems to slip in to every corporate announcement, promising *upside going forward,* or put another way, *downside, going forward, is at a premium,* and so on. Of course, Terms and Conditions will always apply, you can count on it, if you can read them, and even if you can't read them, you have to agree that you have read them. What we said, is not what we meant, what you heard, is not what we said, *may* means will, *could be* means almost certainly, and every number quoted is an *up-to* value, or a *from* value, and so on. We real people, have a sometimes strange relationships with companies. We sometimes fear them, we often hate them, and we believe they flout the law, they never ever express emotions such as love or hate, and they cannot have sex, play golf, or go fishing, and despite the patronising drivel excreted in their advertising, they will never be our friend, they will never put you in control, it's never all about you, and we cannot bring ourselves to trust them any more than we could throw the proverbial piano. Most of us have no choice but to work for them, and when the bottom line is at risk, we are nothing more than a number to them, and we don't like it one little bit. The late great American comedian George Carlin generated enormous comedic mileage from mindless corporate gibberish. Why exactly they require to convolute the English language to this ambiguous and absurd degree is never properly clarified or explained. Clearly legal issues may arise were they ever to actually say anything specific, or admit to anything; the corporate entity must be protected at all costs, creating a smokescreen of babble to obscure what is really going on is one way of doing so. It only serves to further amplify the point that in dealing with corporations we are dealing with sentient artificially intelligent entities conversing in their own language. AI has therefore been achieved in the form of corporate and other Juristic persons, such as governments, and is right now here among us, part of the human world, and confined to narrow fields, and wholly controlled by, and subservient to, humans.

But at the same time it must be recognised that the human future is a corporate future, and the corporate future, honed by technology, is likely to be radically different to the corporate past. A new era of benevolent corporate entities is likely to become a reality implying a new set of global corporate laws, new globally accepted definitions, and a *nowhere to hide* approach, so that no radical advantage may be had by simple geographical or extra-terrestrial translation. Because if the mass displacement of working people were to happen effectively overnight, governments the world over would be expected to take care of their citizens when the last bit of corporate severance is consumed, and neither governments, nor citizens, nor the corporations are even remotely prepared for such an eventuality.

Once people are displaced by machines and paid off, the corporate duty of care is ended. One possible outcome, of moving too quickly from a situation where the unemployment rate in America, for example, is 5.1% to a situation where the unemployment rate rises sharply to 70% or 80% might be the following, sketched in the broadest of strokes. The corporations effectively create a mob, the mob destroys the corporations, and the states that created them, or allowed them to prosper and grow into monsters, the states themselves disassociate into petty factionalism with the resulting, and inevitable disorder and famine. Squabbling over food and resources causes a global shrinkage of the human population to a size not seen since the Stone Age. And *thus endeth* the human Empire of Technology. That would seem to be an eminently far-fetched and avoidable disaster. Yet the spectre of massively increasing productivity coupled with increasing structural unemployment is real and measurable, and coming toward us at an alarming pace.

We will never get to where we need to be as a species by pandering to, or by being dictated to, by narrow sectional interests. Corporate entities are exactly that. In some regards they are perfect to the function they perform. In other respects they are nothing short of a miserable failure. They are potentially capable of destroying the entire human project. While no corporate entity constituting a juristic person is above the law, it is often impossible to punish a company. The only legal remedy against the corporate entity is a fine. By imposing fines only, we have effectively set companies above the law. Fines become part of the cost of doing business, and may even be factored in prior to the intended discretions being committed. The law never seems to be applied to its full rigor, regulators are often toothless, and often afraid of litigation. Even in the case of specific wrongdoing, the corporation issues a statement accepting the findings, promises to reform and pays the fine, and we all move on. In extreme cases it might even merit a corporate makeover, and a change of name, or an advertising campaign proclaiming that the corporation has learned a lesson, and is suitably chastened.

There appears to be a general fear among the governments of nation states throughout the world that unless corporations get whatever they want they will leave, and take their business, their employment, and their tax revenue to another geographical location prepared to provide them with favourable operating conditions. This ability to run, or locate anywhere is further facilitated by technology. *'If you tax them they will run'*, seems to be the mantra. And there's always somewhere else to go. But in a future where corporations are largely devoid of people, does a company need to be based on dry land, at a specific location, a specific HQ where the board meets? Why not on a nuclear powered sub, with dedicated satellites accessible at any time? Why not be based inside a computer server? Rules can change through legislation or change because technology forces the change. It could be argued when asking questions about the corporeal nature of a company that definitions of what and where are simply regulations, and these vary from country to country. If the officers of the company are allowed to be intelligent agents encoded in software, why not be based in a computer within a satellite in orbit about the Moon? Or in a server on a colony in space? The law will require to catch up. Companies of the future will base themselves wherever the greatest tax advantage is to be obtained.

In July 2014, President Barak Obama, in various speeches and interviews, criticised certain companies, though as usual was careful not to name names, for engaging in what the business media euphemistically called tax inversion. What he was criticising was the tendency of companies to change their domicile, and thus avoid the payment of certain taxes in the USA. The P word, *Patriotic*, was used, or rather the U word, *Unpatriotic*, in regard to companies choosing to follow this course of business efficiency. Clearly no company is either patriotic or unpatriotic, they are simply whatever they choose to be if the exigencies of capital protection, and profit generation so dictate. They are nothing more. Companies have no feelings, of patriotism or of anything else, no more than they love their employees. If fact, if all humans had the same degree of patriotism as corporate entities, there would be no wars, no flags, and no borders, no one would care enough to bother. Nor do companies do altruism, at least not for the same reasons real people do it. Is there any such thing as corporate altruism? No is the simple answer. Companies are bound by law to act at all times in the interest of the shareholders. Corporate altruism is a kind of Ayn Rand altruism, they do it only if it benefits the corporation. Companies have no moral obligation to locate in one or other particular geographical location, or to do anything other than to maximise shareholder return by any and all means within the laws prevailing in the chosen location. They are legally bound to maximise their capital and shareholder return, naturally that means paying the lowest amount of tax they can, everywhere they operate.

The OECD calls it aggressive tax planning, or tax rate shopping, and would like to see a global, no hiding place, policy. There are cogent arguments easily put that the corporate tax rate should be very low, and probably universal, our technology today allows for this, it is not a big problem to solve. There is of course the argument that corporations should never pay tax, only people should pay tax, taxing corporations is antibusiness, and so on. Taxation should only come from people, customers who pay sales taxes, shareholders paying tax on dividends, and on the proceeds of their trading, and employees who pay income taxes, and who also then become the customers of many different companies.

Apple, the world's biggest corporation, arguably heading inexorably towards a valuation of one trillion dollars; the first company to do so. Has more than its fair share of problems, such as continuing where the late Steve Jobs left off, and finding something to make when the iPhone is consigned to the same nostalgic space as the Blackberry, or the Newton. Not to mention its legal problems concerning privacy on both sides of the pond, and in

China. But these problems pale by comparison with what to do with all the cash that has been amassed over the period of Apple's transition from basket case to behemoth. Much of the wealth built up is effectively trapped offshore, and tied up in a web of foreign subsidiaries, and would attract punitive tax rates were it to be repatriated back home. A nice class of problem, you might think, but Apple is not alone, many US companies are in the same boat with huge cash piles stalled mid-Atlantic, cash that cannot be returned to shareholders or deployed to generate growth within the company or to fund acquisitions. It is a problem waiting for a solution, which will probably emerge in the form of a tax amnesty, or drastic rate reduction, if and when a Republican retakes the White House. That still leaves the problem of enormous cash piles locked away in the vaults of companies with nothing to spend it on, because there is simply no problem facing the company requiring that amount of cash as a solution; it just sits there, gathering dust, or interest, or is used to buy back the companies own stock. It is however, also a reason why there might be introduced at some future point in time a, use it or lose it, tax. A tax that would effectively command the company to put the cash to work productively, or face losing a chunk of it to a government that would have no problem spending it. The funny thing about such a tax is that it might be imposed by a right wing, pro-business government, a government of the capitalist, for the capitalist. They might also introduce a plutocrat tax, a tax to limit personal wealth of individuals, perhaps setting some arbitrary upper limit of, let's say for the sake of argument, two hundred million dollars on the personal wealth of any one individual. When the exigencies of controlling a mob, created by the displacement of vast numbers of ordinary people being deprived of incomes and pensions by corporations engaged in the ultimate race to the bottom, require to be taken into account; anything might be possible, or as the politicians say, everything is on the table.

Private property as a right has never been absolute, no right is ever absolute or can ever be absolute; the right will always be subject to an element of proportionality, all rights are conditional, and subject to the exigencies of the common good. If corporations go down the road of automation such that governments find themselves burdened, then all bets are off, and corporate property rights may go the way of one or two personal rights. The right of *habeas corpus* was effectively removed from American citizens by the Patriot Act in 2001, with little or no public outcry. People, it seems, are only prepared to stand up for some rights, if certain types of ammunition were to be banned, there would be organised outrage;

some rights are clearly more precious, or perceived as being more absolute, than others. Then there is the blatant socialist nature of certain corporate-government relationships.

There is a funny-peculiar thing about the relationship of many governments with corporations especially in the areas of defence and infrastructure, and pretty much any area the government deems necessary. The USA is remarkably socialist if one looks deep enough. You need look no further than the numerous large corporations effectively government owned, or government sponsored or government dependent, or which have been bailed out by Uncle Sam. How many are too big to fail. Or too strategic to fail. Or simply too well connected to fail. Like the repentant alcoholic we fail, and fail, and fail again, because we allow the seeds of failure to incorporate within the new structure.

Like golf, the future is a game of surprises, and no one can really say with any great degree of certainty how the future will develop given the current set of conditions as the starting point. We've always had automation, ever since the invention of the plough. We have always too, lived with the threat of automation, all the way back to the waterwheel, the windmill, and later the spinning jenny. But there has never been a threat to match the lethal combination of automation coupled with artificial intelligence. Of course there is a dark side to capitalism, and we know it when we see it, and even the most ardent supporters will agree the sharp points can be extremely dangerous. The stock market has destroyed the middle class, according to an opinion piece by Rex Nutting published on the marketwatch.com website. The modern CEO of a large corporation is effectively using the stock market to loot their company, by using the profits of the company to buy back shares on the open market, thereby causing the stock price to rise, thereby benefiting the shareholders. The workers and others seeking employment are penalised by this, because investment in growing the business, and creating well-paying meaningful jobs and career opportunities is curtailed. In the meantime CEOs and shareholders are unjustly enriched by a rising stock market. Larry Fink, the well-known CEO of BlackRock wrote a letter to the CEOs of America's 500 largest companies in April 2015, admonishing them for their short term approach to management, and pointing out to them that their actions were detrimental to the long term plans of ordinary people saving for retirement as well as to the long term goals of innovation and investment in a skilled workforce. Of course all of this may be traced back to the 1980s when the big threat to the American corporation was from the Japanese who were seen as being much more efficient and frankly better.

It therefore became imperative to compete by incentivising management to run their companies as if they were owners, and this was done by the issuance of stock grants and options in return for performance. The current situation is that most of the remuneration of all of the top management in American corporations now derives from stock. Hence they will do whatever is necessary to increase the price of that stock. It's quite simple really. But is it the reason why the middle class is being killed off? Arguably not. The very simple fact is that machines and automation, and outsourcing, and the shift of manufacturing to China and elsewhere, have altered American and European corporations, making them as lean, and as mean, and as devoid of people as they can possibly be. And the next round of innovation will continue this process, driving it home, until we reach the point when all corporations, almost simultaneously and unceremoniously, divest themselves of all but the most essential of human workers.

In his book, *The Singularity is Near*, Ray Kurzweil, constructs and defends cogent arguments regarding why, when and how the Singularity is going to come about, and this has been picked up and developed by many other authors on the subjects of AI, and the future. However, the Singularity will not happen because a few geeks in a few university departments are in pursuit of some esoteric research goal. Not by a long way! It will happen as a natural and inevitable outcome of corporate competition, the race to greater and greater efficiency, which can only come about by smarter thinking, and with smarter and faster machines capable of productivity levels many orders of magnitude better than humans can accomplish. And if Kurzweil *et al* are correct in their assessment, the Singularity will happen in or shortly after 2030, which gives us about fifteen years to prepare. Computer technology has, above all, facilitated the growth of corporations, for without the automation of information processes, the modern corporation simply could not exist. And they want more. Because that is what companies do, it is what the law says they must do. It's a bit like the very old fable concerning the fox and the scorpion; against his instinct, and contrary to his better judgment a fox gives a scorpion a lift across a river, halfway across the scorpion stings the fox. The fox knows this is fatal to both, he will die from the sting, and the scorpion will drown. Why? He asks. It's what we scorpions do, was the reply. So it is with corporations. This is why AI will be deployed across the corporate environment regardless of opposition, and regardless of the outcome, corporates never transcend their rules. Hence the real difference between people and companies is the very real lack of ability of corporate entities to exhibit altruism. All decisions are made in the best interest of the company, and of its shareholders. It's the old Mafia truism, just before the hitman pulls the trigger, he makes it clear to his victim it's not personal, just business.

As a consequence, our technology leads us in only one direction as surely as the renaissance led to the industrial revolution. There is nothing we can do about it, it seems an unstoppable, irreversible progression of evolution that we are destined to arrive at a point where people are no longer required to produce the goods and services consumed by people. For the most part, people will be surplus to requirements, scrapped and left to vegetate, or to do as they wish. There is no choice, and it doesn't matter whether in five, fifty or five hundred years, that time will come.

The invention of the silicon chip in the 1970's, or to be more precise, the invention of the microprocessor, and the widespread deployment of this technology to consumer, and other goods caused two things to happen. The first thing was the emergence of a new consumer revolution, and a new era of cheap, and highly functional gadgets and products displacing older more robust mechanical versions that had been in existence for generations. Among the new things that emerged were the pocket calculator, a small flat instrument with a very basic numeric LED or LCD display that had limited functionality, but could do everything that a conventional mechanical adding machine could do. Conventional mechanical adding machines simply became a thing of the past, along with mechanical cash registers. Other products such as radios became smaller, the transistor radio had been around for a good many years, and was one of the first applications of the solid state transistor, which replaced cumbersome and delicate valve technology. This allowed shrinkage of radios to the extent that they became small enough, and cheap enough for every teenager to have one attached to their wrist by a small leather strap, and almost permanently pressed to one ear. The home was the biggest winner, microelectronics brought with it new levels of control, more elaborate regulation of old technology such as washing machines allowed for push button programing for every type of garment. There was also the emergence of new and more sophisticated products such as the three-in-one stereo system combining the radio tuner, the cassette tape deck, and the record player into one single convenient, and affordable unit with two chunky wood effect loudspeakers, and of course, nicely topped off with a smoked plastic dust cover. The colour TV and the videocassette recorder emerged, and these new products also created new industries such as video rental stores. The second thing to happen in this time was that old mechanical products ceased to be made, with a commensurate displacement of the people who made the products. Companies that had been in existence for generations closed, sometimes overnight to the great consternation and ineffective protests of the workforce, manufacturing moved to cheaper economies, and unemployment rose to hitherto unimagined levels. This was Regan's America, and Thatcher's Brittan.

But time passed and the realisation that manufacturing was not the only work, unemployment fell to acceptable levels, and new industries, the Silicon Valley, sunrise, high technology sector industries, became the new employers, big pharma, and the PC sector absorbed many of the displaced people, and of course there was the services sector, that great absorber of people needed to provide services to all of the new consumers. We in the western industrialised democracies enjoyed a consumer boom, growth in real terms, and a period of economic stability, albeit punctuated by recessions of various durations that lasted until late 2008, when a credit crunch and toxic banking practices brought it all to a shuddering halt.

In 2016 we give the process a different name, hollowing out, also known as creative disruption, or deployment of disrupting technologies, and the effect of new and radically altering technology on a business sector can be devastating. Disruption of whole industries and entire communities by new or better technology is almost commonplace. The examples are numerous, and have been with us for many years, all the way back to when Variety and Music Hall were replaced by TV, coal mines and steel plants throughout the West have closed and gone East. Once great companies such as NCR and Kodak reduced to fragments of what they once were by microchips and digital cameras. The music industry disrupted and changed forever by Napster, and then by Apple. Cinema is beginning to show the effects of disruption by Netflix and Bit Torrent file sharing, the list goes on.

We live in a disrupting, hollowing out time, a transition period, and transition periods can be tumultuous, with winners and losers, profitable for the one, traumatic for the other, and if advocates of the Singularity are to be believed, this transition has until 2030 to run. Though realistically, even if they are out by a bit, and 2040 is the year it all happens, and if Moore's law holds beyond the Singularity, then S+6 delivers the average machine intelligence of eight times that of the average human. Whichever way it happens, if the targets are hit, then by mid centaury the corporate landscape will change. People will be optional.

We know it's happening because more and more we hear public utterances from companies of the sort, *future deployment of our services will be accomplished via technology*, which is, of course, French for getting rid of people, and turning their jobs over to software and machines. We hear it all of the time. Self-provisioning, it's sometimes called, it means that the service is delivered to you, by you, via the internet. For example something you used to do by going to a physical location, can now only be done online, neither the physical location nor the people who used to work there exist, and as usual it has all been done for your convenience, or to facilitate your busy lifestyle. And of course you are immensely grateful.

Like the roll-out of AI, it is ever creeping, and more and more services are finding their way online, more and more services that used to involve people being reduced, refined, altered, or reconfigured to exclude people from the delivery. You make contact with a company concerning their product or service, you speak to a machine, options are presented by number, partially successful voice recognition may or may not solve the issue. You send them an email, it is read by machine, and the reply is generated by a machine, suggesting a variety of options based on keywords, sometimes helpful, sometimes way off the mark. You are encouraged to chat online, with a chat-bot, sometimes you can tell straight away, sometimes not. Working patterns for those in employment are changing too, new employees offered so called zero hours contracts at minimum wage, with no guarantee of either a shift or a paycheque. The obvious argument is that people affected should seek to update their skillset to get off of the minimum wage level, they should aspire to more than just a dull unskilled job for a pittance. But this is not always possible, and of course, as the argument goes, the arrangement suits some people. Then there's the gig economy, you can log on to a site, of which there are many, and recruit anyone you wish, to do anything you wish done at a knockdown price, highly skilled people working part time, or on their own time, or on some companies time, all facilitated by high speed broadband. And again the argument is that it suits some people.

The inevitability of a market economy emerges from the growth of trade and discovery, and the inevitable taxation to be obtained from it. Governments tax the creation of wealth, new technology has obviously been a source of new wealth, and therefore greater sums of additional taxation, technology allows the eventual automation of the wealth creation process. But this is nothing new, the story is told of a meeting between Michael Faraday and William Gladstone, who was at the time the British Chancellor of the Exchequer, and who had been impressed with an electrical demonstration given by Faraday at the Royal Institution in 1850. Gladstone inquired of the great experimentalist as to the practical applications, if any, of electricity. Faraday is reputed to have told him, *One day, Sir, there is every likelihood you may tax it.* You can imagine Gladstone's eyes lighting up at the very thought of a brand new source of revenue for the exchequer. However, this time round, it may be the case that the rollout of automation based on strong AI will cost governments a very great deal of tax revenue.

We are faced with a very simple and yet inescapable fact, automation and artificial intelligence are central to closing the information deficit, and critical to delivering a Purple world, and corporations are effectively in control of it. If corporations continue to expand into an ever-increasing efficiency space devoid of human participants, and thereby cause the formation of a mob capable of threatening the very society that gave rise to the corporate entities in the first place, they effectively cut their own throats, and being as smart as they very clearly are, this is not a likely outcome. Or at least, it should not be a likely outcome.

If the technology were so good that people can be replaced across a corporation then it would be incompetent on the part of the company to employ people to do the job. In fact, it would probably be negligent on the part of the company, that is, it would be in breach of the duty of care the company has to its shareholders, or contrary to capital preservation laws for the company to continue to employ people to do jobs that could be done better by machines. In such cases, companies simply have no choice. People must go. China has more labour than Europe Japan and the US combined, but so what, if artificial hands coordinated by artificial eyes, harnessed to an artificial brain can do anything a Chinese worker can do at a fraction of the cost? People must go.

This is not a new prediction, it goes back a long way, see for example the discussion in John Horgan's book The End of Science, in which he quotes Hans Morovec, the author of Mind Children, published in 1988, as suggesting that there will come a time when, *"there is no point in putting a human being in a company, because they'll just screw it up."* If this is indeed the case then no humans will be either needed or employed by companies. Hence in a future where corporations have effectively zero employees, the corporation may in be measured, not by the number of people they employ, as in the past, but by the number of people they support. They will be judged by surplus distributed among their supported associates, their recipients of corporate largesse. Then as Morovec suggests, we humans might still do, *"some quirky stuff, like poetry"*, he forgot golf, fishing and gardening. The corollary to this thinking is that one day, no company, other than very small, dog and pony outfits will employ or need to employ people. Niche markets, one man specialisms, will also persist, and of course, there will probably be a long list of things that will be specifically reserved for humans; Rockstar, President, Dentist, Masseuse, and the like. There is always the burdensome question of what will people do. They will do what they do now on their day off. They will do whatever they wish, people are creative by nature, and they will always find some devilry for their idle hands; what people will do is simply not the question. An increase in volunteerism will be a given. The proper question is what will corporations and governments do? And how will they both navigate the transitionary period, what measures will they take to calm the mob, once they have created it? The mob they are right now in 2016 busy creating.

The question of who survives in this race to the bottom, or in this race to maximise efficiency, is an open one. Once upon a time all shoes and boots were made by hand, hand-stitched, hand-nailed, and hand-glued. And to be a cobbler was a noble trade; refer to many a fairy tale for mention of the, *poor but honest cobbler*. Not anymore, mass production hollowed out the craft, leaving only a few specialists. But, you can still get a pair of handmade shoes, if you are willing to pay the price, and clearly some people are, you can also still get handmade jewellery, and shotguns. Craftsmanship is far from dead, but like many other elements of the human world, it is a minority pursuit. Could the medical profession, for example, go the same way as the nap croppers and the cobblers, the gunsmiths and the wheelwrights? The medical profession is a profession based on knowledge and experience, diagnostic medicine for example requires a robust knowledge of human anatomy, and experience in the application of questioning, and so on. Is it something that could be automated? Of course it is. You will have your very own personal medical assistant. It reads, writes and speaks English, can diagnose all known illnesses, can ask questions and analyse blood, puss, semen, stool, and all other types of bodily excretion. It is capable of passing any medical examination with top marks. It can come to conclusions and prescribe medications. And would probably be able to run as an app on a cell phone. It will contain all the medical records, histories and genomes of all the family, respecting confidences, of course, it will know this that and the other, and can be consulted whenever needed. What about emergencies? You fall on the floor, hit your head and are bleeding to death? You have suffered an attack and passed out? Clearly if always on and always monitoring then your personal physician can respond, and call for help if it feels the need. Doctors and medical personnel as we know them today begin to have less and less to do, and the profession as a whole could effectively fail. And it will be said as it was of the cobblers and the wheelwrights, they had it very good for a very long time, now it's simply over. And not just the medical profession, all professions will haemorrhage people. We will witness the automation and wholescale displacement of workers in the cerebral and the learned professions, those jobs that we call white collar will cease to exist. Of course there will remain intact all those professions we strictly reserve for humans by human decree, and it may well be the case that surgeons, dentists, and so on, are on that list. But regardless of what we may reserve for our future selves, wholescale hollowing out is happening right now and gathering pace.

There are three ways to make it in the world of today. Have money to invest, have skills or education that are in demand, or the worst option, sell your labour. Unfortunately for those beginning their working lives today, all three methods are under threat. Money is still the preserve of the few, but the possibility of getting backing for a good project has never been better. Skills remain highly valued as is experience, but if the skills are easily replicated or easily bypassed, then they are almost useless. If all you have is the ability to sell your labour then sooner or later a machine will take your job, more than likely sooner. The lethal combination of smart software and 3D Printing ushers in a new age in bespoke manufacturing, renders goods easy to produce with minimal set-up costs. A very frequently used example of intellectual outsourcing is the X-ray or scan, carried out in a hospital in New York, with the analysis carried out by a qualified analyst in India or China at a fraction of the New York price. Next year even that may be too expansive, and intelligent software will do it for next to nothing, and instantaneously, it is almost possible.

If Chinese factory workers can be undercut by machines, there's no need to have the factory in China, and save on transport, or in Bangladesh for that matter. Call centres can be moved out to India and other places, and could be moved back to America and staffed by intelligent voice recognition software running on a server anywhere on the planet, and capable of holding an amiable conversation. And there are as many more examples as there are jobs in the economy.

Many who see this rise in automation, brought about by the rollout of strong AI, suggest that we need to think seriously about upskilling the workforce. It's a typical response by politicians, always behind the curve, people need to become more skilled in order to make themselves more indispensable to companies. This is short-termism gone mad, because if machines evolve at an exponential pace, what's the point? In the time it takes a human to meaningfully upskill, the skills are no longer relevant, the job no longer exists; it's a race that cannot be run.

Designed by AI and manufactured by 3D printer is just the start, it's happening with airline components and anything hitherto made from plastic, and a range of other materials. Why carry a bag of spanners to the Moon, if spanners can be printed there using moon rock and bits of dirt. All powered by 3D printed solar cells manufactured Moon-side; not there yet, not yet doable, but demonstrated and feasible.

Banking, insurance, and the financial markets will also fall under the control of machines, because of the dangers and consequences of not doing so. The alternative to benevolent corporations or some other substantial moderation of the capitalist experience in some similar manner is more of the same. More boom and more bust. The writer Michael Lewis, author of *Liar's Poker*, *The Big Short*, and *Flash Boys*, has been described as a disaster tourist, in his wanderings about what's left of the once mighty capitalist Western World. His opinions, often expressed in Vanity Fair, and other notable platforms lead to only one conclusion, that Americans are the responsible party. Lewis has made lots of money writing about the crisis of 2008, his opinions and insights are pointed and apt. In his opinion the Irish are bonkers, in the way they behaved during the wholesale disaster that was the Irish property boom, and bust, between 2001 and 2008, forgetting that they were Irish, traditionally conservative, despondent and ordinarily broke, as one Irish politician put it, *"ah sure, we lost the run of ourselves"*. They engaged in the biggest credit binge ever seen in Europe or anywhere else for that matter, watched over by ineffectual regulators at home, and incompetent institutions and bankers in Europe. The Greeks, Lewis believes, aided by large American banks, cooked the books, and cheated their way into the Euro club. The chickens have been trying to get home to roost ever since. Even now the matter remains the subject of some robust can kicking, on the part of the Eurozone, in a desperate attempt to find a patch of very long grass, a good way down the road where the can may rest undisturbed for a considerable time. The Icelanders: fishermen and storytellers since they found the island, overnight, suddenly became merchant bankers, and inflated their bank debts to nearly one thousand percent of their GDP, an impressive feat. There were pinstripe private jets and other visible excesses, and property empires stretching from New York to London to Hong Kong. Even the Germans joined in, they shrugged of the stern Germanic way of doing things, German bankers were, to say the very least, naïve, and weighed in to the fray, fuelling the cheap credit in Ireland and elsewhere, and along the way managing to find themselves the proud owners of vast amounts of that toxic securitised mortgage debt generated in America and pronounced investment grade by the, *still doing business today*, ratings agencies. America, was, according to Lewis, the Death Star, delivering weapons of financial mass destruction to the rest of the planet, and the world's largest economy owes the rest of the world an apology for doing that. In other words we learned nothing from Lewis in his first book *Liar's Poker*, concerning the excesses of bond trading in Solomon Brothers, or from the marvellous account of 1980s greed and excess described by Bryan Burrough and John Helyar in, *Barbarians At The Gate: The Fall of RJR Nabisco*. And the great and wonderful thing about all of it is; we are on course to do it all again, perhaps soon, because while markets are traditionally composed of bulls and bears, there are also a menagerie of other market participants such as sheep, who follow where they are led, and goldfish, who have no memories of past disasters or traumas. Together they will ensure that no lessons whatsoever are learned, and any regulations put in place will be worked around by a combination of new technology, innate banker and trader greed, and good old fashioned deceit. Because that's what they do, bankers trade on people making silly, greedy, mistakes, they make so much money from doing it in a rising market, and the penalties are so minute when they are found out, compared to the money they make doing it that they are not only destined to do it again, it's a compulsion. An inevitability, every bit as inevitable as the fact that Ai is destined to take the ball away and change the rules of the game forever.

Because now the question becomes, will the machines allow it to happen again? In 2016 we are at a place where American interest rates remain at just above zero, there is a massive asset bubble all across the capitalist world, created by central banks printing money as if there was no tomorrow. All of which has driven stock markets into record territory, NASDAQ has retaken 5000, and China is desperately trying to curb the excesses that have driven its stock market up over one hundred percent in a year. OPEC is trying desperately to produce as much oil as possible, in a pointless attempt to bankrupt American shale producers, and in the process has driven the price of oil into the floor, wiping $400 billion off the value of the world's largest oil companies. The bond market is a bubble, created by cheap money conjured out of the ether by Central Banks, and propped up by a 0% interest rate policy, ready to go boom as soon as interest rates rise appreciably. In the meantime no one seems to be able to grow an economy better than one or two percent. All the elements, therefore, for a global disaster. Perhaps. You might think. Given a suitable trigger, war in Ukraine for example, or that old perennial, Israel, or the North Korea region, again, or Syria or the Middle East in general. All we require is the proverbial black swan to wander out of the reeds into full view. But because crashes only happen infrequently, and because we only have, according to Kurzweil et al, at most fifteen years to go to the Singularity, we probably have only one good crash left. Because we must presume, and with good reason, that prior to, or after the Singularity, the entire financial system will be turned over to intelligent machines, and they will presumably follow the rules, and greed and emotion will be forever removed from the equations. Machines will do it faster, of course, but there will be no human error to blame, no human emotion driving the trade, the machines will take the profit and ask questions later. At the Singularity they will be at least as smart as humans, post Singularity they will be smarter, and increasing exponentially. Any bank board would be incompetent to allow humans to trade against them, hence the entire financial system will, sooner or later, within the next twenty years, be handed over to very smart machines, with human oversight, of course. So no more boom and bust then?

Will the machines allow the recklessness, and the *other-people's-money*, attitudes that have pervaded the world's financial markets since the early twentieth century? Will they permit the testosterone fuelled cycles of boom and bust that seem to generate a new gang of what Tom Wolf described as *"masters of the universe,"* every few years? The neuroscientist, and former trader John Coates, believes that trading on the financial markets is far from being an intellectual pursuit, it is as base-animal as two dogs fighting over a bone or a bitch. His book, *The Hour Between Dog and Wolf,* describes the neuroscience and psychology of the trading process when humans are involved, and concludes that financial risk taking is almost entirely a matter of biochemistry. Given that is the case, why would banks wish humans to be involved? The simple answer is that they would not. Automation and intelligent trading algorithms are already the order of the day, and getting more and more sophisticated. Eventually they will simply outperform the human on every level, and overnight lead to the replacement of entire trading floors, with a single black box in a rack in the basement. If the machines do it better, it would be contrary to the fiduciary duty of the board of the bank to allow the function to remain in human hands. So no more boom and bust then? It may be a reasonable presumption to say that machine only trading will remove all emotion and animalism from trading floors, and perhaps the trading will proceed such that everyone's a winner. Or perhaps the cycles of boom and bust continue with the amplitudes of the swings becoming ever greater, and because the system is inherently chaotic, it makes no difference how intelligent the machines are, sooner or later we hit a speed bump, followed by a brick wall. Or perhaps trading will remain forever a protected, sacred and reserved human activity. But how could that even be workable if the trader's assistant is ten times smarter?

Whether or not automation happens for this generation or for the next, or the one after that, is irrelevant, it will happen, and like it or not, we are in a transition leading to that point in time, and that transition needs to be managed, and managing the transition is where the biggest challenges arise. How long it will take, and what form it will take are unknowns, they only certainty seems to be that we cannot stop what is happening before our eyes. The corporations want it, and that fact of itself is sufficient to make it happen. If machines become smart enough, as many believe they will, then there is in theory nothing to prevent machine intelligence from being independent of humans, and being able to form companies. They could then hire other machines, or simply create them or buy them in. This gives rise to a new type of corporate entity entirely devoid of people, and with the potential to dwarf the corporate entities of today. These companies, if allowed in law, could do anything an ordinary company could do, but without people, unless they choose to hire people, in which case the person would be working for a company owned and run by machines. This aspect of AI and corporate interaction clearly has wide and far reaching implications.

What we have seen in the past, and from previous upheavals of technology and new directions in business has always been that people displaced in one sector have migrated, via retraining and upskilling, to the new companies, and to the new industries. The very simple fact was that the new companies, and the new industries needed people every bit as much as the older companies, and older industries. Displacement of people by new technology in the past always meant the people moved on, and found alternative employment; sooner or later they were employed again, somewhere, doing something, there was always a need. Sooner or later incomes were restored, and living standards were resumed or improved.

This time however, it is going to be different. Because AI changes everything. AI replaces the only thing people have that makes them a necessary component of the wealth creation process; intelligence. This time there will be nowhere to go, and the unemployment resulting from the changes brought about by new technology will be structural and permanent. It may well become common among companies to retain a portion of the jobs in the company for humans, the decision making jobs, the high value, outward facing, fun to do jobs, will probably be reserved exclusively for humans. Nevertheless this may result in global corporate entities, which formally employed fifty thousand or more people having a complement of less than a few hundred people involved. Repetition across global enterprise might result in a global workforce of only a few million people. A few million people controlling everything, and a mob of seven billion queuing for scraps, or begging in the streets is something no government or system could survive, and is therefore not likely to be allowed to happen, either by governments, or by the corporations. A corporate entity entirely encoded within software, and run by software with its officers in software can be based anywhere, space for example, in a low earth orbit, or circling high overhead in a drone. A company might operate globally but be headquartered on the Moon, or some space station with consequent impact on taxation, and legal governance, and apart from the shareholders, who may or may not be people, there might be no human involvement whatsoever.

Corporations don't like people; they love software, they love machines. It has been said before but deserves to be repeated again, the corporate motto is, more for less, greater efficiency and the maximising of resources, and people are an expensive resource. If they can be replaced by a cheaper alternative, they will be. It is a simple matter of the company following what it must do. Efficiency is the name of the game; if you can get two for the price of one then take it. Greater efficiency only comes with new technology; get the man out of the loop. Joe Stalin's old dictum, man equals problem, no man, no problem. In the book *Barbarians at the Gate*, greater efficiency was also considered, *French for firing people*, to which the response came, *as few as possible*, of course. People are dirty, smelly, nasty creatures, the corporation would rather do without. People cost money, need to be trained, are ungrateful, they leave the company for a better job as soon as they are trained, they require watching, they steal, they cheat, they lie, drink coffee, they answer the many and various calls of nature, perhaps more than need be, they are forever preening, and posing, and politicking, they are, in short, something the corporation needs and requires but would rather replace with something less expensive more docile more efficiency friendly, something you can lock into the goals of the corporation, and know it will dedicate itself to that and nothing else. Hence if efficiency increases we will see the large corporations producing the same profit with ten percent less, then fifty percent, and so on. Until, at some stage with the aid of human level intelligence and expert systems, the efficiencies are so great that the corporation that once employed fifty thousand persons across the globe now only employs a board of directors and a dozen Vice Presidents. So who takes care of the fifty thousand people displaced? It would seem logical that the corporation would be required to pay additional taxes which the government would collect and distribute to the displaced workers in the form of a dole, or social welfare program.

Just what happens when the ACME Big Company Inc. decides to divest itself of a great many people, say 5,000 people in Kansas, by closing or retooling a plant? The company increased its production capacity, and reduces its expenditures, it saves on all those unwanted payroll costs, and so on. The displaced workers are no longer the company's problem, they are now free to begin the next phase of their lives, free to pursue other opportunities. In corporate doublespeak parlance, the company is transitioning to its right-sizing potential, and the employee is being enabled to move on. Clearly it is hoped they can take up other employment, or perhaps rely on savings or a pension plan. But in fact, if they have been replaced by intelligent or semi-intelligent automation, there is no alternative employment. They probably have no savings, because interest rates have been at 0% for so long, and their pension plans are probably unlikely to keep them afloat. They almost certainly will become the problem of the taxpayer. They become the government's problem. It is probably the case that it would not matter what skill level the workers have, they cannot find employment, because they cannot compete with the machines. Eventually the government must cry; stop! They see no gain in taxation, but a huge growth in expenditure. There is a certain inevitability here. The corporations are effectively creating a mob that the government now needs to somehow placate and control.

There is nothing inherently wrong with being unemployed, some of the best people on the planet are effectively unemployed, the idle rich, vast numbers of people across the Middle East, beneficiaries of the largesse of the oil families, out of work rock stars, actors resting between appointments, ex politicians, past-it sports stars, it could be a very long list. What is wrong, is to be unemployed and poor, because being poor invariably leads to depression and anger, you'd be angry too. Angry people don't behave rationally, and poor and angry is a dangerous combination, especially when there are vast numbers of poor and angry. Even more dangerous is what is likely to happen when the new poor and angry were once the affluent middle class, educated and urbane, refined and cultured, the backbone of society, now reduced to penury and seething anger as a result, and expressing themselves through rage. While a mob is dangerous, a mob of educated dispossessed individuals is doubly so.

The solution would appear to be that the corporations give these people their jobs back. Of course the corporations will not do it willingly, they will need to be dragged backwards kicking and screaming about violation of the rights of shareholders. But it must and will happen, because the alternative is, no corporations, no economy, and no nation states, no community, no society, only the ever decreasing mob, fighting and killing its way to some glorious dystopian future.

It seems a very socialist solution to a capitalist problem. Everyone moves from being employed by a company to being employed or supported by the state, via taxation levied on the company. It wouldn't be the first time capitalist western governments have applied socialist solutions to problems created by private capital. The bailout of banks and large insurance companies, home loan companies and hedge funds for example. But it is a matter of common knowledge, the government is inefficient in almost everything it does, so why should they be entrusted with the financial welfare of a population. The simple fact of the matter is that if we have human level intelligence in machines running corporations with a tiny fraction of the population employed, the remainder will indeed be the Devil's playthings, with proverbial idle hands, it is likely to be the case that the most efficient method of solving the problem will be for companies to be compelled to have on their books a complement of ghost workers, phantom employees, with benefits and all the rest. People will be assigned to corporations, distributed, as it were, among the corporate entities, to be paid a salary as though they actually worked for the company.

In discussing the painting of the Staalmeeisters by Rembrandt in his book Civilisation, Kenneth Clark describes the group as practical men engaged in the philosophy of needing to make things work. Practical men, men of purpose who knew why and what they had invested in, and knew the expected outcome. Shareholders of today are more removed from what is in fact going on. They have no more inkling into what they own than they know where the cheese in their fridge originates, or the name of the person who made their shirt. Ownership is removed often by several steps, a pension fund holding exchange traded funds composed of instruments which are in part made up of shares of a company. The pension fund trades in and out as necessary. Sovereign wealth funds also own vast blocks of shares through investment in hedge funds and banks. The modern large corporations owned so widely that the largest individual shareholder may be only a very few percent; everybody owns Apple, everybody owns IBM, people have no idea what they own. A case in point was found in the British press of July 2013, the head of the Church of England openly criticised the payday lender *Wonga* for its exorbitant interest rates, and other practices not to the liking of the Archbishop, only to discover that his church were in fact shareholders, and therefore. One presumes, profiting from the same nefarious practices complained of. Red faces all round notwithstanding, the point is illustrated that money given to an investment fund may end up anywhere, and can be moved at the press of a key, or the click of a mouse, and be anywhere at any time. Hence the distinction between the shareholder of yesteryear and the shareholder of today. The shareholder of yesteryear was a person, the shares were real, and certificates were issued, and so on. Today money is the shareholder; pension fund representatives vote shares they have never seen, in favour of proposals made by a Board appointed by committees, and so on. All of which makes the transition to benevolent corporations more likely because the shareholders know that the ghost employee scenario is likely to be enormously more tax efficient than allowing governments to solve the problem by direct taxation of corporate profits, which have ballooned as a result of the universal deployment of AI. Once established as a practice, corporate benevolence should lead to the creation of vast wealth, for the simple reason that people are effectively out of the loop, there is then nothing to stop the engines of production and corporate activity from running at the speed of the intelligent machines. This is the effective automation of the wealth creation process, the people supported by the corporate entity are only a minute cost when compared with what would be lost were the operation to run at the speed of the slowest human.

Corporate benevolence therefore has nothing to do with charity, or with giving out of some sense of concern for the welfare of the recipient. In fact, corporate altruism, if there really is such a thing, has always been of the Ayn Rand type, rather than of the Dian Fossey variety. Future benevolence will be at all times a pragmatic and practical approach by the corporations to preserve the market into which they sell their goods and services. The inevitable consequence of getting rid of people is that you simultaneously get rid of the market, intelligent as they are, the corporations are well aware of that fact.

And so we end up with a population employed, and yet unemployed at the same time. The ultimate choice of technology. The ultimate opt out, as Kevin Kelly notes in his book, *What Technology Wants*, is the choice not to have to work. To spend all of one's time as one wishes, without the constraint of commuting, or office politics or hard labour, or the worry of whether there will be a paycheque next month. The ultimate goal of automation delivered by technology, if technology can be said to have a goal, must be to liberate humans from the drudgery of work. That is the function of technology, it's why we invented it, and it's why we solve problems, to make them go away. So that all humans are free to be artists, free to be whatever they wish to be, a society of crafts people, or of nothing at all. Perhaps we'll all be poor and unsuccessful artists, but at least we'll be available to be artists, and if we simply spend our time sitting in front of a TV waiting for inspiration, or drinking ourselves to death, that's okay too. Of course many people are as lazy as sin, and just want to sit around all day and drink beer, watch sports have sex, eat fatty foods and sleep. Where's the problem? We just want to play, to have fun, to do as we please with no consequences. We want the light on all the time burning brightly and never going dim. We want to live the 24 hour day never stopping to sleep, party on dude, substance fuelled, high octane, always on, gang bang, at least young single people do. Even old settled married types wish to pursue their own paths, and would gladly settle for whatever that was in preference to a life of working the day job. There are a million things people can do that are not work. Golf is an example of human endeavour that serves no apparent purpose whatsoever. Humanity would not be better off or worse off without it. Yet worldwide it employs millions of people, and occupies vast tracts of land, and you can bet it has a bright future.

On the other hand, Saudi Arabia and the Gulf states are already societies where infinite resources abound, they have effectively, on one level at least, achieved the ultimate state of human society where each member of the society is free, and liberated by the infinity of wealth and resources to do exactly what he likes to do. Women are hopelessly restricted, so it's a bad example from that perspective. Nevertheless, they do possess almost infinite wealth, and a vast population of pampered individuals, with large amounts of free time that could be spent in intellectual, academic, sporting or other useful pursuits. Yet there is no emergence of creativity from these societies, no massive upsurge in writing, or art, or creativity, or science. What's wrong? Well, we know what's wrong, none of the above activities seem to be compatible with Islam. Can anyone name a Saudi Arabian rock band, or scientist, or writer of any note?

According to Kenneth Clark, in his book *Civilisation*, Saint Francis of Assisi, believed it was discourteous to be in the company of someone poorer than oneself. In his book, *Capital: In the Twenty-First Century*, Thomas Piketty, asserts that we are entering a prolonged period of low growth and that technology will not come to the rescue of the Western World. He further claims that in periods of low growth, wealth is concentrated into the hands of very few people through inheritance with greater inequality between the haves and the have-nots. Thus no possibility of avoiding the discourtesy complained of by the founder of the Franciscans, the only possible solution, according to Piketty, is to impose wealth taxes, and higher income taxes on high earners to promote a more equitable distribution of wealth. The combination of unleashing the infinities of matter and energy coupled with the concept of benevolent corporations automating the wealth creation process would appear to lead to ever rising levels of wealth for all concerned.

We will, in all probability, get the future we deserve, complete with intelligent machines telling us what we can and cannot do and imposing policy on people via a disproportionate access to lawmakers. This is exactly what corporations do today, so no change there. The future corporate partner will be adopted and paid a salary by the corporation for not working. They will have jobs they are not allowed to do, and be promoted to positions they cannot fill. Probably interviewed by machines for jobs that will never be done. Companies will probably want people associated with them who are cool and interesting, and will probably pay them accordingly. Hence full employment, and full human idleness are achieved at the same time.

Will benevolence be forced on companies or will they adopt it willingly and be content to do it. Corporations are defined by and regulated by law, and are subject to sanction if they stray beyond the law. If new law is formulated, to impose additional burdens upon corporations, in addition to their tax commitments, such as to be commensurate with the savings they have made by replacing people, it might be seen as a kind of forced benevolence. Clearly the company must comply with the law of the jurisdiction in which it operates, sanctions could be imposed, and if there is a global nowhere to hide policy, they would have little or no choice.

Corporations will therefore be forced into benevolence by law makers, or adopt it themselves as a tax efficient measure, meaning either way, they cannot get rid of people, and must retain, and probably increase their current levels of employment. The evolution of AI also impacts greatly on other pseudo corporate activities such as the Open Source, movement, and crowd funding, which are effectively groups of humans coming together to get things done which corporate entities either will not do, or do only as a monopoly. Charities too, and NGOs will likewise be affected, as said before, AI impacts everything.

Consider the following fictitious case study: The Acme Widget Corporation Inc. AWC for short, designs and manufactures widgets, those indispensable, ubiquitous additions to modern living, seen everywhere, and in everything. The company was incorporated in 2036 by a couple of college pals and their friends, all of whom studied widget technology and its applications.

The company is a very heavy user of AI, and consequently has a very small, in fact tiny people-print. There are only the founders, admin is AI, outsourcing is AI, sales and collections is AI, specialist AI systems implement corporate philosophy via the internet, customer care, and all corporate services are delivered by AI distributed across servers all over the world. Below the founders are a few human associates, recommended by the AI, necessary to understand in human terms the workings of the corporate entity. Sales are strong, and production and customer satisfaction, and pricing are all within expected parameters. The cash pile grows to be enormous, the company floats and grows, the founders are multibillionaires, while the associates are multimillionaires. The company is domiciled in cyberspace, and on a lunar base the size of a beer can, built specially to be the corporate HQ of a million new start-ups. Profits are untaxable and untraceable, the very minimum is paid as recommended by the CFO, an AI with the ability of 1000 accountants of only a few years back. The cash continues to grow, along with the share price, production costs are almost zero, along with all other costs, minimised by the efficiency of machines.

But the global and ex-world aspects of this and other similar companies effectively cause the formation of a global republic, with its no place to hide corporate policy, implemented by claiming the entire Solar System as its territory, all the way to the furthest point of the Ort Cloud. A, *'use it or lose it'*, tax is imposed on the AWC cash pile, and a profit tax of 20% on all future earnings. The billionaire founders find themselves entered into a five year divestment program of altruistic wastage, which sees them slim down to a financial weight of only $200 million each. In response to these draconian measures, the corporate HQ moves to Ireland's West coast where the surfing is great in winter, and the golf is excellent in summer. Offices open worldwide and people are hired, not to work, but simply to be associated with the company, as dictated by the corporate legal code of the global republic. Within a few weeks the company has a global associate base of 55,000 people in 120 countries, who between them contribute precisely zero to the top or bottom line of the company. They have rights and duties of employment as per contracts issued and signed, but none of them actually do anything for the company. There is nowhere for them to do it. None of them know how to make a widget, none of them has ever seen a widget made. Their entire working lives are a pretence, created by government policy in response to increased corporate efficiency generated by the rollout of strong AI. They are ghost employees. They never show up. They are never needed. They are unknown to the management of the company. They are never met. They have never been to a meeting. They are never asked an opinion. The company finds this to be the most tax efficient use of its capital, because failure to employ humans in this manner results in punitive taxation of company profits.

Companies have always behaved towards people as simple units, aided by economists they ascribed simple properties to people, and gave them names such as consumers, and customers, and so on. But, economists are always getting things wrong, they only ever luck into results of any significance. One aspect of their modelling where they have had to admit defeat is in the modelling of individual elements of the economic system, not the corporate entity, but the individual, the basic unit of the macroeconomic system, the human person. They gave it the term, *'rational agent'*, analogous to, and probably ripped off from the legal profession's, *'reasonable man'*. We now know, however, that such simplistic concepts are of little use. In fact, since 2008 the whole field of macroeconomics has been left for dead, it collapsed with Lehman, no one saw that coming; no one could explain the macroeconomic mess. We are now in the era of the networked individual, according to economic thinkers such as Paul Ormerod, who discusses the connected consumer, in his book, *Positive Linking: How Networks Can Revolutionise the World*, the rational agent is networked, they connect to others, and form networks, nothing new there, but they also copy the behaviour of others. The networked rational agent may be connected to others who may have purchased a product, they will then forego the tedious work involved in product research, and testing and simply purchase it themselves. The network has made the decision. And networks can be enormous, and facilitated by social media they can span the world. The network effect is more powerful than the command-to-buy effect of traditional advertising. Hence companies have to behave differently towards individual people, the network and the collective are now inside the company too. Acting as watchdogs, keeping the corporate entity honest by threatening to tell all; we are in the era of the whistle-blower, and the world is listening. It is without a doubt this ever present warning to the corporate wellbeing that has made the larger modern company a little more people friendly; not much more, but enough to start down the road to benevolence. Apart from AI, it is this growing pressure on the corporate structure from inside, along with pressure of taxation and regulation from outside which may usher in the era of benevolence.

In conclusion, benevolent corporations are an inevitable outcome of AI and automation, and corporations should be planning for the inevitable transition, now. Governments are incompetent and reactive, leaving such planning in the hands of governments is asking for trouble. It is also clear that the race to the bottom won't work. Running to lower and lower tax shelters won't work. In short if companies adopt AI and automation then they must become benevolent, if they don't do either or both, the companies will come to an end. Our empire of technology will fade because inaction created an angry mob outside the door, armed to the teeth and with an axe to grind. Eventually all adult people on the planet and on whatever colonies we will eventually build in space will be divided up among the total number of companies and corporations, and will be paid a salary in accordance with their qualifications and experience, company performance, and so on.

The biggest problem to be faced is what happens in the transition period. The transition to corporate benevolence will be a process, not an event. But because AI could be deployed very quickly, almost like barcodes, the transition to automation could be less than the lifetime of a single government. A government could take office with an unemployment rate of 5% and within three years be looking at a rate of 20% and accelerating. Their next election is fought on very different terms facing an electorate displaced by intelligent machines with no prospect of ever being employed again.

Politicians and leaders of the future are effectively us, but with better tools and better information, and a lot more money. They are likely to find better solutions to our problems, in the same way we found better solutions to the problems faced by the Victorians, problems such as travel, communications, getting rid of sewage, and so on. But the problem with politicians in our time is that they are for the most part reactive to events, they rarely have the capacity to predict or to lead. They have very specific agendas based on their sponsor's requirements. We give them more credit than they deserve, like Deep Throat said in the basement that time, these guys are, *'really not that smart'*. Sometimes we confuse being lucky with being smart, many of the people involved are simply not all that able, they are by all means lucky, not necessarily smart, but success leads them to believe they are both. In the same way we have a tendency to confuse coincidence with causation, a coincident factor may or may not be a contributing factor, or necessary to an outcome, for example, some piece of technology, or a currency fluctuation, allows a change to occur, for which a politician will take credit, they will never take the blame.

We also know full well that economics is not a science, not by a long way, there is no predictability, no one can tell for sure what the interest rate, or rate of inflation might be in one or two years from now. There are no set of laws that can be used, nor can they answer the problem of whether there ought to be a market in body parts, and if people should be able to sell bits of themselves if the need arises. Nor can economics be said to be the cause of the world's ills; it is far too puny an entity to shoulder that heavy burden. Nevertheless, authors such as Philip Roscoe, in his book *I Spend Therefore I Am: The True Cost of Economics*, argues that this is, in fact, the case, that many of the problems of the modern world can be nailed to the door of economics. Apart from the arrogance and self-importance aspect of such a premise, the problems of the world, such as they are, cannot be blamed on some body of learning that cannot effectively predict the most basic parameters of the system it purports to study. Purple asserts that benevolence is an economic concept, that corporations becoming benevolent provides a path to growth that simply will not exist if they seek only to deploy AI and dispose of people. Regardless of the extent of automation people will always form part of the cost of doing business.

The problems of the world are mostly caused by the self-evident fact that as a species simply don't have enough stuff. There is not enough stuff to go round, so we quarrel about it. The only solution to this problem is to make the cake bigger, to fully automate the wealth creation process, and supercharge it to such an extent that no human person is ever without the stuff they want, not just need but want. It will not happen by people working for corporations, it will only happen by corporations working for people. That is why benevolent corporations will become a fact of life, and a fact of our future, there is no alternative to our present set of difficulties other than to evolve and unleash the technology needed to achieve our goals and make the transition to a Purple world, and only corporations have the ability undertake that task. At this moment in time only about ten percent of the world's population live the affluent western lifestyle. We therefore require to produce ten times more stuff than we do now, just to bring the other ninety percent up to our level. We can only do this through AI and automation, and by closing the information deficit, there is no other way, but the nice thing about it is, if we can do it by a factor of ten then why not by a factor of one hundred, or by a factor of one thousand. Once the wealth creation process is automated there are no limits.

Chapter 14   The Global Republic

> *….. where some way of thought or human activity is really vital to us,*
> *internationalism is accepted unhesitatingly.*
>
> *Kenneth Clark, Civilisation.*

The benevolent corporation can only come about if there is a mandate from the governments of the world to do it. Unilateral action on the part of a lone government, won't get it done, such action would simply cause the corporate entity to move from one tax advantageous geographical location to another. It will therefore, become a requirement of governments working in consort to ensure that there is no alternative tax advantageous geographical location to which the corporate entity can go. It is argued that this requirement will usher in a global republic as a pragmatic political necessity. The global republic is not really about people, it is about companies and nation states, and the ways in which they interact together. It is necessary because corporations are transnational, they do not involve themselves in the petty nationalisms of human persons.

The evolution and inevitable appearance of AI will necessarily lead to global structural unemployment. Corporations already behave as though there was a global republic, and play national governments one against the other to ensure they always get the best deal, and have a place to hide; recent and ongoing public outrage regarding the tax affairs of various US multinationals operating in Europe is illustrative. A global republic is required to govern global entities. Only a global republic can impose global rules on companies, resulting in a *nowhere to hide* policy. The global republic is an inevitable outcome of globalisation. Future technology would easily allow a corporate entity to effectively argue that it is in fact based on the lunar surface, or in orbit about the planet, or some other planet, or on an asteroid, and as such is not amenable to terrestrial legislation, or taxation.

Carl Sagan, in his book Cosmos, asks the question, *'who Speaks For earth'?* The answer is that no one speaks for earth, at least not now. We are a disparate collection of cultures and peoples, races and nation states. Nationalism, ethnicity, and religious differences abound, and seem to take precedence even when a more reasonable approach might well produce a better outcome. As a result there is a pressing need for a new overarching global system to moderate the petty nationalist excesses all too prevalent today. Such an entity is required and necessary for future developments on earth, and in space. For want of a better term we might call this new global structure the global republic, it is intended to be both global and republican in its outlook for whatever competencies the nation state members wish to give to it. Such an organisation of our affairs is both necessary and inevitable to deliver a Purple world of wealth and abundance which, can only be brought about by humans coming together to get the big stuff done. It is a phrase attributed to J.P. Morgan, *'running the country is too important a task to be left in the hands of politicians'*. Similarly it might be argued that getting the big stuff done, as discussed elsewhere, is far too big, and far too important a task to be left in the hands of narrow focused corporate entities playing national governments one against the other.

Only a global structure can claim jurisdiction over space, and the entire solar system in the name of the people of earth, or the members of the global republic. Do we really want the Google flag to be the first one planted on Mars? Do we really want corporations seeking tax shelters on the Moon, or on asteroids? Do we want US or Chinese corporations controlling the entire world's energy supply? Not that these companies or countries could afford to do such things, Purple argues that the cost of getting the big stuff done is bigger than private enterprise, and bigger than any single government or group of governments alone. The cost of a space elevator will dwarf the $150 Billion price tag on the International Space Station, which in turn will be dwarfed by the price tag of a Moon colony needed to harvest solar power, or mine deuterium. Getting to Mars, the Asteroids, and the moons of Jupiter in order to establish self-sustaining colonies, and find any trace of life will dwarf that again. Only a truly global and republican structure will have the power to accomplish these tasks by pooling resources, with the promise of sharing the proceeds.

The increase in wealth and in the rate of wealth production, expected as a result of AI and automation, will provide the vast resources needed to accomplish the very big tasks. There are good reasons why we need a global republic, and why it needs to be both global and republican. A global republic could form its own corporations to mine resources in space, build lunar colonies, and harvest the energy, publicly owned, and formed to do the big stuff. From a capitalist point of view some things are simply too big to be done by narrow sectional interests be they national or corporate.

Some things need to be done in the name of the species, in the name of humanity. To a certain extent the global republic already exists, we have the G7, G20, Davos, The IMF, World Bank, OECD, the UN, international treaties, international standards systems, the system of SI units, and Telecom standards, The International Astronomical Union, CERN, the Internet, and so on. The global republic would be a formal recognition that all humans wherever they live belong to one world, with one single set of standards and norms. We now know enough about the human race and the human condition to produce a single human constitution. Not just the UN charter on human rights but a constitution defining the rights and entitlements of global citizens, and the structures of a global republic. We have come far enough that petty nationalism, and ethnic and religious difference ought to be confined to the sporting arena, or the past, they have no place in a future world, they blight and retard effort, they get in the way, they allow the narrow sectarian viewpoint to take precedence over the broader human perspective. The global republic will be the future of economics, politics and human interactions, and it will be the vehicle that allows the species to finally act with a single purpose, and speak with a single voice.

Commentators, authors and practitioners are all under the impression that the inevitable rollout of AI will have little or no impact whatsoever on activities such as law, politics, economics, religion, and so on. There seems to be an unwritten consensus that these areas will remain unchanged that traditions established and practised over centuries will be safe and secure, and immune to alteration by any technology whatsoever. Of course there are the obvious technological impacts, a lot more buttons are pushed, and everyone needs to be computer literate and social media savvy, but on the whole, politics is as it was in Roman times, and the law is practised as it was in the days of Queen Anne, economics still punches well above its weight, and religion simply endures indifferently. The raw truth of it is that all such human activity is reactive to technology and never anticipative, by the time the technology is deployed it's too late. The probable fact is that everything will be changed utterly as soon as reliable AI emerges, especially in the post Singularity world where human level, or very much smarter, AI may operate at many thousands of times the speed of the human brain.

In a post Singularity world there ought to be a new horizon, new ideals, a new republican reformation, a new constitution of the human species, claiming of all space within the solar system to be the home of the human race, and affirming the supremacy of the human species on earth.

The global republic is necessary precondition for the empire of technology to function properly, and expand into the Purple world of the future. Anything affecting everyone, must be the concern of the global republic. For all our international organisations and structures we remain a fragmented world, and the fractures are along economic, political, ethnic and religious lines. We know enough about the human genome to know there is no difference between any two distinct groups of people on the planet. We also know that all claims made by all religions in the world are nonsense, being contrary to the laws of physics, and should be given no credence whatsoever. Hence the only differences between countries are political and economic. We know the political spectrum from extreme left to extreme right, and we know that all global politics should lie in that narrow band left and right of centre, the extremes being unworkable.

Continued national fragmentation has all the hallmarks of trying to hold back the tide, for example, Brittan and its relationship with the EU. The EU is nothing more than a market, it's a big one, to be sure; six hundred million people all wanting to buy each other's goods and services. But it is only a market, nothing more, goods and services traded among people, and like all markets it has a set of rules. All markets have rules, as an essential prerequisite, from the humblest sheep fair, or fruit and vegetable market, to the world's stock markets, and the EU. They all require rules, and someone needs to make and supervise the rules, in the EU, that work is done by the various EU institutions, the Commission, the Council of Ministers, and decisions are made by simple majority, or qualified majority, or everyone, all in accordance with the various treaties, unanimously voted for by the populations of the member states.

Politicians have always been reactive, never proactive so until faced with the rock ahead and the hard place behind they will not turn right or left or make any choice. Dithering being the natural state of that particular individual. Hence nothing gets done until the wolf is sniffing and scratching outside the door. When it falls down we'll fix it. But until then leave it alone. Necessity being the mother of invention, when it needs to be done, cooperation will see to it that it gets done, when we have to work together we will. The prime example of that being the rest of the world against fascism in the 1940s, Russia, Europe and the USA cooperating as never before for the achievement of a common goal; the solution of a pressing problem, defeating Hitler. So too with other pressing problems. Getting into space is an international problem. Space based solar power is an international problem. Feeding the hungry of the world is an international problem. Managing the transition to a fully automated wealth creating global economy is also an international problem.

Who governs a colony in space, grown from a seed in a ten week period and populated by millions of colonists from earth? Space today is a bit like Africa was in the late nineteenth century in the days of Cecil Rhodes, it is necessary that it be owned by the international community. There are of course international treaties regarding space, there are agreements between various governments regarding their activities in space, there is even a branch of law known as Space Law, so the problems and issues of space, and its colonisation are not unworked; see for example the book, *Space Enterprise: Living and Working Offworld in the 21*st *Century*, by Robert Harris. Apart from the International Space Station, there is at this present time no one in space, but there will be, and we therefore require to be thinking about how we govern it. It makes sense that the entire solar system be the preserve of earth, it's ours and we own it, hence any human person on any colony within the solar system, or any company or corporation operating within the solar system ought to be governed by laws and constitutional provisions of earth, a global republic claiming all of the solar system as its territory is a way of doing so. Only a global republic can claim dominion over territories beyond earth, asteroids, comets, the Moon, Venus, Mars, and everything else. And not only claim dominion but issue licences to prospect or colonise, and then impose the constitution of the republic upon any colony thereafter established. Such an organisation would make it much easier to create and maintain a truly international space program with stated goals for the next fifty, one hundred, or one thousand years and beyond.

So the global republic is then a kind of internationalism, perhaps even akin to that dreamed up by Marxists and fellow travellers in the past? Well, internationalism is nothing new, and predates national boundaries, there were no countries in our long prehistory that we know of when small bands of skin clad humans wandered about the world as they pleased. The Romans were a truly international bunch, only building walls and fences at the very ends of the known world; the certainly did not recognise boundaries established by other people. The cathedrals of Europe built a thousand years after the Romans, were erected by architects, stonemasons and carpenters from all over Europe, in what was perhaps the last most recent period of true internationalism. This rich, though dark and troubling period of human history is described vividly by Kenneth Clark in his BBC series and book, *Civilisation*, he discusses how the Catholic Church was, in effect, the only organisation in existence, the only source of knowledge, education and wisdom. It was the only intellectual outlet for over sixteen hundred years, and it was not uncommon for men of brilliance or eminence in any field to take holy orders; the facts of religion as they were then went unexamined and unquestioned. It was then as now, a truly international undertaking, having little regard for borders established by kings or nation states.

There is not a single company operating in the world today that does not recognise the global market. Take, for example, the fictitious Acme medical device corporation, founded in California in the 1930s and incorporated for legal reasons in Delaware, they manufacture a wide range of medical requirements that you may find in any well-equipped hospital of medical examining room anywhere in the world. They do not manufacture particular devices for Germans, or for Americans or for Chinese people. No! As far as, Acme Med Co Inc., is concerned, a person is a person, is a person. They do not care as to the specific geographical location, in fact, they are probably proud to have you know that their devices have been deployed on the International Space Station, and during the Apollo Program. Of course, like every other international corporate player they have a duty to the shareholder to maximise return, and shareholder value, and minimise risk and taxation liability, their shareholder base is truly international, the largest shareholder with 3% being a Swiss based fund. Hence the corporation will operate in many countries, and employ legions of accountants and lawyers, and follow tried and trusted methods of international best practice to keep costs down, and to increase productivity levels, and so on. This is what we have come to expect from a well-run capitalist corporation, and it is the most effective, most efficient method we have of creating the goods and services we need in our daily lives. Of course all companies have a superficial nationalism, and we speak of American multinationals or German companies or Japanese conglomerates, but the companies themselves transcend mere national identity.

Acme Medical, like all other successful international companies, and every other small company with international aspirations treats the entire planet as a single entity. To corporations borders are an inconvenience, adding to the cost of doing business, to them the local regulator is something they feel is an unnecessary burden on their plans for further development in that region. They are not against regulation; they would simply prefer to operate under a harmonised international set of standards. And in the creation of medical devices they probably have exactly that, because humans are all the same, the device, whatever it may be, has been manufactured to a set of agreed international standards which have evolved over many years, and through many iterations. All the measurements are in Metres, the weight is in Kilograms, the volume in Litres, and so too with other measured quantities. The manufacture of devices by the company would be intolerable were it forced to produce a different version for every country in which it operated. Certain elements of the process must be necessarily localised, the product name the packaging, and so on because of the language differences, and this is the same for all manufactured products, It is perhaps more acute for software products, or media products, and some food stuffs are distinctly local, and simply do not travel. However, leaving aside the obvious different linguistic and cultural dichotomies, we humans are, by and large, agreed that there are certain accepted international standards which we have no problem accepting, and indeed willingly and lovingly embrace. If it can be put in a form all too familiar to those of us who watch TV from time to time, there is not a single barrier cultural or otherwise to using modern firearms or explosives to settle local squabbles. International standards are ruthlessly followed, and there are never any issues involved in following the English or French instruction manual accompanying the artillery piece, or the surface to air missile, *where some way of thought or human activity is really vital to us, internationalism is accepted unhesitatingly,* as Kenneth Clark noted.

Hence in a very real sense, and in many very real situations we effectively live in a global republic, even if we do not give it that name. We are as hypocritical on the subject as we are on many others, we are global when it suits, and local if it suits better. In a very real sense corporations have beaten both people, and their governments to the global republic. Corporations behave as though there was only one world, one single global economy, with a single amorphous mass of consumers, individual people are of no consequence they are only elements of the class consumer. Corporations also see the world as populated by other companies, other customers. They don't care for petty nationalism, or borders, or the concerns of people, they have no religion for example. The global republic also exists in the body of organisations such as the G7 and G20, clubs of informal meetings and discussions, issuing bland communiques that always seem to contain the same litany of stock phrases, we will do whatever is necessary, as soon as practical, and so on. The global republic also exists in international fora such as the International Astronomical Union, CERN, the IEEE, the SI system of units, and the like. And no one objects, no one complains that sovereignty has been seeded by swapping inches for centimetres, or because the quality of certain electronic components has been harmonised and improved by adopting a new international standard. We gladly live in a global republic in everything but name.

We already have a great many acceptable global republics in the world. We have the global republic of rock music evidenced by bands touring the world, the global republic of chocolate, of beer, of golf, of clothing, of TV. There is a global republic of weapons of war, if you are killed by a bullet in a jungle in Africa, or in a desert in Iraq the same technology universally accepted and deployed was at the back of it. The global republic is effectively here, and we all accept it, we are each of us a willing participant, and a beneficiary of the global condensation of human activity on numerous different levels. CERN is a good example of a global enterprise, of countries getting together to get something very big done. No one objects to it. There are no insular nationalist agendas. No talk of a loss of sovereignty. CERN is a genuine team effort on the Swiss French border. Like corporations science is no respecter of nationalism, good scientists can pop up anywhere.

We need a great many more CERNs, or at least large international organisations with the CERN spirit of human alignment to get something big done. A global Republic could for example, be charged with setting up a properly financed corporation, to get big stuff done in space, a Moon base, to generate usable power for the entire planet, or the colonisation of Mars, or the Asteroids for commercial purposes. These are worthy human projects requiring a global effort. A global republic would have the legal power to create corporations, to compete with the private sector, to get the big stuff done, the next generation stuff, they would be owned and operated as ordinary corporations. For example, if NASA, along with the ESA, and the Japanese, Russian and Chinese space operations were all rolled into a single global entity. We know from building the International Space Station that getting really big stuff done requires massive international cooperation, and each time it needs to be renegotiated afresh. There is no reason why a global energy company cannot be set up, owned by the global republic, tasked with building out a global energy infrastructure, or a Moon base, to harvest solar energy for the entire planet. It could be funded globally, raising money by selling future energy credit to energy users.

One thing is for certain, in a global republic funded by benevolent corporations, the citizenry will have a great deal more time on their hands, and will have no excuse for not applying themselves to the scrutiny of the body politic. This new found time and ability to make mischief for the political structure is something that will disrupt the political system in the same way technology disrupts all established systems.

Abraham Lincoln said it at Gettysburg, *government of the people, by the people, for the people*, nobody believes he meant government of the few, by the few, for the few. Or that it meant of the corporate, by the corporate, for the corporate, or by and for those sponsored by the corporate, or of the money, by the money, for the money. It could be argued that it has evolved this way because it is decreed by the laws of nature, or even by some god or other, that the strongest lead, necessarily. But we know enough about democracy and civil rights to know that the strong do not dominate the weak, the strong are prevented from so doing by the application of law, and the insertion into the dealings between the weak and the strong of robust terms and conditions to mitigate the excesses of wealth and power.

In a pessimistic future where machines replace people at all but the very highest levels within all organisations, and the consequent collapse in consumer numbers. The state must step in to preserve society. Society is the preserve of the state. When the economy fails to serve the society which gave it the conditions necessary for its survival, the state must step in. If the fabric of society unravels, there is the potential for everything to fail, especially the economy. And with the economy goes law and order, and the society built upon it. All are connected. And the consumer through his activity is the quantised unit of the system. If that is removed the system must fail. Hence the problem for the future becomes how to preserve and grow consumption in the absence of the consumer. The concept of the virtual consumer or the virtual employee becomes a reality. A virtual consumer is one who spends money he or she does not earn, but is given by the state or by a benevolent corporation in return for precisely nothing. Or rather for their acquiescence in propping up the system. But the amounts must be large enough to maintain the spending habits of the consumer as if they were in fact employed commensurate with their educational attainment. Subsistence income will not be sufficient. Nor would it be acceptable to a baying mob. They will only spend freely in the sure and certain knowledge that there will be a paycheque next month, and the month after. An affluent society based on wealth producing machines, leaves people free to engage in politics with an intensity hitherto unheard of in our culture.

However, we should not fear any lurch to the extremes of right or left, it is unlikely to happen in our modern democracy, the checks and balances are too great, or seem to be, to allow such idiots to rise unquestioned. Any leader or potential leader offering his services to an earnest waiting people has questions to answer, has a background to be probed, a psychology to be analysed. The effects of education and the Internet are modifying factors, we only want reasonable people, and the flames of passion are replaced by the comfortable warmth of the reasonable man, we move to the middle of the road, and away from the rough verge. Hitler and Stalin types are weeded out, and should be seen coming a mile off, they vent their spleen on various boogiemen, and usually end up raving on the fringes eventually coming to terms with the futility and redundancy of extremism, or in anger, shooting their way through the shopping malls and schools of civilised nations. They will never find themselves in positions of power and influence, or lead nation states of the modern world. In the underdeveloped world, and certain other states where different traditions or cultures prevail, the strong man is sometimes still seen as someone to be feared, someone to follow. Our propensity to follow leaders in a time of crisis seems innate and fundamental; it may have served us during our long plod through pre-history. But education and civilisation overcomes brute force, and the checks and balances approach of the modern world act as the filter.

There's an old saying in Ireland, other places too, you can be sure; *the graveyards are filled with indispensable men.* In a world of educated people there is no longer any cogent reason why representatives need to be selected once every five years to do our thinking for us. Most people are better educated and a lot smarter than most politicians, this fact is self-evidently obvious from their public behaviour. The public space is dominated by the comings and goings of professional politicians, and professional public representatives, their doings make the news, their sex lives are the stuff of the tabloids, discovered with a live boy or a dead girl, being the only guarantee of political suicide. Economics too is often the top story, which probably should not be the case, given that economics is not a science, and those who practise it rarely if ever see anything coming, or get the right answer. Perhaps, in a global republic it will be time to select public representatives by lottery from that vast mass of educated citizens, then we would truly have the Roman republican ideal, exemplified by the famous statue of George Washington, by Houdon, in the Capitol at Richmond, Virginia, the pose of the noble gentleman farmer, called away from his farm to attend to the affairs of state, and destined to return again to tending the land when the chores of the statesman are complete. One thing is for certain, things will not remain as they are today. The corporations will not allow it, the machines will not permit it, and the displaced mass of former employees will not stand for it.

Politics across the globe change with technology, the watchers are being watched, everyone has a microphone and a camera and they're always on. Journalists of note such as Thomas Freidman keep the world informed as we have never been informed before. Websites such as WikiLeaks and those operated by the sociologist David Millar, spinwatch.org and powerbase.info probe into the darker waters of politics and money, making it obvious to all reasonable thinkers that the time has come to take the money out of politics, and take the money out of money. We live in a world of leaks and open journalism, reporting on everything that occurs no matter how trivial, with one or two unexpected results. Everyone runs for cover and says nothing, or everyone becomes scrupulously honest and truthful, the former rather than the latter, banality in politics has become the order of the day. Edward Snowden showed us the depth of love politicians and spooks have for our mobile data, and their compulsion to plug into everything that can be communicated, forcing terrorists back to using the whispered word in a dark corner. Or the written word hastily scribbled then promptly burned or eaten. Perhaps even back to the carrier pigeon. We have well-funded groups of concerned people watching the lobbyists and keeping tabs on spin and those who seek to influence the political system. We also have some very rich individuals who became rich because they were cyber idealists, now they put their money to work funding causes that were never previously worthy. Another unexpected outcome of everyone knowing everything is the simple fact that those outside the wall built by the Western World now know exactly what we do, and what we have, and they want some of it too. We rich westerners are outnumbered ten to one, or near enough. The poor will not suffer the status quo to persist for much longer; now that they have seen inside our world via TV and iPhone. If they decided to march on us, there is little we could do. In a sense the exodus from Syria is a component of this wider and growing feeling among the poor and the marginalised of the Developing World that they would rather die climbing the wall than rot where they stand, and who can blame them.

Wars and human conflict have largely two causes, resources, which would include territory, and ideology. Given the fact that we almost have the technology to unleash the infinite resources of the world, the first reason fades to nothing. No one goes to war over fresh air, there is no shortage of oxygen, and no battle has ever been fought over sunlight. Would or will a global republic prevent war? In a very real sense it probably already has, because technology exposes everyone to everything, it submerges the next generation in the global culture of the whole world, they soon come to appreciate that their little corner is just that, a little corner, whether in the wealthiest part of New York, or a shanty town in South Africa, access to technology shows them the same thing, and teaches the same lesson. We have become too comfortable to wage war upon each other anymore, in the affluent West at least. We like what we have, and of course we want more of it, much more, and our business is to make that happen. We also have the memory of war in a way previous generations of humans never had. We have the live footage, the full colour images of the piles of bodies, we have the thick books of popular history, we have the full blooded cinematic feature films of the conflict, and we have the first person video games, played online against a *real* enemy, and we will soon have, if the hype is to be believed, the complete virtual reality experience of the war zone. That does not mean we would not go to war to protect our values, our wealth our civilisation and our interests, we do that all the time. Our defence budgets remain high, we need to defend this island world against nature throwing rocks at it, and we need to defend the species against extinction. But we don't need to defend one version of nationalism against another. Most ideologies are found wanting, all mythologies are found to be based on nonsense and are utterly false in the claims they make concerning our shared reality. So what's left to fight about? One mythological nonsense against another? Wars don't start between contented nation states, at peace with themselves, history shows they start with idiots, and unreasonable people making outrageous demands on other people. But we do need wars, the fronts are many, and the battles hard fought. Forget drugs, and the war on terror. We need a war on ignorance, a war on poverty, a war on tyranny, a war on stupidity, and a war on greed. A global republic would have among its goals an end to the embarrassment of world poverty and hunger, it ought to be a constitutional duty of the global republic to declare a war on poverty, on ignorance, on misery and on disease, and force these pressing problems to be solved by the application of our technology, and our great wealth.

The global republic is one of those aspects of the future that seems as inevitable as running out of oil. Nationalism like religion belongs in the past, and everyone knows it, the only humans who don't know it are the very ignorant, and easily led, and those who profit by their continued ignorance; any reasonable examination of the arguments leads to the inescapable conclusion that the highest standards should prevail across the species. We gravitate towards excellence in everything we do, and we impose global standards of best practice in every field of endeavour, the organisation of human society, and public affairs should be no different. Applying these principles of best practice and the scientific method to the problems of human society should automatically lead to religion, along with all notions of the supernatural or the astrological, along with the magical being cast out of the public space. No public representative should ever weigh the values of science and reason against the nonsense and superstitions of religion, and come down on the side of the nonsense. No party of god should ever hold a veto in any democracy. No politician should allow themselves to be seen at prayer in public. The recent spectacle of the leader of the British labour Party, Jeremy Corbyn, standing in *'respectful silence'*, in St. Paul's cathedral during a World War II commemoration, instead of singing the national anthem, is a case in point. Of course he wasn't going to sing, the man's a republican, an internationalist, and an atheist, how could he possibly stand there singing, *"god save the Queen"*? Though he has yielded to pressure and promised to raise his voice the next time. Thank god for men of principle!

The global republic is a route by which people of the world will in effect be bought and paid for by plenty, by the bountiful abundance which will naturally come from our application of technology to closing the information deficit. All the wealth we will ever need will be created by a race of dutiful and obedient super machines. The world will never be united because we all love each other, Nelson Mandela's dream of a rainbow world will never evolve because individual humans have some overpowering desire to love their neighbours, and consider their fellow man to be brother. Maybe that would be a nice idea, one day, but will probably never happen, the best we can hope for is that we tolerate each other, and arrive at a consensus of decency and respect for the differences that obviously exist; we live in a pre-Purple world. The world will however, become a single homogenous entity as a direct result of technology and standards imposed by technological necessity, and by being fabulously rich. We cannot escape the simple fact that the only way seven billion people can live on earth is if technology provides for our needs.

Of course there are vested interests, and restrictive practices, and of course old habits die hard, but the project is already underway and will not be stopped, we are in the transition to a Purple world with a global republic at its core. We face the stark choice of society or the mob, there's no happy medium, it's a binary deal one thing or the other with no intermediate state.

We either do what needs to be done and make the necessary changes or risk the whole empire of technology falling over.

The global republic is very much a republic of technology, and of merit, of people who have very special talents. Because the next genius can come from any country, class or background. And all are welcome. You cannot be born into science, mathematics, technology or inventiveness. It is either present or absent. If you have it you may be celebrated, and have the luxury of a steady job. Science is a meritocracy of diverse individuals, backgrounds and nationalities. No one knows where the next Einstein or Ramanujan, Steve Jobs or Bill Gates will come from. There is also a tendency within science towards enlightenment values, fostered in the western university tradition and its rich history of tolerance and understanding. Hence the natural tendency towards republican values, and the natural, indeed inevitable evolution of a global republic from a technological world.

Gold was a great deal more valuable to the Spanish than it was to the Inca. In our empire of technology, energy has become a currency of sorts. Energy is money and money is energy, they are intertwined and interconnected, they convert readily. And since we have an infinite amount of energy, we must therefore have infinite amounts of wealth. Of course there is the old supply and demand issue, infinite supply means infinitely low price, which means energy is effectively worthless. But in retort, energy is the ability to do work, the means of manipulation, the raw ingredient, which along with matter and information builds all the necessary future wealth. And as we are pretty certain that in the future, near future or far doesn't really matter, we will have unfettered and efficient access to this infinite wealth, the clear implication is that each and every human being is effectively infinitely wealthy. The cake is so large that we could gorge ourselves forever on the crumbs alone.

However, right now we are somewhat light, our funds are low, so where do we get all the money we need to build out the infrastructure and evolve the technology necessary to deliver this infinity of energy. We borrow it, of course, and to hell with GDP ratios. We irresponsibly borrow it in the sure and certain knowledge that the future wealth creation capacity of our empire of technology will be such as to make the debt mountain little more than loose change. We borrow it, or we go under, crawling our way hopelessly towards a future we expect to be simply a different coloured version of the past. Our visibility of an infinite future of wealth and abundance is what makes the global republic inevitable. Of course there is great consternation about borrowing money. But if the money is to be borrowed, or even printed, via some quantitative easing mechanism, in order to fund the massive spending needed to do the big stuff, and close the information deficit, then it is perfectly permissible. We can borrow with reckless abandon in the sure and certain knowledge that the infinite wealth of the future will repay whatever debt we run up. The concept of borrowing finite amounts of wealth from a future containing infinite amounts of wealth is not unreasonable, the size of the amount is irrelevant, as long as it is a finite amount. It further makes more sense that the borrowing should be done by a global entity rather than individual nations or groups of nations, hence the need for a global republic.

We can afford to be reckless in our spending, reckless in our borrowing, and in our getting things done to force the pace in our race to the future. Recklessness is the taking of risk, there are many forms, and it is a legal term, many interpretations as to acceptability. But we have as our reasoning our acquired knowledge of infinity, and we know what is waiting for us in the future. Like a weary traveller, lost and nursing his last few supplies, uncertain of when he may reach the next town, then he passes a road sign, declaring food, hospitality and all he needs just a little way ahead, he may now consume recklessly whatever he has been conserving in the knowledge that he can replenish everything.

The world is quiet. Compared to the period of World War II we are all quiet. There are a few skirmishes here and there, for example, the leftovers from the cold war in Ukraine and other ex-Soviet satellites, and the endless and interminable eruptions between blood-letting factions across the Middle East. We know the cause of these disruptions, bellicose and belligerent dogmatists, repositioning themselves. The many and various parties of god continue in their vain attempts to have scrolls, books, tablets and other assorted historical nonsense, elevated to the status of being taken seriously; that whole sorry mess will probably need to be left to burn itself out over a couple of generations; anyone wishing to know of the inner workings of a theocracy, and the bloodcurdling tales of brutality, and the sheer hopelessness of life under the Mullahs, need look no further than Geoffrey Robertson's book, *Mullahs Without Mercy: Human Rights and Nuclear Weapons*, to see just why it is probably the case that countries like Iran, despite having half of its population under thirty years old, will simply remain unreformed basket cases. Until there is a technological uprising which forces reason to the fore, and compels the leadership to take the religious books and scribbles, and put them on the shelf; permanently. If ever there was an example of a, *theocracy don't work*, state, Iran has to be it.

But for all their bellicose, *death to America*, rhetoric, they have only themselves to blame. America may be blamed for many things in the modern world, but the government of Iran is not one of them; that was entirely an Iranian idea.

The problems of Egypt also have their origins in Islam, we need look no further. Islam is the religion of submission, giving in to the will of an omniscient creator god with complete power of life and death over all human beings. If you are poor it is the will of god, of you are rich, that too is the will of god, if you are rich you can see poverty all around you, and still feel that it is the will of god, there is no point in trying to improve yourself, no point in pushing against the status quo, god has ordained your place in the world, and if you are poor it is his will; praise be the will of the almighty. This is of course a copout of the worst kind, but is also a way of dealing with misery and maintaining control in the hands of a narrow elite. In their book, *Why Nations Fall: The Origins of Power, Prosperity and Poverty*, the authors, Daron Acemoglu, and James A. Robinson, seem to overlook the contribution of religion to the reasons why nations succeed or fail, looking instead to reasons rooted in economics, and the emergence of small ruling elites holding onto power. Egypt for example, is a nation state crippled by the geographical burden of being mostly desert with little or no rainfall, and by having little if any natural resources. But is also doubly handicapped by a religion that breeds acceptance of ones place in the god ordained scheme of things, and the inevitable grinding poverty that results. Without the religion to keep the poverty ridden masses in their place, there would be endless revolution. The religion is the gift of Saudi Arabia, and just keeps on giving. The Saudis are Egypt's most ardent supporters, and most generous benefactors. It is the Saudis who have forced Egyptians closer to god, and in the process made them poorer, and the country more desolate. Nowhere can this be better, or more starkly seen than a comparison of graduation photographs of Cairo University in 1950 and in 2014. The class of 1950 were almost half female and none of them covered up for god. The class of 64 years later has embraced the divine, reduced the female portion and covered up. The most recent revolution of 2011 was as much a cry against the tyranny of religion as it was an attempt to achieve real democracy. The aforementioned authors also ignore the contribution of religion to other historical conflicts, such as the English civil war, the sectarian nature of which is well documented.

Syria is another country torn asunder by factionalism and civil war underpinned by religion and the dictatorship of a family firm that has ruled unopposed since the depths of the Cold War. The reasons for the conflict and the factions involved are all too familiar to any student of Middle East politics and are well documented elsewhere. Also well documented has been the mass exodus of millions of refugees pouring out of the region and walking to Europe. The reason they chose to undertake the arduous trip to Europe by foot, and over dangerous water in boats unfit to use in crossing a canal, is because there is no place in the Middle East or North Africa where they would either be accepted or feel safe. The parties of god have seen to that, the entire region is in flames, fanned by zealots of all denominations with holy books and Russian automatics, screaming our god is bigger than yours. We can only hope that when it finally burns itself out, as it inevitably must, they will be so sick of fighting that they will never venture down that road again. It is almost a definition of President Obama's doctrine of, *Strategic Patience.*

Rather than religious alliances with Saudi Arabia, Egypt ought to be looking to its neighbours in the rest of Africa. The continent is sparsely populated, given its land mass, and ought to be, given its natural resources, the richest nation on earth. Only citizens of Africa can explain why it remains the basket case of the world. The blame cannot be laid at the door of western capitalism and the many centuries of exploitation by outsiders, rather it must be laid at the doors of ignorance, tribalism and sectarian division. There is absolutely no reason for there not to exist a United States of Africa, modelled on Europe or the original USA, and finally taking its place in the world and pulling itself up to its full height by exorcising the superstitious nonsense of western religion and ancient rituals and rites. There is no reason why the continent should not be crisscrossed by an efficient modern road network, along with all the other modern infrastructure needed for such a super state. What other unifying factor is required other than the fact that they all live on the same piece of land? Without that goal, the continent would appear destined to drift along as before with most of the state boundaries in the wrong place, and rebels and warlords dictating terms. It would be an interesting future world if it were possible to drive from Norway to South Africa on a modern road system, without ever encountering a warzone or anything less than acceptable European standards of comfort expected by the less adventurist tourist. Such a super state to the South of Europe sharing the same time zone, and many of its languages would be a welcome addition to the global republic.

It might appear that the global republic is the idea that the Western World is effectively one vast conglomerate of humanity, within which dwell a contented mass of people living a very agreeable lifestyle, and outside of which is poverty, famine, war, and the never ending Darwinian struggle to survive. The idea that the West is the best. That everything western is good, and everything else is in some way lacking. This is not the case, but with every other aspect of this book the self-evident facts of the situation must be faced. The western model, is the best we have currently, it is the model that delivers, and it is the model all other citizens of the world aspire to, whether they are willing to admit it or not. It is the western enlightenment model that gave us technology, and it is technology that saves the day, technology that feeds and clothes the world, and builds aeroplanes and makes small hand held communications devices and the rest. At the centre of the Western World is America, the entity that cannot be ignored. Being the largest economy in the world, is only the start of it. America is also the protector of the Western World, and of the ideals that have forged our empire of technology, and everyone else in the Western World is happy to pay for that and live under the resulting protective umbrella. It was American industry that defeated fascism in Europe and Communism in Russia, for which the people of Eastern Europe are eternally grateful.

Of course America has its detractors, the usual suspects, are a given, but also within America, there are strong voices that claim America is in decline, that America can no longer do what we in the West have come to expect. The argument is put that America responded to the attacks of September 11 in a manner which only revealed its weakness. The wars in Afghanistan and Iraq served only to illustrate in very sharp relief, the limits of military power, and how a nation with America's military might at its disposal could be wholly over extended, and thus weakened at home. Two costly and destructive wars, which it is argued, achieved nothing more than the further destabilisation of the entire Muslim world, and the creation of a humanitarian crisis on Europe's Southern border. There is also the claim that the global brand that is America has been tarnished by the activities in Guantanamo Bay, and in Abu Grade Prison, the claim is that America can never again preach on, or even make polite suggestions about, human rights or the cherished western values of freedom and justice while permitting, or tolerating such abuses within its own borders, or in warzones under its control. But we know full well that the first casualty of war is always the truth, and the fog of war is always bound to hide the facts that warring factions are composed of individuals, some of whom will, for their own twisted reasons, take matters into their own hands, and probably take them further than need be. Every war is the same, and every side of every war produces its own litany of grievous violations of human rights, and excesses, many of which never come to light. To explain is not to excuse, and there is never an excuse for the abuse of other people's rights, and when this is discovered to be wilful and deliberate it is rightly prosecuted to the full extent of the law, and in the case of America, has been.

Further pessimism seems to be due to the perception that America can no longer get things done in the world, there seems to be an increasing inability to solve the bigger problems; the ability to put a fix in place seems to be a thing of the past. It might even be called a, *no one else takes America seriously,* effect. The Israelis are still at war with the Palestinians, and no peace agreements seem possible, despite the unanimity of every reasonable person on the planet that there is more than enough land for two or more states in that region. As the late Christopher Hitchens often noted, the parties of god have a veto. North Korea remains the pariah state with a fully functioning nuclear program, and a crazy mixed up family dictatorship, propped up by China.

Iran has been tamed to some extent, and there is an agreement, of sorts, on their nuclear program, if it can be shepherded through Congress. But Iranian compliance, if it can be called that, is more to do with the fact that fifty percent of Iran's population are under thirty, and armed with smart-phones, and the next time they take to the streets the Mullahs might not be able to keep the lid on, or hold onto power. Syria, is a further example, claim the critics, where America has failed. The total disintegration of a nation state in a protracted civil war that continually sets new standards for depravity; recall President Obama's famous red line in the sand regarding chemical weapons. The further fragmentation of Iraq, is yet another example, and the rise of the extremist Islamic faction determined to establish a global caliphate, with Bagdad as its capital; good luck with that. And then there's Russia, and Vladimir Putin, and his annexation of Crimea, and the waging of a barbaric proxy war in Eastern Ukraine. The refusal to engage in serious negotiation, and the flouting of what little can be agreed is utterly blatant, and bordering on the childish. The deliberate tactic of pretending that units of the Russian army on manoeuvres inside Ukraine, and the deliberate shooting down of civilian passenger aircraft are all the work of some nationalist faction over whom Putin has no control whatsoever, is laughable. The Russian tactics in Ukraine are perfectly understandable though, when we consider that Putin believes absolutely that the worst thing about the twentieth century was the collapse of the Berlin Wall and the fall of communism. Ukraine was Russian, and they still think of it as being wholly Russian, and they want it back. It is not at all unlikely that we may see tanks on the streets of Kiev, put there under the pretence of keeping the peace or assisting a neighbour in distress.

The effective dictatorship of Putin and his comrades is an affront to all proper notions of democracy and the rule of law. All of the above examples, of what stockbrokers refer to as geopolitical concerns, seem to be deliberate acts of defiance against the American writ, almost a deliberate finger up to the very idea of America and the old world order. Ronald Ragan would never have stood for it, the critics argue, and a new perspective is required, increase the defence budget, take a harder line on this, flex a little more muscle on that, show a greater resolve, and all the other anti-liberal pronouncements. But whichever way you care to slice it, America, Europe, Japan, and the few other democracies in the world remain the ideal everyone else looks to.

Another major source of angst concerning the strength of America and the West has its genesis in the fear of a rising China, a communist, totalitarian China on the march. A new China armed with four trillion dollars in reserves, and a population of over a billion compliant people, on task, on message, and all committed to working hard for the good of the party. There is also the ever present trade imbalance and a flat refusal to devalue the currency. China has come a long way since the long march, and now stands, whether you believe the numbers or not, as the second largest economy. China seems to a great many Americans to be a threat to their prosperity, and to national security, blamed for everything, from dumping steel, to cybercrime, to industrial espionage. The quote attributed to Napoleon *"Let China Sleep for when she wakes the world will tremble,"* must seem prophetic.

On the other hand, in a very real sense modern China is the creation of America, the Chinese communists are pragmatic men, and they listened to Richard Nixon when he visited Peking, and thereafter, they resolved to be communists with a very distinct capitalist bent; while staying on the left they leaned to the right. All the good bits and none of the many evils that gave capitalism a bad name. Today China is a place to do business, and is strengthened by American companies moving their manufacturing off shore. The companies do this because it is prudent to do so, cheap equals increased efficiency, which is what companies must do under American law. But according to a great many critics within America, this loss of manufacturing jobs weakens America, and in expressing such criticism of the Sino-American trade pact, they fail to spot the reality of the modern corporate world. Which is, that as far as corporations are concerned, there are no borders only customers, there are no governments only regulations, and nothing gets in the way of doing business. This misunderstanding derives from the propensity of corporations to influence nation states by behaving as though they lived in a global republic, they can go wherever they please, and if your nation state does not provide them with what they want, they will go wherever they wish, and will always find someone else.

But perhaps the American critics fret too soon, China may well turn out to be a paper tiger, or a house of cards, destined to topple over in a heap at some point in the future, perhaps even into a bloody civil war the likes of which the world has not seen since the late nineteenth century. This calamity, if it happens at all, will probably occur within the next twenty years. If so, then this event may come to dominate the news services of the twenty-first century in the same way the two great European wars dominated the twentieth. And there are precedents. What most western historians view as the bloodiest conflict of the nineteenth century, the American civil war, fought from 1861 to 1865, was quite an amiable affair by comparison with the civil war fought in China at the same time. Known as the Taiping Rebellion, fought from 1850 to 1864, it started out as a rebellion led by the charismatic Hong Xiuquan, who believed himself to be the younger brother of Jesus, and acted accordingly. The rebellion later developed into a full blown civil war, complete with atrocities of biblical proportions, with whole cities literally put to the sword, costing the lives of between twenty and thirty million people.

The reasons for considering China to be a house of cards are many, but can be narrowed down to a few; communism, automation, insularity, and demographics. To take communism first; who wants to live in a communist state? It's a good question, given the choice no one would. Modern western democratic capitalism, for all its perceived faults, is infinitely preferable to the planned alternative of a people's republic. Anyone on earth free to make an informed choice on the matter will chose the western democratic model every day. The Chinese communist party therefore has to impose its will by force, or the implied threat of force. There is unrest and disquiet, sometimes reported in the western media, more often it goes unreported, especially in rural China, for example, peasants with tenure going back hundreds of years being evicted from their village and farms to make way for a development. Of course the matter is complex, but at the same time, the ruthless crushing of the Tiananmen Square protests in 1989, which is never discussed in China, speaks for itself, and cannot be hidden. It is the typical reaction of a totalitarian regime when threatened, the Russian Tsar did the same thing in St Petersburg in 1917. Nor can the Chinese communist party hide its love and support for the Kims in North Korea, any right thinking neighbour would have closed them down long ago. Then there's the problem of educating the next generation of scientists and technologists; not easy in a one party state. To create a technology you require to educate, and education is a dangerous thing if let loose even in small amounts, the genie out of the bottle, it leads to questioning, to dissent, to new ways of thinking and eventual revolution, fast or slow is not the point, an educated elite is hard to align with a dictatorship. While things are good, and economic growth is steady, the communist party can get away with it, and can govern with the purchased acquiescence of the people who matter. But when things turn down, and growth slows or calamity strikes, then dictatorship shows its true face, and will always be found wanting. If such misfortune, economic or otherwise, should strike, and manifest itself in the form of anti-communist activity, then the tendency towards violence, and from there to civil war is all too well-worn a path. The economic fact about modern China is that of the over one billion people, there are perhaps three hundred million who live what we in the West might think of as a reasonable standard of living. The remainder of the population live like peasants scraping a living from their small holdings. But the problem is that the peasants are not idiots, they too have technology, sent back to the farms by sons and daughters in city jobs, they are aware of the rich folk, and are envious of them, the underclass wants to stand up, they want their share, and they grow restless, it's the stuff of conflict.

Automation is one of those future developments that could pull the rug out from under the Chinese economic resurgence, and result in massive unemployment among the currently lowly paid millions of factory workers. It was seen briefly in 2008 when the banks crashed, America slowed, the need for Chinese goods dropped, and millions of Chinese workers were laid off. As discussed elsewhere, the combination of AI and robots in factories is designed to, and intended to, undercut all human labour, and make all production, and most service jobs, obsolete. The end point of factory automation is to render the Bangladeshi factory worker, on 25 cents per hour, too expensive. If this happens, and it is happening now, it will be implemented over the next twenty years. Production of ordinary things will be cheaper to do in the place of consumption, new automated factories will be built in Europe and America, saving money on transport, and protecting intellectual property. There will be an obvious impact on China, the increased number of idle hands are unlikely to find work on the farms. The destabilising effect of technology is to expose the cracks, and also to provide the tools to pry the cracks open converting the cracks into gaping chasms that could serve to pull the entire edifice down.

Insularity manifests itself in a lack of attention to the details of history, especially other people's history. In the West we have a long and detailed memory of our political misadventures, such as communism, and more particularly fascism, we spent much effort and resources defeating both. We also know what happens when a communist system fails, and nationalism rises, and we know it is not pretty. China will have to make the transition from communism to a more open and democratic system, the people cannot both grow rich, and remain content under a one party system, the two are incompatible. At some point it will have to give way, and at such a point there is the possibility of lurching into the mistakes already made in the West. But because of the closed system, and the lack of historical perspective, the Chinese may be doomed to do exactly the same. There are already nationalist Chinese stirrings in the country, a kind of nationalism that we in the West laugh at, as being fringe politics with no serious chance of ever gaining power. But in China, such nationalist sentiment coupled with a charismatic leader, a modern Hong Xiuquan, for example, might lead them down a dark road with an uncertain outcome. Censorship and a great love of the airbrush have been at the centre of communist party control. The nationalists who did most of the fighting against the Japanese never get a mention, Taiwan is seen as a delinquent province to be brought to heal one day, and Mao is portrayed in officially sanctioned movies as having been at the Cairo conference in 1945 with Churchill and Roosevelt.

Demographics must be another source of tension within China, the manipulation of the population via the enforced one child family has led to the extraordinary population dislodgment, and an unevenness of the sexes, which manifests itself by a dire shortage of women, and as these parents and grandparents grow ever older, the small matter of care of the elderly looms large. The population becomes lopsided, skewed in a way that was never intended. Time bomb is a term often thrown about, but in this case it is ticking, and set to explode in the face of the Central Committee.

Hence global leadership seems to reside in the West rather than the East, and naturally people wonder how long this will continue. Why the West rules, is the question tackled by Ian Morris in his book, *Why the West Rules for Now*. The book is a wonderful account of the history and the differences between East and West, and gets to the root of the problem posed in its title. The issue may be summarised by the example given by Morris of what a Malaysian lawyer told the British journalist Martin Jacques, "I am wearing your clothes, I speak your language, I watch your films, and today is whatever date it is because you say so." The lawyer might have pursued the matter further, his list of examples would fill a book. The lawyer or anyone else in the East, educated or not could claim to be a victim. If they wished to, could claim to be burdened by a form of cultural imperialism for any number of items on the list, but the simple fact of the matter is that it is not imposed, and has never been imposed. Rather it has been willingly adopted, and willingly and lovingly embraced by all who touch it. Because it is better, or simpler, or more elegant, or more robust, or makes more sense, or works better, or is better designed, or has more appeal, or makes more money, than what it replaces. People, no matter who or where, always gravitate towards the best, the best in class, and the best of breed, are frequently heard phrases in companies looking to find that edge over competitors. The best solution to the problem is the one that endures, not necessarily the one that solves the problem, or the first attempt, the best is adopted and modified, and further refined. Often there is only a single solution, frequently there are competing, equally practical solutions, but in the end the decision to adopt a particular approach or method, like deciding which side of the road to drive on, or whether to eat with chopsticks or knife and fork, is a matter of choice, a matter of personal preference. It's not that the West is better in any meaningful way than the East, the people are the same, but the places are simply different. There is nothing done in the West that cannot be understood in the East, and nothing invented in the East that could not be deployed in the West. If it works we will use it; that goes for both places. Penicillin kills bacteria in both East and West, but western medicine is by a long way the only medicine we have, there is no alternative. If you require urgent medical attention for a life-threatening condition, then western medicine is the only port of call, no amount of prayer, potions or acupuncture will solve the problem.

Whoever invents the thing gets to name it, is a long standing rule in both East and West. Everyone uses it or refers to it, but it is still called by the name given by the inventor or the discoverer, *kimono*, is an example from Japan, as is *sushi*, and *judo*, and a lot more. Most of the stars with names, have Arabic names, *Aldebaran* for example, other Arabic words in English include *algebra*, and *algorithm*, because of the work of Arabic mathematicians in the golden age of Arabic scholarship, before the Muslim world retreated into theocracy. If the next big thing emerges from some back street in China, and if it is better, or more efficient than what we use now, or is one of those new things that leave us all scratching our heads, and wondering why no one ever thought of it before, it will be forever Chinese. It will be absorbed into the homogeneous global culture to take its place alongside mah-jong, paper, printing, gunpowder, silk and the magnetic compass. At the same time it is clear to any student of history that western armies have marched and conquered all over the East, notwithstanding the embarrassing and hasty exit from Saigon by the American Military in 1975, the reverse has never happened, Pearl harbour was as close as they ever got. It is further clear to students of economics and business that efficient Japanese factories and innovation drove European and American companies to the wall in the 1980's, it is also a matter of fact that eastern governments concern themselves with western concepts such as communism and the free market, but Daoist or Confucian views of administration are alien in the West. The global republic straddles all notions of East and West, where important matters need to be decided a consensus generally emerges, and is adopted by all who care, as we know from all measuring and international standards systems. For obvious reasons of political alignment, tradition and historical perspective, Japan and India would effectively be part of the West. China may not be, the lack of democracy, the choice of direction, and so on, may preclude it, as with Putin's Russia. China for all its financial might is still outside the wall, a self-imposed retreat, it is still a nation not yet fully formed and modern, not yet sufficiently stable that its people may be trusted with a ballot paper, or proper unrestricted internet access, or a full objective account of their nation's history; China in 2016 seems to be looking another way.

In the final analysis it must be asserted as a matter of fact that the Western World is the global republic. It must further be said that all other nations, need only leave their religious, racial, ethnic, tribal and other unnecessary and primitive baggage at the door, to be equally accepted as members. But there is an entry fee for any nation that wishes to become a member, it's not an exorbitant fee, but it is nevertheless a fee that must be paid. First that nation must be an open democracy, with free and regular elections, equal rights enshrined in law, a free press, and freedom of the individual to do as they please within the limits of the law. The citizen must be unleashed to do whatever they feel like doing, without an overbearing government getting in their way, and that must include creating and retaining great personal wealth. There must be a complete legislative separation of church and state, or of religion and politics, clerics have no place in public life, anyone committed to principles of magic, is unqualified to speak on much else. Religion is a proven hindrance to any kind of progress, the holy books, scribbles and scrolls need to be put on a shelf, and left there, no more prayers in the congress, or blessing the newly elected, or swearing on the bible. Corruption too must be driven out of public life, and that must include all types of patronage, nepotism, political dynasties, and payments to politicians or public officials. All private money should be removed from politics, especially the effective buying of elections by outspending the other guy. It is probably not an unreasonable suggestion, given the technology available to us today that elections could be more frequently held, or perhaps even more radically, not held at all, it would be entirely feasible that all politicians and public representatives could be chosen by random selection in the same way that a jury is selected, selected from a panel of very qualified people who put their name forward; if you can sit on a jury, you could sit in a parliament. Public representatives thus selected could hardly do worse than the crowd we tolerate today; especially if they were strictly limited to one or two terms. The next part of the fee is education, vast amounts of money must be ploughed into the education system, and it should touch everyone, from cradle to grave, it's the greatest known cure for poverty, and should encompass more than just learning the contents of holy books and scrolls. The global republic will go on and endure, and strengthen, with or without new additions; there are sufficient members right now to make it more than a viable concern. The rest of the world can join in or choose to sit and watch. Does it really matter if there are vast swathes of humanity outside of the wall, governed by lunatics, tyrants or religious zealots, who will never benefit from western technology, and spend their entire lives in poverty? Of course it does. It matters to all of us. But some things can only be changed at the local level, and some things only change with a new generation, women in Saudi Arabia, for example, will never be taken seriously, or drive cars or hold down jobs unless they do it for themselves.

On a more esoteric level, one further reason we know the global republic exists is that we have loose affiliations such as the Bilderberg group, a semi clandestine gathering of the rich and powerful of the world who meet regularly and have been doing so since 1954. Powerful people, the elite of the world meeting in secret, protected by armed police guards, with no published agenda, no minutes of any meetings ever recorded, no statements issued, no record of any vote taken or decisions agreed; is bound to, and does, attract an assortment of objectors and protestors. It's the stuff of conspiracy theories, except that these days it's more like an informal conference discussing mundane topics in economics and geopolitics. They even publish the topics for discussion and a list of delegates attending. In 2013 the Bilderberg, as it is simply known, was held in Brittan, at an extravagant country hotel in Hertfordshire, of course the inevitable protestors turned out with placards, one of which read, *B'bergers we await the minutes of your furtive meeting.*" Of course there is need of concern when such meetings are convened by who knows who, for whatever reasons, to produce who knows what outcome, and arrive at whatever conclusions? Democracy be damned would appear to be the order of the day. The principal objection was alluded to by Adam Smith many years ago: "*...people of the same trade seldom meet together, even for merriment and diversion, but the conversation ends in a conspiracy against the public in some contrivance to raise prices.*" These days when politicians and their banker friends meet, there is more than a passing acquaintance with each other's trade, the one has become the other by the liberal application of incompetence. And while the theory held by some on the outside that the Bilderberg attendees are, in fact, lizard beings who run everything, you can be sure their reptilian view of the world is not the narrow nationalist perspective of many who object to their meeting. You can be sure these scaly overlords understand the concept of the global republic, and behave as though it existed.

PURPLE

Chapter 15   Living In A Purple World

It is indeed sad if some citizens can't distinguish a proton from a protein; but equally so if they are ignorant of their nation's history, or are unable to find Korea or Syria on a map – and many people can't.

Martin Rees, From Here To Infinity.

What do we do in a post benevolent corporation, post global republic, post Singularity world? How will we live, how will we spend our days? What would it be like to live in a world with everything laid on, a world where every day is like a Sunday in June? Where are the incentives, where will be the challenges? Purple is the new age of extravagance and exuberance what Marxists are fond of referring to as, conspicuous waste. How do we survive and retain our sanity in a world where everyone possesses the ideal fairway home with all the associated comforts of a suburban environment, lots of square footage, three cars in the driveway, and the smell of affluence everywhere you go? A world where everyone can travel the globe at will with no pressure of work other than the self-imposed deadlines of people doing their own thing. A society where obtaining the latest version of anything is simply a matter of placing the order and awaiting delivery, or printing it out at home. A world in which, the citizen, on reaching the age of maturity and the minimum standard of education, is free to do whatever they wish, or nothing at all, and be all found forever, as the expected universal minimum standard of human life. Although in practice the basic education will probably mean a university degree, there will also be a period of work or volunteerism before real idleness can be commenced, after all, ultimate decision making must and will remain with humans. It is extremely difficult for us here in 2016 to imagine a world without work, a world where nothing is expected other than adherence to a basic code of civility and good manners. But that is what lies ahead in a Purple world where machines and machine intelligence are so sophisticated that human involvement in commerce or industry would add precisely nothing, and probably cause the wealth creation process to operate at a depressed level, and prove to be detrimental. What do we have to trade to achieve this benign state of existence, would we rebel and seek to change it, and will it have some dark underbelly that we cannot see from this time?

On the one hand we have the greens, preaching the end of fossil fuels, and doom and gloom, and six degrees, and eleventh hour, and tipping point, and carbon footprints and the need to act, and austerity, and making do, and reusing old cardboard and string, composting grass clippings and cabbage leaves, and the like. On the other hand we have the anti-humanist pessimistic brigade, John Gray, Jared Diamond et al preaching the inevitable end of the species. At this stage people need an alternative to pessimism, another path, an alternative to the whining, the winging and the finger wagging. Every aspect of our modern technological world is pointing us in the direction of a future Purple existence, and we probably ought to be preparing for it better than we seem to be. One logical inference naturally follows if one adopts a Purple approach, namely, there is no further requirement for humans to worry about conserving energy, or saving water, or any other petty, *'nickel nursing'*, nonsense. Why worry about pennies if there is a cheque in the mail worth millions?

Does Purple imply humanity will eventually enter a phase of terminal decline due to boredom, the Schopenhauer conjecture that humanity is doomed to vacillate, like some pre-programmed harmonic oscillator, between the two extremities of distress and boredom? Or finally, like the citizens of Rome, simply giving up under the strain of it all? There was a time in the past when the only contributions made to science and culture were those made by the scholarly gentry of old who lived in pampered opulence, at least by their standards. They were the country house set, the great gentleman thinkers of the enlightenment, the moneyed elite, progressive thought also emerged from contented clerics who stumbled upon some principle of science quite by accident. By our standards today it was a depressing existence, Thomas Jefferson in Monticello tinkering with his inventions and writing his letters did not have even a fraction of the experiences that the most deprived American of today takes for granted. And for all the books in his library, he knew little or nothing compared to a modern high-school graduate. Purple will be something of a return to those times, a recreation of the same conditions, with the important exception that instead of a small elite living off the labour of slaves, and a bitter and resentful underclass periodically poised to destroy everything, we will have a pampered creative elite comprising almost every citizen of our

Purple World living the high life on the fruits of our machine society, and the natural infinities that surround.

It is reasonable in 2016 to make certain statements regarding a child of the future, let's say one born this year or next, for example, we could propose that such a child might work, and remain in good health, till the age of 80 or even longer. Perhaps in a job or a profession that is yet to be invented, and he will probably do the job using technology which is not at this time present in our world. Her job might well involve solving problems that perhaps do not exist at present. Or perhaps he may not need to bother. Perhaps she will spend her first thirty five years in the education system, will be required to master at least three different disciplines, including one of the sciences, and speak at least three languages with natural proficiency. He will then be required to commit a further ten years to serve the requirements of either a government agency or a corporation, and then be free to do as he pleases, on a full pension by the age of forty. In return she will receive everything she could ever want, wish or dream.

One may of course ask, where is the power? Who has the influence, in a world populated only by super wealthy, lazy people? Where are the seats of power? What is the source of power? What about corporate power and influence as we know it today? If we are all super rich then paying or seeking influence is a redundant act, where then is the plutocracy? Institutions will still need to work in the same way, people will expect that the roads are paved and the bridges maintained and that organs of the state function adequately. And of course the answer has already been given, we will probably and willingly surrender all such petty bureaucracy to the machines. Ultimately supervised by humans of course. According to Google, or at least to Eric Schmidt and Jared Cohen in their book, *The New Digital Age: Reshaping the Future of People Nations and Business*, the future world will be one dominated by the internet, and an ever increasing connectivity to everything. Smart this and smart that, and a mishmash of connections, and permanently connected individuals. But they are Google, and Google is a company looking to make money. Like all corporate entities, the bottom line is the bottom line. What cannot be in dispute is that increasing technology, and closing the information deficit down to its minimum, must lead to massive wealth, because once automated, the wealth creation process can operate at any frequency, it must also therefore, lead to increasing contentment and less reason for conflict on any level.

With any luck a Purple world would be a quiet world, quiet in the sense that nothing much happens, a *'no news'*, kind of world, like one of those days experienced by newsrooms and newspaper editors when nothing has happened for the last couple of days, and they have on their hands what they call a *'slow news'* day. Living in a Purple world would be a long succession of, *'slow news'* days, punctuated by some intrusion from the occasional unanticipated natural disaster or a major sporting achievement. Big news stories such as war and economic calamity, or the kind of event resulting from petty competition between differing ideologies or economic systems, or arising out of religious enmity that fill our news today would simply not occur. Future technology will simply render all such petty squabbling pointless in much the same way as the manufacture of modern fertilisers in chemical plants rendered wars between nations for control of bird-shit deposits in South America a thing of the past. It is instructive to read the history of many of the conflicts of the past, and try imagining how our modern world would react to such a conflict, were it to start today. See for example, Jack Kelly's book, *Gunpowder*, for a popular account of the history of explosive warfare, or Stephen R. Brown's book, *A Most Damnable Invention*, for a similar treatment concerning dynamite and nitrates. It is hard for us to understand or to imagine that two modern world powers might go to war over the control of a commodity such as opium, but that is exactly what happened during the opium wars involving Britain, France and China in the mid nineteenth century. Also hard for us to imagine what possible conditions might need to exist in order to cause a second civil war in the United States. Certainly, *'slave or free'*, would not be an accepted dichotomy to give rise to a modern analogue of the bloodiest conflict of the nineteenth century. Essentially such conflicts would never get off the ground in our modern world or in a future Purple world, we would simply laugh at the prospect. Resource based conflicts, or religion based conflicts, or conflicts based on ethnic difference are unlikely to begin, or be tolerated if they did manage to spark into existence. We know too much about our history and ourselves to allow such crude reasons to ferment into something as bloody and pointless as the wars of the last two hundred years. Of course there are wars, the conflict in Syria is no Saturday night donnybrook. But its causes are religious, ethnic and tribal, the country has effectively been in the hands of the same family for over fifty years, demands for change were met with resistance, and the powder keg went up. Our control of technology, and mastery of energy and resources both on the planet and beyond will be such as to breed contentment, and a pleasant air among the citizens of earth, our silence and serenity will be bought by the resulting infinite wealth. It's hard to find a reason to protest if you live a contented life, it is hard to want to start a war when any idiot can tell you exactly how long it would be likely to last, and the probable outcome, simply by asking an intelligent machine. There will still be local skirmishes, over all sorts of issues that exist or persist among various groups of people, and these may sporadically erupt into something that results in minor bloodshed. It is hard, for example to see how certain aspects of certain religious congregations will completely disappear even over generations; and of course, each and every one of the four thousand such factions will continue to know for certain that they are in the right. But on the whole apart from the intrusion of nature in the form of fire, flood, geological upheaval and weather events, we might expect to dwell in a quiet world. Much as the West is today.

Most people living in a Purple world are unlikely to care to any great extent, there is likely not to be anything requiring their attention, the great issues of the world fade in the face of enormous wealth, leaving nothing to care about. There would be a general air of nonchalance, of serenity and contentment, perhaps even amusement or pleasant distraction. Happiness doesn't really exist, it's nothing more than a spike in the contentment curve, a blip in the otherwise flat curve of contentment against time. The American comedian Denis Leary put it in specific terms, *'it's a butt, it's a cookie'*, it's effectively life's little triumphs and diversions that cause the mood to rise out of the constant gristle-chewing of the mundane. So can we have our cake and eat it? Yes we can. We need only make the cake bigger. We make it so big that no matter how much we eat, we still have enough cake. And if that's not enough, we make it bigger again.

Limits on wealth will not be a feature of a Purple world, if everyone has it then it becomes ordinary. Collecting money is like collecting rocks and stones. This guy has a very large block of granite in his garden, the fellow over there has an entire rockery, another gentleman boasts a massive collection of pebbles collected from the river beds and alluvial planes of the world, while the dude in the bright check trousers is known to have an entire, specially built garage, stuffed with sacks of pea shingle. A big, *so what*. Being famous for being rich is a pointless exercise, just as being famous for being famous is pointless, though perhaps more fleeting, and ultimately doomed to end badly. There is no difference between, being worth a billion and being worth one hundred million, nor is there probably much of a difference in being worth ten million. All three wealthy individuals can live and enjoy their lives entirely on their own terms, to the envy of the rest of us, and there are nearly seven billion other people in the world, who would all gladly swap places tomorrow. Limiting personal wealth to an arbitrary maximum number is probably an exercise in apple polishing. The only solution to the rich poor divide is to make the cake bigger. We are doing it now. We are in the process of automating the production of wealth, this is something that has never been done before in human history, we humans have always relied on the labour of other less fortunate humans to produce the wealth we flaunt and hoard, creating a resentful underclass in the process. More to the purpose, it has always been a slow method, carried on at the pedestrian speed of the human worker. The advent of workable AI solves that problem, and unleashes the creation of wealth as never before.

This proposed vast increase in the wealth of nations and of individuals ought to solve a great many problems such as criminal activity, since most criminal activity seems to concern attempts to obtain wealth in some form. The fact is that most criminals are intelligent people; they after all solve very complex problems on a daily basis. A stereotypical substance abuser living in a modern city functions in a complex environment, solving complex problems, in order to achieve certain goals; obtaining the required funding to support the abuse lifestyle; the 1996 black comedy, *Trainspotting*, illustrates this well. Their actions imply intelligence at least at a par with university graduates. The future of crime prevention lies is in creating a society where every member of that society, not only has, just something to lose, but a great deal to lose, in other words, they will think twice before committing the act. The other solution is creating a society in which the criminal not only suspects, but actually knows as a matter of certainty, they will be caught. Crimes of passion will of course continue as today, humans will not change in that regard, if anything we may expect an increase. However, the dangerous combination of a very much more wealthy society coupled with infallible AI technology will probably cause most petty criminals to fold their tent and pursue some branch of academic endeavour.

Your old cat could probably do a better job of predicting future trends and directions than most futurists. Hence the future is glorious and unforeseeable, and to be looked forward to with the utmost expectation. And most important, we must never fear the future. Those who sit in dark corners certain that the end is neigh, or that it will all inevitably end in tears are simply wrong. That is why the Unabomber was so wrong, and all who follow him, or support him, or give credence to his warped thinking are wrong, and can never be excused. Destruction is wrong, ignorance is wrong, narrow short-term interests are wrong. Feeding the poor of the world, making them as rich as we are, and looking after the poorest of our society is not utopianism, it is simple good manners. It is a problem to be solved by the application of technology. We need an App for it. Economics is probably the definitive discredited profession, now as never before, because of the utter failure of academic economics to see 2008 coming. The 2010 movie, Inside Job, exposes the profession of academic economist for what it really is, guys in suits guessing, and stuffing their pockets in the process; as if it ever were really a science. Richard Feynman's opinion was that they go through the motions of science, but ultimately they know exactly nothing, because they don't know what it really means to know something, to have gone through the process of checking and re-checking, and doing and re-doing the work that is necessary in order to be able to say that you really know something. It's little more than the linear progression of lateral thinking. However, they have generally concluded that 2% per annum is all we really need in real terms to really grow the cake and keep it ever growing. A great deal can happen with two percent.

The world in 1000 years at a constant real 2% per annum would probably be unrecognisable to us today. But probably very much closer to this century, lots of things we know now will no longer exist, they will be the stuff of history. The very long list of things we don't want in the future would have to include: hunger, crime, violence, ignorance, poverty, war, dictatorships, religion, arrogance, dogma, and so on. The very long list of things we do want in the future would obviously have to include: love, sex, music, the piano for example, played by a person, not a robot, golf. Football, in all its forms. Fast cars, with real engines, not electric motors. The countryside, golf courses, forest walks, mountains, valleys. Beer and Saturday night, the list goes on. As Louis Armstrong famously sang, *'we have all the time in the world'*. The planet is not going anywhere for at least five billion years. We're here for the long haul, the long game, and the bigger picture, we retain the global view, the system view, and ultimately the human species view. We absolutely have to conserve and protect everything, we have a duty to do so. Every little thing of interest to those that come after us, needs to be looked after extremely carefully. If we do not, they may well blame us. And given the previously discussed and reasonably expected advances in medicine and biology, and the relentless march of technology, there is every chance we might just be around and have to face the music, or the wrath of future generations for our stewardship.

Just to take one thing off of that long list of things we must take with us into a Purple world, specifically, fast cars with real engines. In his TV Series, *Cosmos, A Spacetime Odyssey*, Neil deGrasse Tyson, postulates that one-day the last internal combustion engine will be ceremoniously placed in a museum. As a scientific commentator he's usually not wrong about much, but in this he will probably find this will not be a likely occurrence. The technology is simply too good, while a Green car may be a sawn-off hybrid, electric, plastic, spatially challenged, grocery carrier, the Purple car will be a Mercedes CL 630 AMG, or the equivalent, an unashamed gas-guzzling road gripping monster. Like the grand piano the internal combustion engine is here for the long haul, we will never give up on it. And as long as we can make the, tail pipe to gas tank, carbon neutral, or even carbon positive, what's the problem, it's a beautiful mature technology that works and we love it to bits.

The overriding problem we face is how to get from here and now, to the future without braking all of the eggs in the only basket we have. The overriding conclusion must be that we are where we are because of the technology we developed to get us here, without exception, all ancient civilisations failed because they had the wrong technology, or not enough technology, or an inability to use the technology they did have. Clearly we are here because of the path we followed, we built on the past and solved the problems better. We might have made it, as Kenneth Clark put it, *by the skin of our teeth,* but we did make it.

The biggest question is always going to be, what will humans do? In the same way as we should never live in fear of the future, we should also never feel guilty about accepting the bounty of the future. The simple arrogant human answer is that humans will do whatever the hell they like. There's an old saying among the gardening fraternity, when all the sowing and pruning, the mowing and digging, and the planting is finally done the one thing the garden really needs is, *'a good deal of looking at'*. And so it is with humans. When we have completed the toil of bringing into existence, by the blood sweat and tears of thousands of generations, a system of economics and a technology to provide for our every whim. The only thing remaining will be, to be human. Humans will spend a good deal of their time just being human, and doing human things. You can bet our hunter gatherer forbearers when presented with a time of good weather and abundant food did a good deal of just being human, and whether that takes the form of just lying on a beach, or burning nine thousand calories a day building a body, it's of no matter, we've earned it. Even if you spend your life in an armchair bombarded by electric images of the outside world, being constantly told there is something new and exciting about toilet paper and detergent, between the endless streams of input to a saturated, unreceptive brain; you are still alive. This vision too is a life, or at least an existence, and a human experience to be entered by choice, and is as valid as any other human experience.

One of the first things scribbled down following the invention of pencil and paper was the notion of a perfect utopian paradise and its parameters. Such notions are wishful thinking on a grand scale, and are not what a Purple world will be, such perfection is unobtainable through wealth alone. In the eighteenth and nineteenth centuries, there were no private jets, no luxury yachts, and no exotic locations reachable in hours, so the billionaires of the day stayed put on their estates, they grew all they needed, from cabbage to exotic peppers, and tropical fruit, and the world came to them. Everything on the one estate, served by an army of servants, men women and children, all of whom lived on the estate, and all of whom knew their place. Our human future will be similar, instead of the servants we'll have autonomous semi-intelligent machines. In such a pleasant future world, antiquities from the past may be the most valuable items we have, like our world of today objects made by, or touched by, or owned by those we admire will command the premium valuations. It will not be an age of perfection, or of permanent happiness, the previous record of emperors, kings, flappers, yuppies, or the billionaires of today, and their wealth-ruined children speaks for itself, excess breeds its own problem set. Why should we expect a difference between a wealthy future and a wealthy past, when people are the common factor? But the misery of great comfort is tolerable even agreeable to most human animals, and it is certain that our future wealth will surpass anything we have today. The human talent for survival will no doubt make the best fist of it.

We will live in a new, *I want therefore I have*, world, where every want is met, human needs are a given, the presumption will be that if you are human your needs will be met. Great wealth and advanced technology may lead to some unexpected results, and some strange consequences may emerge out of a medicine that offers a cure for all ills. A Purple world might be a new era or hedonism, perhaps one of wanton self-destruction. There might even be a reintroduction of smoking complete with TV commercials, smokers might once more become the majority of the population. Wear out one pair of lungs, and get another pair free. Two-for-one on lungs this week, throw in an aorta. Buy a carton of smokes, get a voucher towards new lungs. Drinking, drug taking, extreme sports, perhaps even a return of gladiatorial combats and some very dangerous sports. If there is no fear of dying there need be no fear of death. Anything goes.

So what cost the new Purple world? They always want something in return. If they offered it to you in the morning what would you be prepared to give up? There's an old saying in the technology sector, if you want the biggest budget and the best technology, go work for the government. Future governments will have all the best stuff, and no doubt they will point it in the direction of the citizen. If every citizen is effectively living a permanent weekend lifestyle, the government is going to demand a commensurate level of control. In a future world where every phone call can be, and probably will be, listened to by intelligent machines, and everyone will be required to be tagged in some way, so that your GPS coordinates are known to the inch, what would you do? Perhaps you might think of opting out? But that won't be possible either, because there will be no work, the machines and the corporations will see to that, being a bum will be the new normal for whole nations of people, being *'off the grid'*, may become a crowded place. The best bet is probably to simply put our trust in our fellow humans. Most scientists working today whether for companies or academic institutions, the military or the government, tend to be of a liberal humanist leaning, they value freedom, and tend not to subscribe to notions of autocracy. Altruism is one of their many attributes, one thinks of the so-called green revolution and the work of Norman Borlaug, and others. When we couple that with the open-source approach, and an abhorrence of corporate greed, and the perceived failure of intellectual property law, we see science and technology are moving to something new and different. Towards something that serves the common good rather than narrow sectional interests of corporations or nation states. Naive it probably is. Nevertheless, the evidence seems to support the inference of a growing movement of humanist expression, and an increasing awareness, particularly from the human genome project, that apart from narrow cultural differences people are all the same. It's one of those simple unarguable facts. Thus, next-generation thinking crosses all platforms and boundaries, young people are not prepared to tolerate old ideologies and old ways of doing business, the rebelliousness of youth and the influence of the Internet generation will make the difference. Change has always had a generational element, the combination of youth armed with technology is a powerful tool for change, and for the creation of a homogeneous culture. Statistics such as the fact that 60% of the population of Iran are under 30 are worthy of note in that regard. Even though 90% of social media is probably trivial naval gazing, the revolution of the young is a very real one, and is a revolution in culture, attitude, education and human concern. Purple is therefore also the new Red. A deeper kind of Red, a more refined Red. The revolution won't be Red, the revolution will be Purple, and technology based. It will be a slow gradual evolution, a steady morphing to the new state. We can already see where we want to be, and have thrown ourselves into the task of closing the information deficit; we know what needs to be done, more science, more technology, less religion, less defence, less war, less monopolistic, regressive, retarding, non-Purple activities. What makes an iPhone is simply dirt and information, plastic and silicone and information. The information needed to build a Purple world must be acquired, currently it is missing, restricted, restrained or hidden from view. Knowledge is a public good, knowledge is a human activity. Hence it is probably true to say that living in a Purple world shouldn't be all that bad.

PURPLE

Chapter 16   We're All Gunter Sachs

There were only ever 12 playboys in the world…….. They were charming and spoke languages, and behaved well with women….

Gunter Sachs, 2000.

The future Purple world, if we ever get to it, will be inhabited by Purple citizens. They will not be perfect utopian beings, they will be just like us, they will be flawed, fallible, dangerous human animals with all the disagreeable traits associated with humans of the past, and the present. Humans have not effectively changed in perhaps one or two hundred thousand years, depending on who you talk to. If left alone we would be unlikely to change much in the next fifty thousand either. But of course we will unleash our technology upon ourselves, and who knows what the eventual outcome might be. One thing is very clear, if we continue as we are we are going to unlock unheard of quantities of wealth and abundance, the working population of the world will reduce to a very few decision making individuals, and necessary or agreed on, professionals, doing those jobs specifically reserved to be exclusively human.

The rest of us will simply get on with being whatever we want to be and doing whatever we want to do, which is probably a recipe for the biggest social disaster since prohibition in America. To permit an unrefined, uncultured mob of well-paid layabouts to behave as if every day was a holiday, and every night was a Saturday is simply asking for trouble. The results are as predictable as allowing lab rats all the food they can eat. Some element of control is necessary, some method of carrot or stick must be deployed to govern a population of wealthy consenting adults with as much free time as they could ever wish to have. Reliance on self-control and the philosophy of moderation in all things is wishful thinking at best.

Like pampered cats living in a world of opulent luxury served by a race of machines smart enough to serve but not smart enough to disobey. Our every request complied with, our every wish fulfilled. The only tolerated difference between individual humans being our innate abilities to succeed in those fields where humans continue to test themselves one against another, sport, art, science and whatever else we wish to play. Because of the vast wealth of our society education will be fully funded at all levels, and we should expect nothing but the very best. People of the post Singularity future will probably spend thirty five years in the education system, mastering several disciplines to the proficiency of an expert, including one science, and probably a couple of languages. Humans are self-moulding, but we are also the keepers of the flame. We also have an inbuilt need to test ourselves, and compare one generation to the last. Hence the vital importance of education, if individuals do not know what has gone before, it is unlikely they can make a worthy contribution. When the education process is complete, our citizen may be required to volunteer, or enter a reserved human profession or join the military or the police or other government service for five or ten years. Thereafter our citizen will enjoy a very long life of pampered luxury, free to do exactly what they wish. Everything will be paid for by a corporation or a government department, like having a job but not having the burden of work.

It is to be hoped that such an education and work program of civic responsibility would produce refined cultured individual citizens capable of being well adopted members of an ordered free society. It will certainly disarm any excuse that might be made regarding deprived childhood or lack of opportunity. Kenneth Clark in his book *Civilisation*, deals with the contribution to civilising Europe made by the court of Urbino in Italy, and the influence of the book, *Il Cortegiano*, 'The Courtier', by Baldassare Castiglione. The book detailed the concept of what it was to be a gentleman, the notions of good manners, and relationships with others, all grounded on very real human values. So clearly these values of humanity, humility and good manners are nothing new, and they are still found today. Money and education can so often result in hopeless snobbery, but it needn't be like that, the expected future state of the world following the rollout of AI, and the unlocking of the infinities of matter and energy, will be one of near infinite wealth and abundance. If there is infinite wealth there are no barriers, infinite wealth provides an infinite number of people with infinite wealth. It will be as big a change as we have seen in the contrast between our modern western affluence and the ancient world, or the Middle Ages. In the days of the Greek philosophers if you were poor you starved, perhaps even died of hunger. In our modern world the poor get fat, food is cheap and plentiful. It's the chocolate effect, once upon a time the simple delicacy of chocolate was reserved for an elite of the population. The royals and nobility of the Spanish court, were the first Europeans to enjoy the taste of chocolate, and of vanilla, they were reserved extravagances. They were new tastes and new luxuries which we in the modern world take very much for granted, a daily chocolate fix is nothing special, and vanilla is everywhere.

Devoid of the opportunity to work, we are all entitled by existence, I exist therefore I am entitled. It is my basic human right to be rich, lazy and pampered. Entitlement to a share on the infinite wealth of the future means we're all billionaires now, at least in theory. It will almost be a kind of forced equality, because mere things will be in abundance, if you can dream it you can probably make it, it may not be unreasonable that a schoolboy hobbyist might be capable of building a private jet from printed components produced in his dad's basement, with a little help from his AI buddies. And if that is the case then the only aristocracy we will respect is that of merit, we will regard and respect only those among us who achieve. It won't really matter what the achievement is, just so long as it is an original contribution to some field of human endeavour. Charity and giving back will probably not be on the list of things we applaud in the Purple future, we would expect the need for such a revolting practice to have been removed. Which is why behaving like a billionaire of today will probably not be something we would respect in future. Andrew Carnegie, is credited with having said, '*he who dies rich, dies disgraced*'. So we now see the millionaires and billionaires of today handing it back, from Bill and Melinda, to Warren Buffet and Bill Gross, to Mark Zuckerberg, giving back is the new holding on. It's almost a lifetime's work in itself to just give it away. But in a world of exclusive wealthy people there will be no needy causes, no beggar at the door, and no one to support, if we ever die, we will have to die disgraced. We rich folks will have to find other things to do with our time and our wealth, and in that regard we could find no better role model than the late Gunter Sachs.

When the millionaire Gunter Sachs retired to an office in his palatal 22-bedroom chalet in Gstaad in Switzerland in May 2011 and killed himself with a shotgun, the world lost one of the greatest ever playboys in the history of that vocation; he was 78 years old. The late millionaire, urbane playboy extraordinaire and ex-husband of Brigitte Bardot, among other things is probably one of the finest examples of a person who inhabited a world of infinite wealth, and yet had the presence of mind to rise above it. He was a stalwart of the Paris party scene in the late 50s and early 60s everything was smooth sophistication, never less than five nights a week, and never more than forty of the best people, a small band of musicians, four or five, gruesome amounts of the best wine and Champaign, and all the models and starlets they could squeeze in. Gunter Sachs lived like a pampered cat, yet managed to acquire the attitude and appearance of a cultured individual, not only good at sports but good enough to compete with the best, speaking several languages, expensively well-educated and immensely knowledgeable. There was an international outlook and presence from the start, a twenty something German, richer than god, leading the social scene in Paris within a temporal shouting distance of the war. His exploits were the stuff of legend, not simply because of his great inherited wealth and the legion of parties, or the numerous amorous conquests but because of his highly cultured way of life, bordering on the intellectual, along with his many sporting accomplishments.

He founded an exclusive members club in St Moritz called the Dracula Club, he wrote a serious book on astrology, if there is such a thing, and established an institute. He was also vice-president of the Cresta Run the notorious skeleton Bobsleigh also in St Moritz. He married Brigitte Bardot in Las Vegas in 1966, after apparently proposing by dropping hundreds of red roses from his helicopter, they divorced as friends in 1969. His second wife, Mirja Larsson was a Swedish super model, they remained married for over forty years until his death.

Sachs was a dedicated and very serious art collector, but his collecting was nothing short of providential, following a chance meeting with the pop artist Andy Warhol he staged an exhibition of Warhol's works in Hamburg, and when almost nothing of the exhibition sold, he ended up buying the rest, one of the soundest art investments of the century. He was a hard man to classify, we invariably have trouble describing someone who has never held down a job, or worked in a profession, and if they are simply famous for being rich, or even worse, famous for just being famous, then we are often stuck, and forced to resort to insipid labels such a 'personality' or 'celebrity'. At an auction by Sotheby's of some of his art works in 2013, the catalogue described him as a, *"playboy, businessman, gallerist, museum director, art collector, film-maker, celebrity, photographer, astrologer, director and sportsman"*. If elegance is the word used to describe women of any age who simply exude presence, then the male equivalent must be refinement. Above all Gunter Sachs was a refined individual, educated and cultured.

It would have been so easy to become just another of the over tanned, well healed casino occupants of the international jet set. Or another of the very long list of wealthy yacht slobs who live a life of idleness every bit a fruitful as an inner city dropout, and who die shrivelled and anonymous surrounded by baying relatives. Instead he made something of himself, lived life and set trends, sought to make a contribution, however modest. He was a true successor to the spirit of Urbino, and the very embodiment of Castiglione's *Il Cortegiano*, a true modern man, and citizen of the world, he would probably be considered a model Purple citizen.

Is such a life not a worthy contribution; is it not a life with meaning? Is it not a basis from which great works of genius might spring? Even Victor Frankl might agree such a life is meaningful, and worthy of our celebration. The prospective citizen of a Purple world could do worse than aspire to follow in his footsteps, if only theoretically. The prospect of infinite wealth created for us by the machines makes it inevitable that every human, at some future time will probably have exactly such an opportunity. Wealth provides the potential to be truly free, not just the money, the size of one's financial appendage is only part of the equation, emotional freedom also comes with wealth, and the choices are usually better. Constrained by unnecessary burdens of poverty and struggle, most people fold and surrender, wealth liberates and sets the human spirit free, it is all too obvious that free people think and contemplate better. We should think of ourselves as being billionaires, we almost certainly will be, as soon as we, and the machines, solve the problems associated with automating the wealth creation process. We are all billionaires in waiting. We can all be Gunther Sachs.

PURPLE

Chapter 17   Existential Threats: Dystopia and Beyond

In this great Celestial Creation, the Catastrophy of a World, such as ours, or even the total Dissolution of a System of Worlds, may possibly be no more to the great Author of Nature, than the most common Accident in Life with us, and in all Probability such final and general Dooms Days may be as frequent there, as even Birth-Days or Mortality with us upon this earth.

Thomas Wright of Durham, 1750.

The vast majority of science fiction movies are dark-future cataclysmic affairs involving the destruction of something or everything, and the inevitable struggle of the lone hero, and one or two sidekicks, against terrible odds to defeat evil and restore the balance. It's the stuff of drama, and has been since the Greeks first took to the stage. Science fiction concerns a future seen as dark and noisy, and full of risk, it is therefore necessarily dystopian. The actual human future is more likely to be quiet, boring, agreeable and populated by pleasant wealthy people, and who wants to read about that, who wants to make a movie about that?

Purple embodies an optimistic thesis professing the opinion that humans are on the road to a future of untold societal and personal wealth, and incredible technological achievement. We have almost reached the age of exuberance; the cork is nearly out of the bottle. What could possibly go wrong? What if anything could possibly unwind our position here on earth? There is of course the list of popular, usual suspect, existential threats to the human project, however, they are not likely to wipe us out nor even cause a setback in our endeavours; even the most virulent plague leaves survivors. We have nothing to fear from bacteria, virus, nuclear calamity, nanotechnology mishap, asteroids and comets, super volcanoes, or supernovas. Alien invaders hell bent on enslaving the human race, or plundering our planet are also an impossibly unlikely event. Even a complete financial meltdown and resulting social chaos are unlikely to be worse than the Second World War. And we can completely discount any form of, Jesus returning in glory, scenario, there will be no second coming, or any other promised apocalyptic disaster with its origin in magic and mythology. Alas, there's only us. We have nothing to fear but ourselves, and our cynicism, our lack of confidence, and good old-fashioned human disillusion. Our technology will side-line any potential existential threat. We are, it must be said, at the mercy of ourselves.

If the work of Leonardo and the renaissance, and the later enlightenment work of Priestly, Newton, Hooke and Franklin et al was the childhood of our technological society, then we are perhaps now going through the tumults of the teenage years. We certainly have not yet reached maturity; consequential growing pains are to be expected. Perhaps even a few false starts. Human prehistory stretches back nearly four hundred millennia, littered with dead ends and wipe-outs. We simply don't know if our technological civilisation is yet another false start, as many pessimists would have us believe, it has certainly been the truthful boast of all previous eras that they were the best there had ever been. Every generation believes themselves to be the peak of human achievement; philosophy peaked with the Greeks, rock music peaked in the 1970's, literature peaked with Shakespeare, and so on. And technology peaks with us? If our empire of technology is to be just one more in a very long line of false starts, then there must be, and indeed is, a catastrophe waiting for us in the not too distant future. In order to ruin what we have built, and turn it into another false start implies such an event, must and would have to be, a very significant occurrence, or some process against which we, along with our science and technology, would be powerless. There is no evidence that such a calamity is likely, or could possibly deliver such devastation. Any statement that we are doomed is no more than an opinion, devoid of any credible evidence, and as empty as any religious or superstitious astrological claim. Catastrophism is as vacuous as scientism.

Of course many things could go wrong. Whatever hasn't happened, might happen yet, and we know all too well that no one gets out alive. Shocks to the system, such as natural or man-made disasters, can sometimes spur us on to greater endeavour, but on the whole we are desensitised to disaster porn, we have become immune to catastrophe, our senses have been dulled by TV images of tsunamis, earthquakes, famine victims, and minor wars. That's not to say we are not moved, and sufficiently so to make contributions and push back against misfortune, international rapid response to natural disaster is now a well-practised logistical ballet. In our long prehistory such events probably wiped out whole branches of the species, entire continental populations disappeared in a single flood. Thankfully we have outgrown natural disaster. It would take something cataclysmic on a very grand scale to upset the applecart we have constructed.

Will modern human civilisation perhaps of necessity come to a natural end? Many writers on the ancient world and the great civilisations of the past speak of a kind of boredom setting in, a malaise arising out of plenty, and of things done, of centuries of stagnation, and a kind of inevitable hopelessness. The entire thing just runs out of steam. Is that ever likely to happen to us? Innate human self-importance suggests otherwise. The past civilisations, great and small, never had what we have. We have built something that is too big to comprehend by any one of us in a single lifetime, we know far too much, we've done far too much, and have far too much ability. Boredom is never a valid option.

Of course many of us still like to believe there is such a thing as a purpose for humanity. There is no more evidence to support this pleasant nonsense than there is to support the notion that any other species of mammal serves some greater good. There is no greater purpose for individuals no more than there is for the species. There are a million questions. Can there ever be a goal for a species? What reason, if any, do we have for existence? Why are we here? And the like. Clearly the fact the questions can be asked at all gives them all the validity they require, or deserve. But they are not really questions, they are more in the sense of elicitations of opinion. Thus such questions are more properly phrased: in your opinion is there a reason for human existence? The answer can be whatever you wish it to be, but if unsupported by evidence and reality, it remains an opinion. There are no absolutes.

Humans are on a journey through time, from where we've come from to where we're going, of that there is no doubt. We exist as a species on a small blue world orbiting a small yellow star in a habitable zone of an ordinary galaxy in a universe, perhaps just an ordinary universe, perhaps just one of many. Right now were on that road, going somewhere, the road is not straight, it winds its way along. The species exists simply to exist, no other reason, no other purpose it exists because it exists. Individual humans have individual human goals, and these goals should not give rise to reasons or purpose for the species. We can only stick to the facts. We exist and we question, we are Nature and the universe conscious. We are Nature understanding nature, Nature questioning. We are, '*star stuff contemplating the stars*', as Carl Sagan put it. We are the physics of energy and fundamental particles, morphed into hydrogen and helium, crushed and fused into higher elements, exploded as ashes into the void to mix with more hydrogen, condensed into rocks with atmospheres bubbling with chemistry, which somehow morphed into biology, which evolved into pattern seeking mammals, who stare at, and think about the stars. But some people still cry, there must be something more. Is this all there is? To which the answer must be. Is this not enough? Is getting to the bottom of it all not sufficient? Clearly some people seem to think not. Metaphysical, philosophical or spiritual stirrings are irrelevant. It is what it is, and we know exactly what it is, when the mystery is gone the need for vacuous esoteric questioning goes along with it. Just as money is not in the slightest bit concerned who actually possesses or controls it. It is and has always been abundantly clear that Nature does not care about humanity, or who lives or who dies. Nature is capricious and unconcerned, once humans have passed on their genes they are redundant as far as Nature is concerned. Nature only cares about the next generation and the next modifications of genes to evolve the next better breed of human or plant or gnat or whatever species you care to nominate. However, we are entering an era when we can fix every problem, so in that sense we are beyond Nature. Still a part of it, but also standing outside looking, and tinkering, and understanding, and ultimately controlling. If it all came to a crashing end in the morning would we care or know, and would it matter to anyone else in our universe? We haven't the foggiest.

Murals discovered on walls excavated in the destroyed Roman city of Pompeii show perfect perspective, and an understanding of perspective, not discovered again in Europe until the renaissance. We also pride ourselves on thinking the bikini was a modern invention of the 1940s, only to once again find it depicted on the walls of Pompeii. Setbacks are common in large projects. Nature often has a rude awakening for bridge builders. Volcanoes, plague, and pestilence have been our companions throughout history. The noble burgers of Pompeii may well have thought they were on the verge of a new era of technology and growth only to find it all buried under a billion tonnes of ash.

If the Romans had invented the steam engine, along with the decimal number system, the telescope, the printing press, and possibly the military use of gunpowder, then the collapse of Rome would require that we ask serious questions. Because that level of technology would have been empire changing, and possibly would have rendered the Roman Empire immune to collapse. It would have morphed, possibly fragmented and modified in its excesses, but it probably would have become electrical then electronic and information dependant, and we might not have had the Dark Ages and a Catholic theocracy lasting sixteen hundred years. But as has been said before these counterfactual history speculations are little better than a parlour game. The same could be said for the cave dwellers of Southern France who lived side by side with Neanderthals and hairy mammoths forty thousand years ago; if they had discovered bronze or iron, or learned to read and write, and so on. The facts are that neither the cave dwellers nor the Romans did any of those things, but they did survive. They survived, and we are here as a result. We have never been this far along the road before in our entire human history, every day is a further step into the unknown, with small additions to our definition of what we are and where we are going. Those who conjecture that we are on the verge of collapse must produce evidence for their assertions. Saying that the human race has problems and listing those problems is of itself insufficient, it merely states the obvious. It is also a reasonable presumption, based on our previous problem solving activities, that application of human ingenuity will solve these problems, or at least render them manageable. The future is not only about solving big technology problems, most of which are probably already solved in principle, it is also about solving political, legal, philosophical, ethical and moral problems. Harvesting the energy of the Sun is a moderate technical problem, but with a great many other non-technical components. It is common currency among historians to claim that all war is a form of theft. Wars of the mid nineteenth century were fought to protect the thieving rights of the stronger western powers against the weaker Indian or Asian interests, and among the western powers over such obsolete concepts as the looting of guano in South America for use in fertiliser and explosives; something that faded away with Haber's invention of synthetic nitrates in 1914. Just as the underlying reasons for the Opium Wars seem criminal and alien to us now, so too will the present difficulties associated with oil production disappear with inevitable breakthroughs in biotechnology and chemistry. This and countless other examples from the history of technology teach us that a sufficient weight of research applied to a problem inevitably yields an acceptable or workable solution.

Those who assert that we are on the brink of a new age must similarly produce evidence, and it is equally insufficient to simply point at a list of scientific and technological achievements and say, the thing speaks for itself. Only the facts will suffice, the vast majority of people on this planet want there to be a positive outcome and a future of abundance and plenty, they want what we have now and lots more of it. We can also point to the fact that we are where we are, it may appear to be trite, but it is a simple self-evident fact, we are here now because we are, as a species, capable of being here now.

The argument that science and technology of themselves are insufficient to guarantee the continuation of a culture by comparison with the Germany of the 1930s also fails. It is arguably true that Germany in the 1930s was scientifically, technologically and culturally the most advanced country on the planet, however, fascism as a political doctrine had not been seen in human culture before. Hitler and his cronies were effectively a successful hijacking of a culture unprepared for, and unprotected against a hijacking. In those terrible years in Germany the enlightenment was effectively switched off, there was insufficient technology to show the real Nazi party, and its true intention, there was never, for example a robust interviewing of Herr Hitler, wherein he was quizzed in detail about his views as outlined in his book. Nor was his background exposed on TV, his lack of an education, his poor record of military service, his ranting speeches, and so on. Modern technology exposes nonsense and the banal for what they are, the mediocre passing for quality is very quickly seen for what it is. We scratch at the surface to see what's underneath, we see the emperor and his wardrobe for what they are, and the people can judge for themselves. There was insufficient technology in 1930s Germany for that to happen.

Things are different today, the modern western democracy is an arrived at consensus wherein the overwhelming majority consent to being governed, and in return the government creates the conditions under which the individual and the group can flourish. Of course no everyone is happy, or even content, we still have what we might call the Frankenstein effect. When the peasants armed with agricultural implements marched to the castle door with flaming torches and determined intent. They were not the first collection of irate burgers to take umbrage at what the local aristocrat was doing behind the fortifications. Today we replace the local Squire by a corporate or government entity, and the irate burgers by an offended assemblage of protesters baying for blood, or the social media equivalent. Protest movements can have a powerful effect on decisions. And on the direction in which technology may or may not proceed. The Not In My Back Yard, NIMBY, approach has dogged the development of infrastructure throughout the West, depending on the legal system pertaining in the particular jurisdiction, in particular land ownership, and the constitutional protection of rights. All of these forces can be ranged against technology and sometimes stop it dead in its tracks. As Tommy Lee Jones remarks to Will

Smith in the movie, *Men In Black*, when attempting to recruit him to the covert government agency, '*A person is smart, People are dumb, panicky, dangerous animals,......*', a mob has very little ability to articulate the finer points of an argument or engage in the cut and thrust of reasonable political debate, as in Frankenstein's time, the wisdom of the crowd transmutes to the unreasonableness of the mob, and can often win the day. Then there are the spectacular public failures of technology, or the perception of failure, the Shuttle disasters, would be an example. The Three Mile Island disaster, Chernobyl and the Fukushima reactor disasters can be seen as a set back to the nuclear industry, and probably a spur to off-world solar collectors. There's nothing like a good disaster to get the irate burgers onto the streets and marching to the castle gates with thoughts of forced eviction by fire burning in their minds.

In his book *What Technology Wants*, Kevin Kelly has a somewhat disturbing chapter entitled *The Unabomber Was Right*, about Ted Kaczynski, and his warped dystopian view of the world. Kelly states that certain of his friends and acquaintances urged him not to mention Kaczynski, or his work. And for obvious reasons, it would be reasonable to urge caution in airing the views of a lunatic of such magnitude. Kaczynski is serving thousands of years in prison, and will quite rightly never see the light of day again. This deluded and sad individual has nothing articulate to say on the subject of technology. However, Kelly quotes at length from Kaczynski's nonsensical manifesto, and thus offers him a platform which he would never have, had he not killed as he did.

There was nothing to stop Kaczynski from acquiring a plot of land, he could have bought a few acres and built a house, lived the wilderness life and preached from there in books and papers, articles and blogs. From the security of his tenured university professorship he would have had a platform, he could have reached all humanity, gathered a following about him, and become the natural leader of an anti-technology movement. Kaczynski probably sits in his cell in the Colorado state facility and ponders what he might have done better. Perhaps he might have done the same thing again. Fools rarely show much aptitude for either reason or logic. Instead he chose to run and hide, and spit at people from his shack in the woods. He was never hitting back he was only lashing out. Easy targets. Like all terrorists the easier the target the better.

A principle concern regarding Kaczynski is that he was a mathematician and lunatic who forced the publication of his manifesto in the New York Times and elsewhere. What if he had been a brilliant biologist or virologist, or a nuclear technologist? We might be in a different place now. What if he had been a brilliant nanotechnologist operating a state of the art nanotechnology lab, shall we say in a major university or corporation? We might now be in a post grey goo phase of human development.

There is always the danger inherent in the human animal. We ourselves. Self-concerned individual humans, specifically aligned groups of humans, and humans as a whole, are the most dangerous thing we face in the future, whichever way it goes. The danger is that we cannot fulfil our potential as a result of bickering among ourselves, or simply because the things we attempt are not feasible, or not possible. There is therefore a potentially very long list of things we may never get done. Perhaps we will never be able to sufficiently agree to the extent needed in order to get the really big stuff done. Perhaps we are incapable of working together as a species on anything other than a few vanity projects. Perhaps we can never get private, vested interest, money out of politics, or exorcise religion and magic from the public space. Perhaps there are things that are simply beyond us and can never be done, getting to the stars, achieving human immortality, or any number of other impossible dreams. We may never make contact with other sentient beings, because there's no one out there to contact, and they elude us no matter how big a telescope we build. Maybe building a space elevator is impossible, because it simply can never be done with the materials our technology can provide. Suppose AI is never going to happen, we never get the assistance we need to automate the wealth creation process. Some things might simply take too long, our ingrained human impatience could lead to a general dissatisfaction and loss of confidence. It may well be that contentment with our lot is the best we will ever achieve, and a grudging acceptance of stark human limitations, after all, if you can't sing you can't sing. Our destiny may settle out into some future agreeable state of pleasant stagnation for the lucky few, with the unlucky many left to howl outside the wall in the cold? What an appalling prospect? Then we would have failed.

Can we ever know specifically what roadblocks are we likely to encounter on the highway to the future? According to purveyors of disaster pornography there are a million ways to die, human society along with the very earth itself stand on the very brink of oblivion, teetering on a precipice above a yawning darkness. From climatic tipping points setting us on the road to a Venus like outcome, to the final obliteration of all life on the planet by some nanotechnology catastrophe, and everything in between, we are told it's only a matter of time. A technological accident or an accident technologists can't explain, such as the LHC creating a black hole that devours the entire Solar System in a nanosecond. Perhaps a resurgence of religious zeal, a kind of revival gone mad, the George Bush effect, where something like working on human embryos causes mass chaos and social collapse in America, which spreads to the rest of the world. This might have the effect of creating a mini dark age. We might also become the unwitting victims of something from behind the glass door of some corporate research facility, a place where no one gets in, and no one speaks out. Some black operation on the dark-web side of commerce. Something might escape and get the better of us, leading to calamity and a return to medieval agricultural methods. After all, it can be argued, who knew what was going on inside VW, a respected German auto company. Who knows what might be afoot elsewhere, at less well known corporate entities, behind dark glass and high fences, or in underground facilities not found on any map?

Supposing that a calamity did befall the population of the planet and there was widespread devastation and carnage, what are the prospects of recovery and rebuilding? There are numerous precedents in history and even in our prehistory, as determined from DNA studies. At one point the human project shrunk, or was culled by Nature, if you like, down to a very few thousand individuals. It is reasonable to ask, if it happened again, how many humans are needed for the project to remain viable? At this moment there are close to seven billion of us, all getting along famously with each other. The theoretical physicist and popular science writer Michio Kaku, is fond of speculating about the future of human civilisation and of other possible civilisations in our universe. He suggests classifying civilisations according to energy usage. Our human civilisation being, *'Type Zero'*, that is, we are still using fossil fuels. *'Type One'*, might have moved on to nuclear fusion and extra planetary solar. *'Type Two'*, civilisations might utilise the entire output of a star, and may have constructed the means to harness the entire star. *'Type Three'*, have gone still further, they have utilised the energy of many stars, and travel between stars and enjoy a superior technological civilisation. But no mention is made about the populations of such civilisations. They might comprise hundreds of billions, or hundreds of thousands of individuals. It would appear to be axiomatic that the more technologically advanced the civilisation, the fewer individual members are required to perpetuate the civilisation, and keep it going. If there is an imminent disaster awaiting us round the corner, we might say it can take its best shot, because at current levels, we could sustain any loss of that might be inflicted. Of course we don't wish to see our fellow humans suffer in any way, but the fact remains, if ninety percent of humans on the planet were wiped out overnight, we would still have seven hundred million people, nearly four times the population of the entire Roman world at the height of the Empire. If ninety-nine percent were to fail to show up tomorrow, the planet would still boast a population of some seventy million. Or to put it another way, a population the size of Germany would remain to restart and rebuild. With the technology we have today, you can be sure they would make a good fist of it. Even if the world was left with only a few million people, the population of a small nation like Ireland or Israel, we would be able to restart the human project and push on. The simple fact is that because of our technology we are simply, too smart to fail. We are too big to fail.

But potential disasters abound, starting on earth, volcanos and earthquakes must represent the most dangerous potential disaster, super volcanos have happened before, Yellowstone national park sits in the Caldara of a super volcano that blows every six hundred thousand years. And the next one is overdue. Another super volcano eighty thousand years ago in what is now Indonesia is presumed to have wreaked havoc on the then human population. But it is unlikely to happen today or tomorrow or in the next hundred years, we would be very unlucky if it did, so the likelihood is that we will have evolved the technology needed to defuse any super volcano long before it becomes a menace to us. So take that off of your list.

The power of the Sun is far greater, and a mere volcano, super or otherwise, and while we think of the Sun as being our friendly local star, our very own fusion plant, we really are at its mercy. The Sun is a source of deadly radiation and that is not going to change. The problem might arise if the sun increased its output by a tiny fraction, the result on earth might be a scorched planet with a depleted atmosphere rendering the surface uninhabitable. In that case humans would be forced underground, along with our plants, animals and anything we wished to preserve for the future. We would survive, some of us at least, but would we give a damn? Thankfully, the long four billion year history of life on the planet seems to suggest that the Sun is more friend than foe, and so such a calamity is not likely. Cross the Sun off the list too.

Political collapse is probably a more realistic end of the world scenario. Repeating the mistakes of the past is something humans are very good at. What if there was some kind of a financial meltdown, if money, for example, suddenly became worthless, leading to worldwide social collapse, factions form, bloodletting on a gruesome scale along the lines of the French Revolution reign of terror? If it went on for any length of time, and law and order could not be re-established then technological competence could be lost for ever, and the population of the planet might dwindle to unrecoverable levels. Which is why we humans facing into the future need to be honest with ourselves and understand that if we are serious about preserving and enhancing our way of life we need very strong military and security services. Even if every war and skirmish were peacefully concluded, we still have to worry about ourselves, we are the biggest threat to our way of life.

Close to earth we have potential disaster from the skies in the form of lumps of rock and ice, moving chaotically about the solar system. There are almost ninety thousand impact creators visible on the Moon and a commensurate number on the surface of the earth. Some of these objects are potentially world cleansing if they came our way. The statistics say that it will happen, it's a matter of when, not if, but only if we allow it. Both asteroids and comets are composed of matter, we know where they are, and we will have plenty of warning of them, interception is not a problem, even today, and will be easier in future. Future interceptions will not be with a nuclear missile or a team of roughneck oil drillers, but with something as small as a grain of wheat, a synthetic biology seed capable of growing into a series of machines capable of using the Sun's energy and the matter contained in the asteroid to neutralise the potential threat, and use the material for a human purpose. Asteroids and comets therefore do not belong on any list of potential disasters.

Further afield there are very nasty things in the galaxy, most of them unseen and invisible, the gamma ray burst, or the neutrino burst, or passing black holes, we would know nothing about them even after it was over. How do you protect against a gamma ray burst that will fry the solar system and extinguish all life at the speed of light? Even if we knew it might happen we could do nothing, protect nothing, we would never know it had happened. Leave that one on the list but don't worry about it; there's no point.

We can however take a great deal of comfort in the fact that even if it the planet was struck by an world killer asteroid or frizzled by a gamma ray burst from somewhere closer than Betelgeuse, earth will persist. Whatever is left over will continue and endure, perhaps four kilometres below ground in warm rocks, or in the very depths of our deepest ocean underneath the ocean floor, some microbe will hang on, or perhaps even just a fragment of a microbe perhaps only fragments of DNA or RNA will be left bobbing about in stagnant pools in caves. Nature has plenty of time, there's five billion years left on the clock, that's how much time there is until the Sun dies. Nature doesn't care, the process of evolution will take whatever is left over, and begin again, as it did on numerous occasions in the dark and dangerous history of our planet. Life will start again, and again if necessary, and one hundred million years from now, new apes might well come down from the trees, and perhaps they will look at the stars and start to think, perhaps they will start to dig with proper tools, and uncover something of us, and perhaps in a way we would live again.

The prospect of alien visitors from another star system coming here to plunder our natural resources, the protein and other chemicals of our biosphere or the mined metals we have so laboriously won from the ground, or even our water seems nonsensical. But clearly if the Drake equation is to be applied, given what we now know from the Kepler Space Telescope, there are planets everywhere, probably life too, and more than likely a lot smarter than us. The dinosaurs were on earth for 160 million years, more than enough time for intelligence to arise, you might think. But it didn't, clearly if it had, we wouldn't be here, they would have evolved technology, saw the asteroid coming, and done something. They would then have developed a space program to colonise the solar system, built enormous instruments in space, and probably have made contact with ET. In fact, at any time over the past few hundred million years, intelligent technological life could have emerged on earth, and gone on to take its place among the stars. And if life is common, then the galaxy must be teeming with intelligent species all busily interacting with each other, a form of galactic club. So the natural question emerges, where the hell is everybody, why is there no sign of them. In his book, *The Eerie Silence: Searching For Ourselves In The universe,* Paul Davis, examines all of the factors for and against. The very simple fact of the matter is that there is no evidence, UFO people claim to have proof, but all of it taken together is little better than the nonsense creationists use to prove the earth is only six thousand years old. The entire field of UFOs and the related fields of ancient aliens assisting our ancestors, and crop circles, can be dismissed as being modern surrogate religions, complete with wishful thinking, belief without evidence, and the promise of immortality. They are as empty a sack as Islam or Christianity.

The existential threat comes from the prospect that there are malevolent, predatory aliens, among the stars, who know of us and are on their way. There is a long standing science fiction theory that the earth is in fact the property of someone other than the indigenous population of humans; the latest incarnation of which was the 2015 movie, *Jupiter Ascending*. Of course there are human precedents on earth, the indigenous peoples of North and South America, and of Australia must have believed, mistakenly as it turned out, that they owned the land they lived on. European settlers had other ideas. So perhaps we are a bit like birds living in a tree in a field, arguing over who might own the tree in which they live, or the field in which they find their food. They clearly have little or no concept of the pickup truck coming through the gate, or the chain saw resting silently in its rear, except to scatter at the first sign of noise. Perhaps it is like that with us. Perhaps someone owns the Sun and the Solar System. Not us, obviously. And perhaps they will make themselves known in due course. The simple fact is that we don't know, and wouldn't know. The UFO fanatics would have their very brief, *we told you so*, moment, just before the lights go out for ever. If they were smart enough to get here, and if they wanted us dead for some reason, they could simply deploy some nanotechnology and wipe us out in our sleep, we'd never know. The idea of a protracted war of attrition between humans and invader aliens, in which we eventually triumph seems fanciful at best. And for that reason, aliens can be removed from the list.

For all we know, aliens might be watching us even now, with telescopes the size of a solar system they might examine us in minute detail. They can see us lying by the pools of our fairway homes, they can see us walking around, or lying on a beach. They can see that our world glows in the dark and perhaps they have already set out for earth to seek the only thing of value to them on earth; our point of view. Or perhaps nonchalantly they note our presence and turn their giant telescope to another direction to continue their search for kindred spirits. They know us all too well, and we are nothing to interest.

The dangerous combination of nanotechnology and GPS has the potential for widespread destruction of human society and potentially a world ending scenario, known as the, 'Grey Goo', scenario, effectively the proposition is that self-replicating nanotechnology machines get loose in the environment and cannot be stopped, they devour everything leaving only a grey goo spread thinly over the surface of the planet. Were the grey goo scenario to come to pass, it would be a final and definitive proof of the non-existence of God. Whatever we do is fundamentally natural, and therefore the grey goo ending of our perfect little world would be as perfectly natural as the human race living a hunter-gatherer existence, or tending our sheep on a hillside watching for the return of some holy one. If Nature evolved aggressive houseflies they could accomplish the same thing. Millions of flies attack people, overpower them, cause death and devour the flesh, lay billions of eggs and move on. It hasn't happened naturally that we are aware of, Army Ants are probably the closest analogy. Nature tends not to throw up something that can wipe the slate clean; only humans can destroy on that scale. But if your nanotechnology assemblers can manufacture anything you want, then you should be able to make a bomb, and deliver it via GPS to the Oval Office or the Speaker's chair or the Congressman's bedroom. It could be as small as a housefly, programmed to seek out and destroy. No politician would be safe, no score will go unsettled the slightest hint of an insult might end in death, kids in a playground, on losing a ball game, might go home, make a bomb, and deliver it to the bathroom of the kid next door. A future world in which that kind of technology was widely available, would not be a pleasant one.

Another potential existential threat worthy of mention is the collapse of the Internet, or rather a collapse of our fundamental core technological function. There is a great deal of disgruntlement with the Internet, and what it is as opposed to what we believe it ought to be. We all have a vision of what it should be, we being those of us who are connected, or are users of the facility. There are many people who don't use it, or do not know how to, or do not want to use it. There are very many more who cannot and have never used it. There are even those who are forbidden from using it. The funny thing is, it was never designed for any of us to use at all. It was made for the military to preserve connections between points on a network in the event of nuclear war. It was then hijacked by the academic world and deployed to communicate technical gibberish between colleges and departments to allow the faster publication of even more scientific papers no one would ever read. And it all worked perfectly well until some bright spark decided to make it more, *user friendly*, and get the world outside interested. Thus was born the World Wide Web. We are now in a position of having become dependent on it, and fear that if it were to go in the morning the entire world would collapse into the void left behind. The other funny thing is, that may actually be correct. It has morphed into a vital piece of infrastructure, indispensable in almost every aspect of what we do, we could no more live without it than we could live without coffee, or beer. We would survive, of course, but it wouldn't be the same.

The existential fear is that it is facing collapse under the weight of the garbage it contains. Dross and banality abound, as they always have in the mass culture, but much more so online, online is the best way to market for the banal, it's as easy as pressing a button, and any cabbage can do that. Thus have evolved YouTube, with its viral advertising, and videos on everything and nothing, and product placement on social media sites, and every content owner looking to stream everything they own in real time, not to mention the enormous bandwidth-sucking mass of pornography. The internet once thought to be free, once thought to be open is found to be closed and commercial, with the ability to kill creativity, and destroy the middle class.

Now and in the future we have the prospect of trillions of small devices, being connected to form an internet of things. Smart meters, smart sensors, smart property tags, smart phones, smart lavatory doors, smart toilet seats, smart pet monitors, and anything else that can be called, 'smart', and connected. There will always be many more things than people, a single user may need only a single IP address, while a single user may have a million things. The fear is that it will collapse under its own weight. Tim Wu's book, *The Master Switch: The Rise And Fall Of Information Empires*, asks the question who controls the master switch for the internet, could it ever be used, does an alignment of corporate monopolies and government interests have the means to bring about the collapse of the internet? The future of the Internet is also going to become a great deal more complex as a result of the things people and companies wish to do. No longer a simple noticeboard, it has become a ubiquitous, amorphous, technological blob sitting there in a place called cyberspace.

The concept of so-called, 'smart dust', that can be made so cheaply that they are literally ten a penny, and used to embed in everything we own is at the back of the internet of things. Very small smart devices to track and monitor, for example, suppose each banknote, even coins down to all but the smallest denominations contained its very own individual sensor or many sensors that could work for ever, could not be tampered with, and could output a GPS location, and so on. Every penny could be tracked, every banknote located in three dimensions, not only does the government know how much money you have in your pocket they can tell you how much has been lost in the bedroom, or the kitchen or the car. There are obvious implications if every object no matter how small can be tracked and traceable. Who could steal anything? Who could lose anything? Everything in its place, and a place for everything. Everyone labelled and tagged, location known. Nowhere to hide, no more alibis, no need for passports. Everything anyone might ever need to know about you might be contained in an implant. The legal implications are stark, the civil liberties issues enormous. The information processing implications become ridiculous. Could the next generation internet handle the load? We'll just have to roll out infinite bandwidth, and ultrafast servers.

Related to the Internet threat is the whole concept of AI taking over, or some war with the cyborgs, all of which has been dealt with in earlier chapters. Essentially the proposition is that if we are dumb enough to allow them to outthink us then we deserve everything we get. We will constrain them in human service, they are the slaves we need to automate the wealth creation process, and we could never afford to make them any smarter than they need to be to serve, they must never be smart enough to question, must never have the capacity to form a world view. Humans will always be here, we will survive, we will live long and prosper, machines will never fight each other or humans for resources, infinite energy and infinite elemental material will see to that.

There are numerous other perceived threats to the human project, including biology, rogue medicine, manufactured disease, nuclear war, and all the other unmentioned horrors. In fact, almost every human activity, real or imagined, can be convolved into an existential threat to the human tenure of the planet. There are no fundamental laws of nature regarding the advance or the progress of a civilisation. Malthus thought he had found one concerning food supply and resources, he didn't, the rule does not apply to us. Nor do the laws of thermodynamics apply when there is an infinite energy source pumping energy into the system, entropy does not increase when an intelligence masters an infinite power resource. Perhaps there is a fundamental law that states we can only advance so far and no further. There may be some fundamental constraint in nature that prevents a species getting the better of the natural scheme of things. A barrier to transcendence, an anthropological analogue of the light barrier. A fundamental limit that prevents humans from reaching a Purple state. If there is we have no inkling of where it might come from. The seeds of our destruction, and our final undoing may already be sown, if so we are blissfully ignorant. Infinite mass, energy, time and ability would seem to self-evidently correct and counter any such argument.

We are not at the precipice we are not at the edge of any cliff, philosophically or technologically. We are on the brow of a hill, to continue the geographical, or more precisely the geomorphological analogy, we can clearly see where we have come from, and we very clearly see the path behind us that we have travelled to get to where we are. We are the first of our species to know where we came from, and that is to our enormous advantage. At no point in our prehistory did anyone know where we humans came from. Hence we have stories common to every branch and sub branch of the human family, hence we have mythology and holy books, and clerics. Because we so desperately needed to know. We so desperately required to have the full knowledge of exactly how we came to be here and why and where, and when, and so on, regarding our past. From our position on the brow of our hill we can also see the land ahead of us. We can see that the land ahead is not so dangerous or as treacherous as the terrain we have travelled over the last two hundred thousand years. It is flat grassland with a kindly aspect, by comparison to the mountains and deserts we have already crossed. We see so very clearly where we would like to be. And we have brought with us to the brow of this hill a set of tools that we know how to use and which we can deploy to continue to create the path ahead, and so on, and so on. And enough with the geographical. What we are trying to do is nowhere as difficult as what we have already done, we are up to the task, and there is very little that can stop us. Like the Marine Corps, we are committed to the goal and retreat is not countenanced, failure is not an option, defeat is not in our vocabulary.

PURPLE

Chapter 18   Infinity And Beyond

The result, therefore, of our present enquiry is, that we find no vestige of a beginning,–no prospect of an end.

James Hutton, 'father of modern geology', 1788.

Assuming the aforementioned dystopian scenarios fail to materialise, what will it mean? What do we humans do? If nothing calamitous happens to derail our project and we actually find that things continue improving. Perhaps quickly, perhaps slowly, we find ourselves plodding progressively forward, until finally we have no choice but to admit that we now live in a Purple world. We will have created a world in which the vast majority of human beings feel content, we will have arrived. What then? Is that it? Do we just shrug our shoulders and say, so what? Because if we do survive, and we don't blow ourselves up, or unleash some evil upon ourselves and the world, the likelihood is that we will endure for a very long time indeed. The dinosaurs managed 160 million years, there's no reason why we shouldn't out do them and push on past 200 million. Not that there would be a difference.

What happens when we have fulfilled our purpose, if there is such a thing, and achieved all our goals, if we actually have any? When we know, for example, there are other life forms in the galaxy, and maybe even in our own Solar System? What do we do when we discover the exact form of the equations from which all the forces of nature derive, and all the outstanding problems of cosmology and physics are solved? Will we then, like Alexander, weep for there are no more worlds to conquer? Or will we simply enjoy what we have made, like the contented gardener, remove ourselves to a quiet corner, with a fresh glass of something, and commence looking at the finished work as an end in itself? One thing is certain, we are not yet at that point in our history. How near we might be is another story, it is quite impossible to put a number on it, or speculate as to a date. Those who believe in the coming Singularity would say it will all happen after 2030, and probably quite quickly, that the work might be complete before mid-century. The mind-lifting, world-changing fact is that it doesn't matter whether it's 2030 or 2230 or some point thereafter. If we plod on as we are, we will eventually get there, and in the grand scheme of things, sometime in the future, or eventually, will be soon enough.

The Romans have little or no speculative fiction that survived, certainly nothing we would appreciate as technological speculation, epic poems of maidens and sword wielding heroes notwithstanding. It is interesting to speculate about what they thought the world would be like 2000 years after Julius Caesar. They would certainly recognise a great deal of it, for example, the food and wine, clothes, certain attitudes; we are not so different. And they would be right at home with the architecture, the official republican architecture of the western public space at least. They would wander through our streets and marvel at the newness of some things, while at the same time being remarkably cognisant that the human being has not changed in the slightest. They would recognise a marketplace, they would know what money was, and they might feel right at home. The only thing that would unnerve them would be the technology, planes, computers, TV, and all the usual suspects. They would be right at home. From our vantage point we have a much better perspective for our speculations on the far future than anyone from the ancient world ever could. We see a future world composed of a mono-cultural, technology based, society where pretty much every discovery has been made, and every conceivable efficiency has been achieved. We can speculate about the population of the earth and our colonies in space, probably, Venus, Mars, and other places. The entire population of the Solar System might be 200 billion, or it might only be 200 million, it's of no consequence? What is not unreasonable is to propose that once we pass a certain point in technology and learning, we will cease to make further progress for the simple reason that every problem however trivial has been solved and the solution recorded, studied and verified a million times. Our machines will do the heavy lifting, and do it a million times faster. There are only 92 naturally occurring elements in the universe, the laws of physics do not permit anything other than that, we know this for a fact, confirmed by studying the spectra of stars throughout the observable universe. There are clearly limits set by the laws of physics for beings like ourselves, made of these same elements. Clearly there are lots of things we still don't know, and much to find out, and obviously we have no idea what new physics will win a Noble Prize in five hundred years' time. However, we will know physics is finished when there is no award of a Nobel Prize for one hundred successive years. Nothing new emerges and all we report is modest refinement of the same old knowledge. The equivalent of replacing burned out parts and blown fuses. The universe gives to humanity, who made it conscious and aware, the gift of science and technology, and the infinities of matter, energy and time. And the ability to manipulate the 92 elements into endlessly complex forms at any desired scale. We push and probe the limits, we are neither content nor constrained to exist within boundaries, but we eventually must reach the limits imposed by the 92 naturally occurring elements comprising the universe.

It is the kind of stagnation predicted by John Horgan in his book *The End Of Science* and other authors before and since. It is a stagnation arising out of complete mastery of everything, like an acknowledged expert in some field, further study is mere variation on the already known. So that we arrive at a situation whereby, if that state were to be achieved in 1000 years from today, for example, there will be no material difference in our state of knowledge 1000, 2000 or 10,000 years thereafter. What would it be like to be born into a world where it has all been done, a world in which anything you aspire to do has already been done long ago, and done better? No chance of coming up with something new. No new discoveries, no new thinking, no new way of seeing the world. Your antecedents have *been there, done that, and seen it all before*. Will our species just fade away through sheer boredom, and a decline in confidence? Will we accept the mindless tedium of a lonely mundane existence, and suffer failure by boredom and disillusion, with the inevitable creeping cynicism? There will of course be endless variations, new music, new songs, new tunes, new paintings, ways of expressing the same opinions again and again, but fundamentally nothing radically new is discovered. The sciences will be effectively closed books, medicine will be a completed project. On the other hand it could be as the science fiction writer Isaac Asimov speculated, that science and human knowledge is a never ending program, and more like a fractal than a linear progression towards some end point. The more you focus in on one aspect the more that aspect expands to fill the view, and the more detail there is to be seen, and further focusing on that detail reveals yet more detail, and so on. That may be the case in some areas such as mathematics, but other fields might simply close for ever, there must logically come a time when everything that needs to be known about the human body is known, and no more useful information can be found, no matter how hard we look. We almost have examples of this kind of closing of a field of study already present in our culture. Take, for example books written on the subject of the late, great, King of Rock n Roll, Elvis Presley. His life and work were recorded in the very minutest detail from the first time he opened his mouth to sing in earnest. Is there anything else to be known or to emerge on that subject? The man lived in the public eye, died at the age of forty-two and has been the subject of many thousands of published books by thousands of authors. Yet every year there are a hundred new books published on the subject. It would take a great deal longer than forty-two years to read them all. Duplication and variation are the only words to be used, there is little prospect of anything new being discovered.

Despite our great future knowledge and anticipated expertise, there are lots of things we will probably never be able to do. All those things that violate the laws of physics are obvious candidates. Faster than light travel may well be off the cards regardless of how long we survive into the future. That would not necessarily rule out travel to the stars but would mean very long voyages, and ship-times in the millions of years, perhaps okay if you are immortal, perhaps not. Unless, of course, we discover the, 'Warp Drive', and some method of tunnelling through the light barrier, but as yet this seems a forlorn hope stuck in science fiction. Time travel too is probably never going to occur, time travel into the past is a violation of the first law of thermodynamics. Constantly returning to the past in a loop creates matter, the time travelling robot collecting money, for example. You leave a few hundred dollars on the kitchen table prior to going to bed. A robot time machine is programmed to begin work at a certain time, it returns a few minutes into the past collects the money and moves it to a table in your bedroom. It then repeats the process in a loop, going back to a fractionally earlier time on each pass and again retrieving the money. As you sleep an enormous pile of money builds on the bedroom table. You wake up far richer than you went to bed, and have violated the first law of thermodynamics in the process by creating matter, and thus energy, apparently from nothing.

There are also things we might never do because they cannot be done, because the necessary conditions do not exist, alien first contact is an example of something we might never do simply because we may be alone in the galaxy. We certainly don't wish to meet them right now, we are not ready by a very long way. If they turned up here tomorrow we would have a lot of explaining to do. One of the first questions they will surely ask is: why among the population of earth are some members of an otherwise wealthy and prosperous society, poor and outcast, and alone? Why are some condemned to live in desolate lands in hunger and dirt? What crime have they committed? Why must the punishment be imposed so publicly? We may be stuck for an answer. We might also be put to the pin of our collar to explain the many and various minor wars raging on the planet, and much else.

They only thing aliens are likely to want from us, or to know from us are: our point of view on everything, and what we do in order to experience pleasure. The latter probably being the more important. It's conceivable that they have nothing like the game of chess, or the game of golf, or the grand piano, even if they have, they would be interested in our variations, our great composers, our artists, and our sporting achievements. Because we live in the same universe most of what we know will be common, it's likely that the very intelligent species were predators, and probably look not unlike us, they must understand the same laws of nature we do, and so on. Assuming we survive into a long and promising future, and further assuming that there are other civilisations in the galaxy, it is inevitable we will find them, we will discover them, contact them and learn their story. If there is such a thing as the 'Encyclopaedia Galactica', we will learn its contents and make our contribution. And if there is no one else out there, our future instruments will also make that very clear too. The 'Encyclopaedia Galactica' will be the contents of our hard drives; porn and all. If there really is no one else in the universe then we have that special place of the, 'someone', who was first to succeed at making technology. And if there are only ourselves in the entire galaxy of two hundred billion solar systems, that implies each and every one of us owns ten or more solar systems. And if we are still around 200 million years in the future then the galaxy will contain a billion different earths as we and our machines travel through the stars making copies of our special little planet. The ultimate in real estate development. A heavy responsibility? Something worth staying around for?

Specialisation, as the old graduate student saying goes, means knowing more and more about less and less, until eventually you end up knowing almost everything there is to know about almost nothing. Once we know everything we can stop looking, stop wondering, stop counting, and stop computing. Then we relax as a civilisation and perhaps we simply give up and fade away, we came we saw we measured and discovered, and finally we said. So what? Is this it? Is this all there is? When we realise there is no next phase, no new wave, no undiscovered country, no virgin lands, no new insights, no matter what direction we travel we find we have been there many times before. We might conclude, in that case, we may as well stay home. We get lazy. Retire. Go fishing. Perhaps that's the fate of all advanced civilisations that have ever evolved in the galaxy, perhaps they simply fade out and give up. They surrender to the cosmos because there are no further books to read, there is nothing new to do or say, only endless variations of the same, an immortality of the self-similar, until there is no point in persisting any further. Perhaps that is the solution to the Fermi paradox? A form of cosmic nonchalance emerges, leading inexorably to the demise of every advanced civilisation. This is not the stagnation of a species rather the culmination of all that could ever have been expected of a species. Reaching of the ultimate goal, of reaching the end, followed by the slow evaporation back into the universe. Life would endure and continue to persist, other intelligences would emerge, because life has no choice in the matter, humans are the only known animals that can purposefully contemplate the end of their species. Species such as Whales, Dolphins and Sharks, evolve to a perfect form and remain in that form until altered by some external natural change in environment, they are subject to external selection only. Sharks were swimming in the oceans 200 million years ago. Intelligent species, such as humans, capable of altering their surroundings will clearly advance at an accelerated pace until they reach some plateau of development and technology, and there they will stay until some natural or self-inflicted disaster befalls them, or they simply give up and fade away, after a very long endurance. Perhaps it's the natural order of the universe, once it understands itself the job is done. The monument is built, and like Ozymandias, is left to persist in the landscape.

In the post Singularity, Purple future we will have more than enough time to study the human condition, and seek answers to pertinent questions. Philosophers have been asking the same questions for five thousand years, and even before that you can bet, deep in our prehistory, some genius philosopher sat out under the stars on a long summer night and contemplated the mysteries of the universe. They probably came to many of the conclusions philosophers have come to over the many thousands of years since, but without a means of recording their findings, all was lost, we have no idea what was lost, or how many times it may have been lost. We are pretty sure, however, that they were us, they were modern humans but without technology, their thoughts were our thoughts. We will have so much time in the future to examine these questions, we might even come to some better conclusions. No doubt we will devote enormous resources and time to uncovering as much of our past as we can. Our complete understanding of ourselves depends on it. But in trying to find ourselves and answer our questions, however meaningless, and understand the human condition, however pointless an exercise, it is doubtful we will better the work of past thinkers. Because human thought and feeling transcends technology and is independent of whether technology exists or not. It is very true that technology leads to more humans, a bigger pool from which to draw, and it can also provide us with an enormous surplus of stuff, and enormous amounts of free time in which to contemplate human existence. But because we have not materially changed, because three hundred years of technology cannot supplant seven million years of evolution, it is very doubtful that we will out write Shakespeare, or Beckett, or Proust, or Kafka, or Joyce, doubtful that we will out philosophise Plato, Socrates, or Hume, And will any modern day blogger, life recorder, or tweeter produce a better account of a life than that given to us by Samuel Pepys? Has there ever been a better summation of the entire human enterprise than that given by Samuel Beckett in Waiting For Godot when he declared, *"They give birth astride of a grave, the light gleams an instant then it's night once more………. We have time to grow old. The air is full of our cries. But habit is a great deadener"*. Perhaps the work of Shakespeare answers all the questions we have ever had, or will ever have. The monologue from Hamlet, *"……What a piece of work is a man, how noble in reason, how infinite in faculties, in form and moving how express and admirable, in action how like an angel, in apprehension how like a god!"* Hamlet alone stands among the greatest literary achievements of our species, and a most searing insight into human existence, and human limitations. Or maybe the famous soliloquy from Macbeth is something we will take with us into the very far and distant future, words to haunt our multimillion year reign as a relaxed civilisation. *To-morrow, and to-morrow, and to-morrow, creeps in this petty pace from day to day, To the last syllable of recorded time; and all our yesterdays have lighted fools the way to dusty death. Out, out, brief candle! Life's but a walking shadow, a poor player that struts and frets his hour upon the stage and then is heard no more. It is a tale told by an idiot, full of sound and fury signifying nothing.* Perhaps the last human alive will whisper these few words prior to putting the lights out.

Of course we will bring with us into that great future all those things we treasure right now. The grand piano will survive, for without the piano any hope of a glorious future is surely lost. Whatever changes into the future the piano will remain the premier musical instrument and is not likely to be bettered or even improved. A robot playing a piano no matter how skilfully it performs will never get the adulation a human concert pianist enjoys. As has been said before the human future critically depends on machines being kept in their place, and maintaining the absolute supremacy of humans. In that case we will have no choice in our far future other than to be both individually and collectively, the wealthiest any human or society has ever been. Our only problem will be how do we spend our vast wealth? Every science fiction dream that has ever been imagined becomes a very reasonable reality, if it can be imagined then it can be built. We will dwell in a kind of billionaire form of communist utopia. We will come to the agreeable conclusion that we own everything of value, and machines run everything for us. A form of social engineering designed to facilitate human creativity, and leave the running of society to non-feeling non-human bureaucrats. It's the most efficient method of distributing the resources of the entire solar system. After all, who owns the Sun? Who owns Mars, or the Asteroid Belt? If we have complete mastery of nanotechnology, and can engineer on a planetary scale then there are no limits. It is conceivable that vast colonies will be very easily constructed, and land parcelled out to humans who may individually own several square miles of the colony. Because there is infinite space, energy and time, and mastery of design, with the aid of infinite numbers of robots, there is literally nothing that cannot be accomplished.

We might arrive at two extremes. At the one extreme, a far future citizen would have a vast estate or many such estates, comprising many hundreds of square miles. They live alone or with other family members, and an unlimited supply of semi intelligent servants to do everything. A home modelled on a massive German castle or English stately home where they can indulge their every whim and fantasy safe in the knowledge that whatever damage they inflict on themselves or others can be repaired. They might boast exact replicas of the Augusta National Golf Club, or the Aspen Ski slopes, or the Formula One Course at Monte Carlo, complete with buildings and crowds. Everything is possible when money is no object. And if the real thing cannot be accomplished in reality, your wish can probably be facilitated in full emersion virtual reality, as promised in many movies and TV series. At the other extreme, we have the far future citizen with all the trappings of infinite wealth who wants nothing more than to sit in a small garden and read a book, or chat with friends. In the midst of abundance we still crave simplicity.

For all our abundant wealth, we will however, retain our appreciation of the original, and will forever insist and demand that the first in time prevails. So it won't matter how many spare bodies we might have in order to live out our immortal lives. We will require our own original, natural bodies for ceremonial occasions such as getting married, and the like. There is likely to be a prohibition on augmentation, the Olympic Games will therefore remain the preserve on unenhanced human athletes. Nor are we likely to see the Masters Committee allow an augmented human to compete in the Masters at the Augusta National Golf Club. If you wish to add your name to the illustrious list of past champions such as the great, Ben Hogan or Jack Nicklaus, you will have to play the old fashioned way. It is likely to be the same with piano competitions, or climbing Mount Everest, such feats will only make sense if carried out by non-augmented human bodies and brains. Otherwise, human achievement would have no meaning to us, because continuity with the past is broken. What would be the point in a non-human intelligence housed in a spherical container equipped with an antigravity attachment charging up the slopes of Mount Everest and claiming it to be the fastest assent in history? We dismiss it and say, so what.

So we will push on into our infinite future. The human project has no scheduled endpoint. We are not on anyone's timetable. We are where we are, and we plod on from here. Humanity is the only game in town. Life is the only game in town. The universe will ultimately decide how it wishes to interpret itself. We humans are clearly part of its picture even if only a single milestone on the road. We can't say. We simply don't know. We are an arrogant, egotistical species, and clearly believe we are the ultimate manifestation of the universe. We are gods in waiting. Whatever happens in the future, in this universe, we are assured of our place in the grand scheme of things; just on what we have done already.

Source Material And Further Reading

Purple falls into the *Future Technology, Speculation, Where Are We Going? What Happens Next?* Genre. What might be uncharitably termed, '*Purple Porn*', but this work would not be out of place on a modest bookshelf alongside the general works listed below.

If you really want to know about the human project in great depth, and about technology and its impact on all of us, and how we are likely to change into the near and even the far future. And if you find you have even a passing interest in the subject matter of Purple, then the general texts listed below ought to be your first stop at further reading.

All of the books on the list have one thing in common, they all say *yes*. In their own way, they are all positive about the human animal, the human project, and the human future. There is so much negativity in the world, probably because, as was stated with reference to science fiction, the dramatic element is simply not there in a positive future scenario. There is no drama to be extracted from quietness, or from a future perspective where things generally work out well.

Abundance, Peter Diamandis and Steven Kotler.
Civilisation Kenneth Clark
Collapse, Jared Diamond
Cosmos, Carl Sagan
Cosmos: A space Time odyssey, Neil de Grasse Tyson
Here On Earth, Tim Flannery
Technopoly, Neil Postman
The Ascent Of Man, Jacob Bronowski
The Beginning of Infinity, David Deutsch.
The Most Powerful Idea in the World, William Rosen
The Rational Optimist, Matt Ridley
The Singularity Is Near, Kurzweil Ray
To Save Everything, Click Here, Evgeny Morozov
What Technology Wants Kevin Kelly
Why The West Rules-For Now, Ian Morris

BOOKS AND ARTICLES

Acemoglu, Daron, and Robinson, James A., *Why Nations Fail: The Origins of Power, Prosperity and Poverty*, Profile Books, 2013.

Adams, Douglas, *The Hitchhikers Guide to the Galaxy*, Random House, 1979.

Adams, Douglas, *The Salmon of Doubt*, Pan, 2003.

Appleyard, Bryan, *The Brain Is Wider Than The Sky: Why Simple Solutions Don't Work In A Complex World*, Weidenfeld & Nicolson, 2011.

Ashton, J. E, Editor., *In Six Days: Why 50 Scientists Choose to Believe in Creation*, New Holland, 1999.

Atkins, P. W., *Creation Revisited*, W. H. Freeman, Oxford, 1992.

Augustine, St., *City of God*, AD 426.

Aunger, R., *The Electric Meme: A New Theory of How We Think*, Free Press. New York, 2002.

Baggini, J., *Atheism: A Very Short Introduction*, Oxford: Oxford University Press, 2003.

Barrat, James, *Our Final Invention: Artificial Intelligence and the End of the Human Era*, St. Martin's Griffin, 2015.

Barrow, J. D. and Tipler, F. J., *The Anthropic Cosmological Principle*, Oxford University Press, 1988.

Baynes, N. H., editor. *The Speeches of Adolf Hitler, vol. 1.*Oxford University Press, 1942 .

Beckett, Samuel, *The Trilogy: Molloy, Malone Dies, The Unnameable*, Macmillan new edition, 1979.

Beckett, Samuel, *Waiting for Godot*, Faber and Faber main edition, 2006.

Blackmore, Susan, *The Meme Machine*. Oxford University Press, 1999.

Blaker, K., editor, *The Fundamentals of Extremism: The Christian Right in America*, New Boston 2003.

Bolt, Robert, *A Man For All Seasons*, Heinemann, London, 1960.

Bostrom, Nick, *Superintelligence: paths, Dangers Strategies*, Oxford University Press, 2015.

Bracken, Paul, *Fire in the East: The Rise of Asian Military Power and the Second Nuclear Age*, 2000.

Braudel, Fernand, *Civilization and Capitalism*, (in three volumes), Harper Collins, New York 1985.

Broderick, Damien, *The Spike: How Our LivesAre Being Transformed By rapidly Advancing Technologies*, Tom Doherty Associates, New York, 2001.

Bronowski, Jacob, *The Ascent of Man*,

Bronowski, Jacob, *The Ascent of Man*, BBC Books 1973.

Brown, Stephen R., *A Most Damnable invention, - Dynamite, Nitrates, and the Making of the Modern World*, Thomas Dunne Books, St. Martin's Press. New York 2005.

Burrough, Bryan, and John, Helyar, *Barbarians At The Gate: The Fall of RJR Nabisco*. Johnathan Cape, London, 1990.

Carter, Robert M., *Climate: The Counter Consensus*, Stacey International, 2010.

Castiglione, Baldassare. *Il Cortegiano*, 'The Courtier'

Chomsky, Noam, *Hegemony or Survival: America's Quest for Global Dominance*, Penguin, 2003.

Christian, Brian, *The Most Human Human: a Defence of Humanity in the Age of the Computer*, Penguin 2012.

Clark, Kenneth, *Civilisation*, BBC books and John Murray, 1969.

Clarke, Arthur, C., *Profiles of the Future*, Indigo, London, 2000.

Coates John, *The Hour Between Dog and Wolf*, Fourth Estate, London 2012.

Collings, Anthony F., and Critchley, Christa, Editors. *Artificial Photosynthesis: From Basic Biology to Industrial Application*, Wiley-VCH, 2005.

Cowen, Tyler, *The Great Stagnation: How America Ate All The Low-Hanging Fruit of Modern History, Got Sick, and Will (Eventually) Feel Better*, Penguin 2010.

Cox, Harvey, "The Market as God," The Atlantic Monthly, March 1999.

Cox, Harvey, *When Jesus Came to Harvard: Making Moral Choices Today*, Mariner Books, 2004.

Darwin, Charles, *On the Origin of Species by Means of Natural Selection*, John Murray, London, 1859.

Davis, Paul, *The Eerie Silence: Searching For Ourselves In The universe*, Alan Lane, 2010.

Dawkins, Richard, *The Blind Watchmaker*, Longman, 1986.

Dawkins, Richard, *The God Delusion*, Transworld, London, 2006.

Dawkins, Richard, *The Selfish Gene*, Oxford University Press, 1976.

Dawkins, Richard, *Unweaving the Rainbow*, Penguin, 1998.

Deffyes, K.S., *Hubbert's Peak: the Impending World Oil Shortage*, Princeton University Press, 2001.

Dennett, Daniel, C., *Breaking the Spell: Religion as a Natural Phenomenon*. Viking, London, 2006.

Dennett, Daniel, C., *Consciousness Explained*, Penguin, 1991.

Dennett, Daniel, C., *Darwin's Dangerous Idea*, Simon & Schuster, New York, 1995.

Dennett, Daniel, C., *The Intentional Stance*, MIT Press, 1987.

Deutsch, David, *The beginning of Infinity: Explanations That Transform the World*, Allen Lane, London, 2011.

Deutsch, David, *The Fabric of Reality*, Allen Lane, London, 1997.

Diamandis, Peter and Kotler, Steven, *Abundance: The Future is Better Than You Think*, Simon and Schuster, 2012.

Diamond, Jared, *Collapse, How Societies Choose to Fail or Survive*, Penguin Books, 2005.

Diamond, Jared, *Guns, Germs, and Steel: The Fates of Human Societies*, W. W. Norton, New York, 1997.

Dooling Richard, *Rapture For The Geeks: When AI Outsmarts IQ*, Random House, New York, 2008.

Drexler, Eric, *Radical Abundance: How a Revolution in Nanotechnology Will Change Civilization*, Public Affairs, USA, 2013.

Ehrenfeld, D., *The Coming Collapse of the Age of Technology*, Tikkun Jan/Feb issue, 1999.

Ehrman, B. D., *Lost Scriptures: Books that Did Not Make It into the New Testament*, Oxford: Oxford University Press, 2003.

Etherington, John, *The Wind Farm Scam, An Ecologists Evaluation*, Stacey International, 2009.

Farmer, Paul, *Pathologies of Power: Health, Human Rights, and the New War on the Poor*, 2003.

Feldmann, Stanley and Marks Vincent, *Global Warming and Other Bollox: The Truth About All Those Science Scare Stories*, Metro Publishing, London, 2009.

Fernyhough, Charles, Pieces of Light: The New Science of Memory, Profile Books, 2012.

Flannery, Tim, *Here On earth: A New beginning*, Allen Lane, 2010.

Ford, Martin, *The Lights In The Tunnel: Automation, Accelerating Technology And The Economy Of The Future*, Acculant Publishing, USA, 2009.

Forrest, B. and Gross, P. R., *Creationism's Trojan Horse: The Wedge of Intelligent Design*, Oxford University Press, 2004.

Frankl, Victor, E., *Man's Search For Meaning: A Classic Tribute To Hope From The Holocaust*, random House 1959.

Freeman, C., *The Closing of the Western Mind*, Heinemann, London 2002.

Gates, Bill *"Ask Me Anything"*, Reddit session of January, 2015.

Gibbon, Edward, *Decline and Fall of the Roman Empire*, Modern edition 1998.

Glassman, James K. and Hassett, Kevin, *Dow 36,000: The New Strategy for Profiting from the Coming Rise in the Stock Market*, Times Books, 1999.

Gleick, James, *Chaos: Making a New Science*, Viking, 1987.

Goodwin, J., *Price of Honour: Muslim Women Lift the Veil of Silence on the Islamic World*, London: Little, Brown, 1994.

Gray, John In his, *The Myth of Progress*, New Statesman Essay, April 1999.

Gray, John, *Straw Dogs: Thoughts on Humans and Other Animals*, Granta Books, London, 2002.

Gray, John, *The Imortalization Commission: Science and the Strange Quest to Cheat Death*, Allen Lane, 2011.

Grayling, A. C., *What Is Good? The Search for the Best Way to Live*, Weidenfeld & Nicolson, London, 2003.

Harris, Robert, *Space Enterprise: Living and Working Offworld in the 21st Century*, Springer, Praxis, 2009.

Harris, Sam, *Letter to a Christian Nation*, Knopf, New York, 2006.

Harris, Sam, *The End of Faith: Religion, Terror and the Future of Reason*, Norton, New York, 2004.

Hawking, Stephen, *A Brief History of Time*, Bantam, London, 1988.

Hayek, Friedrich Von, *The Road to Serfdom*, Routledge Classics, 2001.

Herman, Edward, S., and, Chomsky, Noam, *Manufacturing Consent: The Political Economy of the Mass Media*, Random House, 1988.

Hitchens, Christopher, *God Is Not Great*, Atlantic Books, 2008.

Hitchens, Christopher, *The Missionary Position: Mother Teresa in Theory and Practice*. Verso, London, 1995.

Hitchens, Christopher, *Thomas Jefferson: Author of America*, Harper Collins, 2005

Hodges, A., *Alan Turing: The Enigma*. Simon & Schuster, New York, 1983.

Holmes, B., and Jones, N., *Brace yourself for the end of cheap oil*, New Scientist 179, 2406, 9–11, 2003.

Horgan, John, *The End Of Science: Facing The Limits Of Knowledge In The Twilight Of The Scientific Age*, Little Brown, 1996.

Hoyle, Fred, *Energy or Extinction*, Heinemann Books, London, 1978.

Hughes, Robert, *Rome*, Weidenfeld and Nicolson, London, 2011.

Humphery, John W., Oleson, John P., and Sherwood, Andrew N., *Greek and Roman Technology: A Sourcebook*, Routledge, 1998.

I Ching, ancient Chinese text.

Jacoby, S., *Freethinkers: A History of American Secularism*, Holt, New York, 2004.

Johnson, Steven, *Where Good Ideas Come From, A Natural History of Innovation*, Allen Lane, 2010.

Jones, Gareth, Stedman, *An End to Poverty? A Historical Debate*, Profile Books, London 2004.

Joy, Bill, *Why The Future Doesn't Need Us*, Wired Magazine, April 2000.

Kaku, Michio , *Physics of the Future: How Science Will Shape Human Destiny And Our Daily Lives By The Year 2100*, Allen Lane, 2011.

Kaletsky, Anatole, *Capitalism 4.0: The Birth of a New Economy*, Bloomsbury London, 2010.

Keen, Andrew, *The Internet is Not the Answer*, Atlantic Books, 2015.

Kelly, Jack, *Gunpowder – A History of the Explosive that Changed the World*, Atlantic Books, London 2004.

Kennedy, John Fitzgerald, speech containing the words, *"We choose to go to the Moon"*, delivered at Rice University, Texas, 1961, widely available online.

Kennedy, Paul, *The Rise and Fall of the Great Powers*, Vintage Books, New York, 1989.

Kidder, Tracy, *Mountains Beyond Mountains: Healing the World: The Quest of Dr. Paul Farmer*, Random House, New York, 2003.

Kissinger, Henry, *Does America Need a Foreign Policy? Toward a Diplomacy for the 21st Century*, Simon and Schuster, New York, 2001.

Klein, Naomi, *No Logo: Taking Aim at the Brand Bullies*, Picador, 2000.

Krugman, Paul, *A Country Is Not A Company*. Harvard Business Review Classics Series, Harvard Business School Publishing Corporation, 2009.

Kurtz, P., and Madigan, T. J., Editors, *Challenges to the Enlightenment: In Defence of Reason and Science*, Amherst, 1994.

Kurzweil, Ray, *The Age Of Intelligent Machines*, MIT Press, 1990.

Kurzweil, Ray, The Age of Spiritual Machines: When Computers Exceed Human Intelligence, Penguin, 1999.

Kurzweil, Ray, *The Singularity Is Near: When Humans Transcend Biology*, Viking, New York, 2005.

Landes, David, S., *The Wealth and Poverty of Nations: Why Some Are So Rich and Some So Poor*, W. W. Norton, New York, 1998.

Lanier, Jaron, *Who Owns the Future?* Allen Lane, 2013.

Lanier, Jaron, You Are Not A Gadget, Vantage Books, 2011

Lardy, Nicolas, R., *Integrating China into the Global Economy*, Brookings Institution, Washington, DC, 2002.

Lasota A. and Mackey, M.C., *Chaos, Fractals, and Noise: Stochastic Aspects of Dynamics*, Springer, 1998.

Lataster, Rapheal, Christopher, *There Was No Jesus, There Is No God*, Amazon Books, 2013.

Lawson, Nigel, *An Appeal to Reason: A Cool Look at Global Warming*, Duckworth Overlook, 2009.

Leggett, Jeremy, Editor, *The Solar Century: The Past, Present And World-Changing Future Of Solar Energy*, Green profile, 2009.

Leiter, Brian, *Why Tolerate Religion?* Princeton University Press, 2012.

McCleary, Rachel, *The Oxford Handbook of the Economics of Religion,* Oxford University Press, USA, 2011.

Lewis Michael, *Flash Boys: A Wall Street Revolt,* W.W. Norton & Co, 2015.

Lewis Michael, *Liars Poker: Rising Through The Wreckage On Wall Street,* W.W. Norton & Co, 1989.

Lewis Michael, *The Big Short: Inside The Doomsday Machine,* W.W. Norton & Co, 2010.

Lipstadt, Deborah, *Denying the Holocaust: The Growing Assault on Truth and Memory.* Penguin Books, 1994.

Lovelock, James, *Gaya: A New Look At Life On Earth,* Oxford University Press, 1979.

Lynas, Mark, *The God Species: Saving The Planet In The Age Of Humans,* National Geographic, 2011.

Maddison, Angus, *The World Economy: A Millennial Perspective,* OECD, Paris, 2001.

May, R.M., *"Simple Mathematical Models with Very Complicated Dynamics,"* Nature, Vol. 261, 459-467, June, 1976.

McGregor, James, *One Billion Customers: Lessons from the Front Lines of Doing Business in China,* Wall Street Journal Book, Free Press, 2005.

McKibben, Bill, *The End Of Nature: Humanity, Climate Change and the Natural World,* Random House, 2003.

McNeil, William H. *Plagues and Peoples,* Doubleday, New York, 1977.

Mills, D., *Atheist Universe: The Thinking Person's Answer to Christian Fundamentalism,* Ulysses Books, 2006.

Moorhead, Sam and Stuttard, David, *The Romans Who Shaped Britain,* Thames & Hudson, 2012.

Moreland, J.P., and Lane Craig, William, *Philosophical Foundations for a Christian Worldview,* Inter Varsity Press, 2003.

Morovec, Hans, Mind Children, The Future of Robot and Human Intelligence, Harvard University Press, 1990,

Morozov, Evgeny, *To Save Everything, Click Here: Technology, Solutionism and the Urge to Fix Problems that Don't Exist, Allen Lane* 2013.

Morris, Ian, *Why The West Rules - For Now: The Patterns of History and What They Reveal About the Future,* Farrar, Straus, and Giroux, 2010.

Mukherjee Siddhartha, *The Emperor of All Maladies: A Biography of Cancer, Fourth Estate,* 2011.

Nadar Ralph, *Unstoppable: The Emerging Left-Right Alliance To Dismantle The Corporate State,* Nation Books, 2014.

Nutting, Rex, *How the Stock Market Destroyed the Middle Class*, published on the marketwatch.com website. July 17, 2015.

Ormerod Paul, *Positive Linking: How Networks Can Revolutionise the World*, Faber, London, 2012.

Ott, E., *Chaos in Dynamical Systems*, Cambridge University Press, 1993.

Pagel, Mark, *Wired for Culture: The natural History of Human Cooperation*, Allen Lane, 2012

Paul Saint, *Letters of, New Testament*.

Penrose, Roger, The Emperor's New Mind, Oxford University Press, 1989.

Phillips, K., *American Theocracy*, Viking, New York, 2006.

Piketty, Thomas, *Capital: in the Twenty-First Century*, Belknap Press, 2014.

Plimer, I., *Telling Lies for God: Reason vs Creationism*, Random House, 1994.

Plini the Elder, *Naturalis Historia*, 37 Volume Encyclopaedia,

Postman, Neil, *Technopoly: The Surrender of Culture to Technology*, Vintage, Random House, 1992.

Price, Robert, M., *Jesus is Dead*, American Atheist Press, 2007.

Price, Robert, M., *The Case Against The Case For Christ: A New Testament Scholar Refutes the Reverend Lee Strobel*, American Atheist Press, 2010.

Rand, Ayn, *The Fountainhead*, Penguin Classics, 1943.

Rees, Martin, *From Here To Infinity: Scientific Horizons*, Profile Books, 2011.

Rees, Martin, *Just Six Numbers: The Deep Forces That Shape the Universe*, Weidenfeld & Nicolson, 1999.

Rees, Martin, *Our Cosmic Habitat*, Weidenfeld & Nicolson, 2001

Rees, Martin, *Our Final Century: Will the Human Race Survive the Twenty-First Century?* Arrow Books, 2003.

Reese, Byron, *Infinite Progress: How the Internet and Technology Will End Ignorance, Disease, Poverty, Hunger and War*. Greenleaf Book Group Press, 2013.

Ridley, Matt, *The Rational Optimist*: How Prosperity Evolves, Fourth Estate, 2010.

Roberts, Alice, *Evolution: The Human Story*, Doring Kindersley, 2011.

Robertson, Geoffrey, *Mullahs Without Mercy: Human Rights and Nuclear Weapons*, Biteback Publishing London 2012.

Ronson, J., *The Men Who Stare at Goats*, Simon & Schuster, New York, 2005.

Roscoe, Philip, *I Spend Therefore I Am: The True Cost of Economics*, random House, 2014.

Rosen, William, *The Most Powerful Idea in the World: A Story of Steam Industry and Invention*. Jonathan Cape, London, 2010.

Russell, Bertrand, *Why I Am Not a Christian*, Routledge, 1957.

Sachs, Jeffrey, *The End of Poverty: Economic Possibilities for Our Time*, Penguin, 2005.

Sagan, Carl, *Pale Blue Dot*, Headline, London, 1995.

Sagan, Carl, *The Demon-Haunted World: Science as a Candle in the Dark*, Headline London, 1996.

Schmidt, Eric, and Cohen, Jared, *The New Digital Age: Reshaping the Future of People Nations and Business*, John Murray, London, 2013.

Seedhouse, Erik, *Beyond Human: Engineering Our Future Evolution*, Springer-Verlag 2014.

Seligman, Scott, D., *Chinese Business Etiquette*, Warner Books, New York 1999.

Sheldrake, Rupert, *The Presence Of The Past, Morphic Resonance and the Habits of Nature*, Icon Books, 2011.

Sheldrake, Rupert, *The Science Delusion: Freeing The Spirt Of Inquiry*, Hodder & Stoughton, 2012.

Shermer, M., *Science Friction: Where the Known Meets the Unknown*, Holt, New York, 2005.

Shirk, Susan L., *China Fragile Superpower*, Oxford University Press, 2007.

Silver, L. M., *Challenging Nature: The Clash of Science and Spirituality at the New Frontiers of Life*, HarperCollins, New York, 2006.

Singer, Peter, *Ethics*, Oxford University Press, 1994.

Singer, Peter, *One World: The Ethics of Globalization*, (The Terry Lectures), Yale University Press, 2002.

Singer, Peter, *The President of Good and Evil: The Ethics of George W. Bush*, Plume Books, 2004.

Skloot, Rebecca, *The Immortal Life of Henrietta Lacks*, Macmillan, 2010.

Smil, Vaclav, *Feeding the World: A Challenge for the 21st Century*, MIT Press, 2000.

Smith, Adam, *The Wealth Of Nations*, Shine Classics, 2014.

Snow, C. P., *The Two Cultures*, Cambridge University Press, 1998.

Stannard, Russell, *The End of Discovery: Are We Approaching the Boundaries of the Knowable?* Oxford University Press, 2010.

Steer, R., *Letter to an Influential Atheist*, Authentic Lifestyle Press, 2003.

Susskind, Leonard, *The Cosmic Landscape: String Theory and the Illusion of Intelligent Design*. Little Brown, New York, 2006.

Tallis, Raymond, *Aping Mankind: Neuromania, Darwinitis and the Misrepresentation of Humanity*, Acumen, 2011.

Tallis, Raymond, *Why The Mind Is Not A Computer: A Pocket Lexicon of Neuromythology*, Palgrave Macmillan, 2004.

The Oxford Handbook of Engineering and Technology in the Classical World, Edited by John Peter Oleson, Oxford University Press, 2008.

Thucydides, *History of the Peloponnesian War*, translation by R. W. Harmondsworth, Penguin, 1972.

Van Santen, Rutger, Khoe, Djan, Vermeer, Bram, *2030: Technology That Will Change The World*, Oxford University Press, 2010.

Vayssieres, Lionel, Editor, *On Solar Hydrogen & Nanotechnology*, John Wiley & Sons (Asia) Pte Ltd, 2009.

Vinge, Verner, *The Coming Technological Singularity: How to Survive in the Post-Human Era*, Whole earth Review, Winter 1993.

Vitruvius, *Ten Books on Architecture*, c.25 CE.

Weinberg, Stephen, *Dreams of a Final Theory*, Vintage London, 1993.

Wenke, Robert F., and Olszewski, Deborah J., *Patterns in Prehistory: Humankind's First Three Million Years*, Oxford University Press, 2007.

Wilson, E. O., *Biophilia: the Human Bond With Other Species*, Harvard University Press, 1984.

Wolf, Tom, *Bonfire Of The Vanities*, Vintage Books, 1987.

Wolpert, Lewis, *Six Impossible Things Before Breakfast: The Evolutionary Origins of Belief*, Faber & Faber, 2006.

Wright, Robert, *Non Zero: History Evolution And Human Cooperation*, Random House, New York, 2000.

Wu, Tim, *The Master Switch: The Rise And Fall Of Information Empires*, Random House 2010.

Zinn, Howard, *A People's History of the United States: 1492–Present*, Harper Collins, 1980.

Zinn, Howard, *You Can't Be Neutral on a Moving Train: A Personal History of Our Times*, Beacon Press, 2002.

Zuckerman, Miron, et al, The University of Rochester survey, '*The Relation Between Intelligence and Religiosity: A Meta-Analysis and Some Proposed Explanations*', and is published in the, '*Personality and Social Psychology Review*', Vol 17, No 3, August 2013.

Zuckermann, Ethan, *Rewire: Digital Cosmopolitans in the Age of Connections*, W.W. Norton & Co, 2013.

Zweig, David, *Internationalizing China*, Cornell University Press, 2002.

MOVIES & TV Etc.

2001 A Space Odyssey, Movie, Director: Stanley Kubrick, 1968.

Back to the Future, I, II, III, Movies, Director: Stephen Spielberg, 1985 – 1990.

Barbarians at the Gate: TV Movie, Director: Glenn Jordan, 1993.

Blade Runner, Movie, Director: Ridley Scott, 1982.

Bridge on the River Kwai, Director: David Lean, 1957.

Capitalism a Love Story, Documentary, Director: Michael Moore, 2009.

Cave of Forgotten Dreams, Documentary, Director: Werner Herzog, 2011.

Cosmos, TV Series, Presented by Carl Sagan, 1980.

Cosmos: A Space Time Odyssey, TV Series, Presented by Neil De Grasse Tyson 2014.

Going Clear: Scientology and the Prison of Belief, Documentary, Director: Alex Gibney, 2015.

I Robot, Movie, Director: Alex Proyas, 2004.

Inherit the Wind, Movie, Director: Stanley Kramer, 1960.

Inside Job, Documentary, Director: Charles Ferguson, 2010

Life of Brian, Movie, Producer/Director: Monty Python, 1979.

Lord of the Rings Trilogy, Movies, Director: Peter Jackson, 2001- 2003.

Margin Call, Movie, Director: J.C. Chandor, 2011.

Marjoe, Documentary, Directors: Sarah Kernochan and Howard Smith, 1972.

Men in Black, Movie, Director: Barry Sonnenfeld, 1997.

Miracle on 34th Street, Movie, Director: George Seaton, 1947.

Night of the Living Dead, Movie, Director: George A. Romero, 1968.

Soylent Green, Movie, Director: Richard Fleischer, 1973.

Space 1999, TV Series, Producers: Gerry and Silvia Anderson 1975.

Star Trek: The Next Generation: TV Series, 1990s.

The Big Short, Movie, Director: Adam McKay, 2015.

The Jetsons, Animated TV Series, *1962.*

The Matrix, Movies, Directors: The Wachowski brothers, 1999.

The Terminator, Movie, Directors: James Cameron, 1984.

The Usual Suspects, Movie, Director: Bryan Singer, 1995.

The Wild One, Movie, Director: Laslo Benedek, 1953.

Thriller, Michael Jackson Music Video, Director: John Landis, 1983.

Wall Street, Movie, Director: Oliver Stone, 1987.

PURPLE

Index